Nuclear Magnetic Resonance

Atta-ur-Rahman

Nuclear Magnetic Resonance

Basic Principles

With 290 Illustrations

Springer-Verlag
New York Berlin Heidelberg Tokyo

Atta-ur-Rahman
H.E.J. Research Institute of Chemistry
University of Karachi
Karachi, 32
Pakistan

Library of Congress Cataloging in Publication Data
Atta-ur-Rahman
   Nuclear magnetic resonance.
   Bibliography: p.
   Includes index.
   1. Nuclear magnetic resonance spectroscopy.
I. Title.
QD96.N8A88   1986   538′.362   85-27633

Typeset by Polyglot Pte Ltd, Singapore.

9 8 7 6 5 4 3 2 1

ISBN-13: 978-1-4612-9350-7   e-ISBN-13: 978-1-4612-4894-1
DOI: 10.1007/978-1-4612-4894-1

Preface

Nuclear magnetic resonance spectroscopy is presently going through an explosive phase of development. This has been brought about largely on account of the advent of Fourier transform NMR spectrometers linked to powerful microcomputers which have opened up a whole new world for structural chemists and biochemists. This is exemplified by a host of publications, especially on new pulse sequences, which continue to provide new exciting modifications for recording two-dimensional NMR. Moreover, NMR is no longer confined to structural chemists but has moved firmly into the area of medicine as a powerful nondestructive body scanning technique.

With this background, I felt that there was need for a text which would provide a fairly comprehensive account of the important features of ^{1}H- and ^{13}C-NMR spectroscopy in one book, as well as make available an up-to-date account of recent developments of new pulse sequences, with particular reference to 2D-NMR spectroscopy. Since this book is written for students of chemistry and biochemistry as well as for biology students who have chemistry as a subsidiary, it was decided to avoid a complex mathematical treatment and to present, as far as possible without oversimplification, a qualitative account of ^{1}H- and ^{13}C-NMR spectroscopy as it is today. I hope that the book satisfactorily meets these objectives.

I would like to thank a number of students of my research groups for their help in the drawing of diagrams. They are Mr. Mohammad Iqbal Choudhary, Miss Kishwar Jahan, Mr. Sohail Malik, Miss Anjum Muzaffar, and Miss Khurshid Zaman. I am particularly grateful to Mr. Siraj-ud-din Nizami, Lecturer, Department of Chemistry, for his help in the drawing of figures. I wish to thank my wife, Nargis, for her assistance in the preparation of figures and her enduring patience over the innumerable evenings which went to the writing of this book. Finally I wish to record my indebtedness to the numerous research papers, reviews, books, and monographs which I consulted, all of which are acknowledged as references or as recommended reading at the end of various chapters.

I dedicate this book to my parents, Jamil-ur-Rahman and Amtul Subhan Begum, who lit a candle in me many many years ago.

Karachi, Pakistan   Prof. Atta-ur-Rahman

Contents

Contents

Chapter 1

Chemical Shift in ^{1}H-NMR Spectroscopy

1.1 INTRODUCTION

The phenomenon of nuclear magnetic resonance is based on the fact that nuclei of certain elements possess a spin angular momentum and an associated magnetic moment. When such nuclei are placed in a magnetic field, they can adopt one of a number of quantized orientations, each orientation corresponding to a particular energy level. The orientation with the lowest energy is the one in which the nuclear magnetic moment is most closely aligned with the external magnetic field while the orientation with the highest energy is the one in which the nuclear magnetic moment is least closely aligned with the magnetic field. Nuclear magnetic resonance involves transitions between these energy levels (or changes in the orientation of the nuclei) with respect to the external magnetic field. These transitions may be induced by the absorption of radio frequency radiation of the correct frequency, which is measurable on a recorder in the form of an NMR signal of the nucleus. Pauli first postulated such magnetic properties of nuclei in 1924[*] but it was not until 1946 that the first NMR experiments were independently carried out by two groups, Bloch[†] at Stanford University and Purcell[‡] at Harvard University. These experiments have ushered in a new era in which nuclear magnetic resonance spectroscopy has become well established as a powerful tool for structure elucidation, and more recently has also been increasingly employed to obtain sectional pictures of human organs.

It is a fundamental fact of physics that a spinning charged body produces a magnetic moment. Since a nucleus is positively charged, if it then has a spin angular momentum, P, its spinning will result in the rotation of the positive charge which may be compared to a current flowing in a circle. This would produce a magnetic field parallel to the spin axis, and the nucleus would have a

[*] W. Pauli, *Naturwissenschaften* **12**, 741 (1924).
[†] F. Bloch, W.W. Hansen, and M.E. Packard, *Phys. Rev.* **69**, 127 (1946).
[‡] E.M. Purcell, H.C. Torrey, and R.V. Pound, *Phys. Rev.* **69**, 37 (1946).

magnetic moment, μ. In quantum mechanics, the angular momentum P is given by the relationship

$$P = I\frac{h}{2\pi} \tag{1.1}$$

where I is the spin quantum number and h is Planck's constant. The spin quantum number is quantized and in different nuclei it may have values of 0, $\frac{1}{2}$, 1, $\frac{3}{2}$, etc. Thus if I is zero, the nucleus will not have any angular momentum. Examples of such nuclei are ^{12}C and ^{16}O. This results in a great simplification of the proton nuclear magnetic resonance spectra of organic molecules as no interaction is observed between the protons and ^{12}C or ^{16}O nuclei to which the protons may be attached. The spin quantum number I is related to the atomic number and mass number as follows:

Atomic number	Mass number	Spin quantum number
Even or odd	Odd	$\frac{1}{2}, \frac{3}{2}, \frac{5}{2}$, etc.
Even	Even	0
Even	Odd	$1, 2, 3$, etc.

The above relationships emerge from the fact that the protons and neutrons in the nucleus possess spins. Protons will thus form pairs with other protons in the nucleus with opposite spins. Similarly, neutrons will "pair" with other neutrons in the same nucleus but with opposite spins. In nuclei which have even numbers of protons and neutrons, all the spins will be paired and the spin number I will be zero. However if there is an odd number of either protons or neutrons, the spin quantum number I will have a quantized value of $\frac{1}{2}$, 1, $\frac{3}{2}$, etc. If the sum of protons and neutrons is even, I will be zero or a multiple of 1. If the sum is odd, I will be an integral multiple of $\frac{1}{2}$. The nuclei with which the organic chemist is most frequently concerned are ^{1}H and ^{13}C, both of which have a spin quantum number of $\frac{1}{2}$. Other elements which may be of interest are ^{19}F and ^{31}P which also have $I = \frac{1}{2}$ while deuterium, ^{2}H, and ^{14}N have a spin quantum number of 1.

When placed in a uniform magnetic field, the angular momentum of a nucleus is quantized and it will adopt one of $(2I + 1)$ orientations with respect to the external magnetic field. In nuclei such as ^{1}H or ^{13}C where $I = \frac{1}{2}$, there will be $(2 \times \frac{1}{2}) + 1$ orientations, i.e., two orientations. In the lower-energy orientation, the nucleus will have its magnetic moment μ aligned with the magnetic field while in the higher-energy orientation it will be aligned against the magnetic field. In each orientation the nucleus will have a particular potential energy equal to $\mu B_0 \cos\theta$ where B_0 is the strength of the external field while θ is the angle between the nuclear spin axis and the direction of the external field. Nuclei with $I = \frac{1}{2}$ have a symmetrical distribution of charge around the nucleus.

In nuclei such as ^{2}H or ^{14}N which have $I = 1$, there will be $(2 \times 1) + 1$ orientations, i.e., three different orientations, each having its characteristic

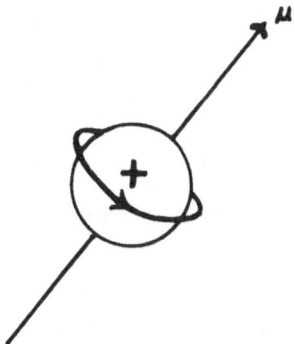

Figure 1.1. The spinning positively charged nucleus produces a magnetic moment μ.

potential energy. Nuclei with $I > \frac{1}{2}$ have unsymmetrical distributions of charge, which result in an electric quadrupole moment, Q.

As mentioned earlier, electromagnetic radiation of the correct frequency can cause transitions between adjacent energy levels. The relationship between the electromagnetic frequency ν and the magnetic field strength B_0 is governed by the Larmor equation

$$\nu = \frac{\gamma B_0}{2\pi} \tag{1.2}$$

where γ is the gyromagnetic ratio, and is a characteristic constant for a particular nucleus (Figure 1.1). This is the basic mathematical equation for NMR.*

Now let us consider a nucleus such as a proton (^{1}H) with a spin number of $\frac{1}{2}$. When placed in a magnetic field, it behaves like a tiny magnet, with the difference that while a magnet can adopt any number of orientations with respect to the external field, the tiny spinning "proton magnet" is allowed only two orientations, the lower-energy orientation being the one in which its north pole is aligned with the south pole of the external field. The energy difference ΔE between the two energy levels is proportional to the external magnetic field. Since the nucleus is spinning on its axis, the external magnetic field causes it to "precess", i.e., instead of the spinning axis remaining stationary, it undergoes a circular motion such as that exhibited by a spinning gyroscopic top before it is about to topple, i.e., the two ends of the spinning axis trace two opposite but circular paths (Figure 1.2). Nuclear magnetic resonance experiments are concerned with the measurement of this precessional motion (Figure 1.3).

---

* The equation can be simply derived, and its derivation is reported in numerous standard texts of NMR (see list of recommended books at end of chapter).

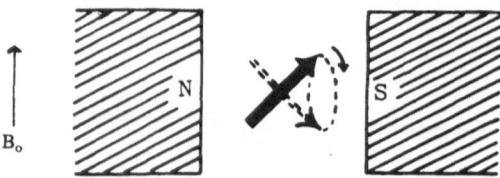

Figure 1.2. Precessional motion of the magnetic dipole when placed between the poles of a magnet.

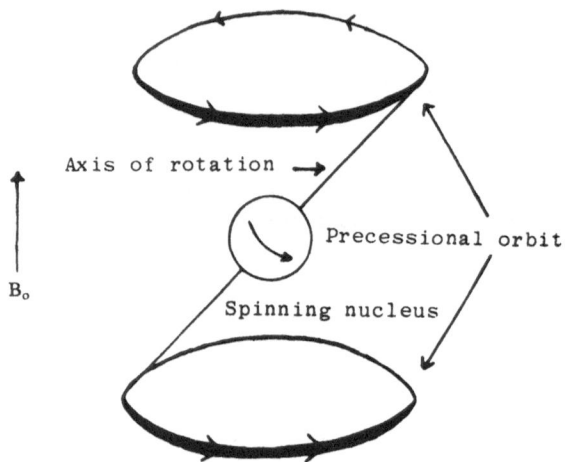

Figure 1.3. Precessional motion of a nucleus in a magnetic field.

In order to cause transitions between the two spin states, a second radio frequency field is applied perpendicular to the original field. When the value of this electromagnetic radiation reaches the precessional frequency of the nucleus, absorption of energy will occur. It is important to note that the frequency of the electromagnetic radiation required to induce transitions from one nuclear spin state to the other is *exactly* equal to the precessional frequency ω_0 of the nucleus. The precessional frequency ω_0 is directly proportional to the applied magnetic field B_0 and also to the gyromagnetic ratio

$$\omega_0 = \gamma B_0 \qquad (1.3)$$

Thus with increasing values of the applied magnetic field, B_0, there will be increasing differences in energy (ΔE) between the two spin states (Figure 1.4), and since the precessional frequency of the nucleus is proportional to B_0, the nucleus will precess at increasing frequency values. In order to bring it into "resonance," correspondingly higher frequencies of the applied electromagnetic radiation will therefore be necessary. The advantage of observing resonances at higher frequency values of the electromagnetic radiation (which

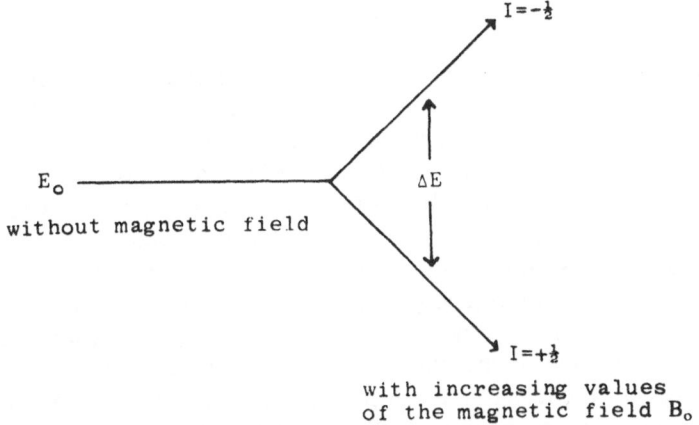

Figure 1.4. Increasing difference in energy levels of the two magnetic states with increasing value of magnetic field B_0.

is generated from a suitable oscillator) is that it results in increased dispersion of signals and a greater sensitivity. One can therefore induce resonances in different protons by keeping a constant value of the applied magnetic field and gradually increasing the oscillator frequency. As the precessional frequencies of the different nuclei are reached, resonance will occur and NMR signals will be observed. Alternatively, and this is more convenient in practice, one can keep the oscillator frequency constant at, say, 60 MHz or 100 MHz, and vary the magnetic field. With increasing values of the magnetic field, the nucleus will precess at correspondingly higher frequencies and when the frequency of precession reaches the value of the oscillator frequency, resonance will occur, i.e., energy will be absorbed by the nucleus at its "chemical shift" and it will be tilted away from the direction of the applied magnetic field, causing a change in the impedance of the oscillator coils which can be measured on a recorder.

As mentioned above, for a nucleus with spin quantum number $I = \frac{1}{2}$, only two energy levels are possible. When I is 1, there will be three energy levels, but only transitions between adjacent energy levels are allowed, i.e., from $m = +1$ to $m = 0$, or from $m = 0$ to $m = -1$; transitions from $m = +1$ to $m = -1$ are not allowed (Figure 1.5).

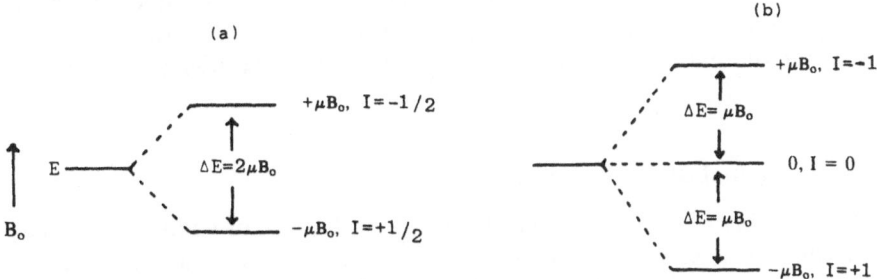

Figure 1.5. Magnetic energy levels for nuclei (a) with spin $\frac{1}{2}$ and (b) with spin 1.

1.2   RELAXATION PROCESSES

The theory of electromagnetic radiation tells us that by absorption of energy, transitions of the nuclei will occur to a higher-energy state while by emission of energy, resulting in downward transitions, the nuclei will relax to the lower-energy state. If the two spin states of a set of protons, for instance, were equally populated, then the total number of protons undergoing transitions from the lower spin state to the upper spin state would be exactly equal to the total number of protons undergoing transitions in the opposite direction, and no absorption or emission of energy would be observed and there would be no NMR signal. Fortunately, the population of the upper- and lower-energy spin states is not the same, there being a Boltzmann distribution of the nuclei with a slight excess in the lower state. In practice what this means for a 100-MHz instrument is that if we have one million nuclei in the lower-energy state, there will be 999,987 nuclei in the upper-energy state, leaving a population excess of 13 nuclei in the lower state. In order to detect this slight excess of nuclei, one has to have a very sensitive detection system. Thus, when a solution of an organic compound is placed between the poles of a magnet and irradiated with a suitable radio frequency, the nuclei simultaneously undergo transitions in two opposing directions. Absorption of radio frequency results in transitions to the higher-energy state while relaxation processes cause transitions from the upper to the lower spin state. Since the probability of upward transitions is slightly higher than that of downward transitions due to the greater population of the ground state, a *net* absorption of energy will occur. With this absorption of energy, the population of nuclei in the ground state will decrease while the population in the higher-energy state will increase till there is no further excess in the population of the ground state, resulting in the weakening of the intensity of the NMR signal, and finally in its disappearance. A "saturation" state is then said to have been reached. In order to reestablish the Boltzmann excess of the ground state, there are a number of mechanisms involving radiationless transitions by which the nuclei in the upper energy state can relax back to the lower spin state. The result is that while the Boltzmann excess of nuclei is somewhat diminished, it is not totally eliminated and an equilibrium state is established at an intermediate value, so that an NMR signal can continue to be obtained.

There are two main types of relaxation processes by which nuclei in upper energy states can relax to lower energy states. These are *spin–lattice relaxation* (or longitudinal relaxation), T_1, and *spin–spin relaxation* (or transverse relaxation), T_2.

1.2.1   Spin–Lattice Relaxation (T_1)

The assembly of molecules constituting gases, liquids, or solids under study are together referred to as the lattice. In gases or liquids, the molecules are undergoing rapid rotational and translational motions, and in those molecules which have magnetic nuclei, these motions will result in corresponding magnetic fields of varying magnitudes which may be considered to be built up of a number of oscillating components. A magnetic nucleus precessing under the influence of an external magnetic field will also be subjected to the fluctuating fields generated by the lattice, and whenever the oscillating component of the lattice field becomes correctly oriented and its frequency becomes exactly equal to the precessional frequency of the magnetic nucleus, exchange of energy from the nucleus in the higher-energy state to the lattice can occur. This energy is transferred to the lattice in the form of translational or rotational energy. The energy of the magnetic nucleus is thus converted to the thermal energy of the molecular system comprising the lattice. Spin–lattice relaxation, T_1, is therefore directly responsible for maintaining the Boltzmann excess of nuclei in the lower-energy state. The more efficient the relaxation process, the smaller will be the value of the relaxation time T_1. Spin–lattice relaxation is dependent not only on the type and rapidity of the molecular motions of the molecule in the lattice but also on the gyromagnetic ratio of the magnetic nucleus. In solids in which the molecular motions are very restricted, T_1 values are very large on account of the inefficiency of the spin–lattice relaxation process. T_1 relaxation times may be used with great advantage in structure elucidation of organic compounds as T_1 values can vary in different compounds, providing a valuable insight into the chemical environment of the nuclei being observed.

1.2.2 Spin–Spin Relaxation (T_2)

The field generated by a precessing nucleus will contain a static component, which will be aligned in the direction of the applied field, and a rotating component, which will be moving at the precessional frequency in a plane perpendicular to the applied field (Figure 1.6).

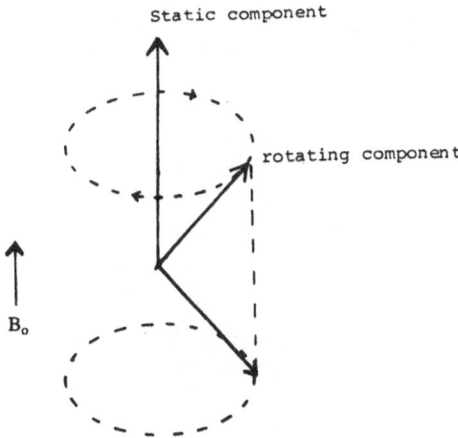

Figure 1.6. Static and rotating components of the field generated by a precessing nucleus. The static component is aligned with the applied field while the rotating component rotates at the precessional frequency in a plane perpendicular to the applied field.

If another nucleus is located near the first nucleus, it will experience the effect of the static component as a variation in the main field. A "spread" in the value of the main field will, therefore, be caused since the total field experienced by a neighboring nucleus will be the sum of the main field and the static components generated by the local field. This will result in a broadening of the resonance signals.

The rotating component of the magnetic vector, which is in a plane at right angles to the external field, is correctly set up to induce a transition in a neighboring nucleus provided that the neighboring nucleus is precessing at the same frequency. This would result in a mutual exchange of spin energies between the two nuclei and it would also cause a broadening of the resonance signals.

1.3   CHEMICAL SHIFT

The frequency at which a given nucleus comes to resonance is dependent not just on the magnitude of the applied field and the gyromagnetic ratio of the nucleus, but also on a third factor—the molecular environment of the nucleus. Thus for a given nucleus, say protons, it is observed that the atoms in different chemical environments give NMR signals at different characteristic values of the applied field. This means that when the NMR spectrum of, for example, ethyl alcohol is recorded, the signals for CH_3, CH_2, and OH protons will not appear together as one signal but as three distinctly different groups of peaks. One important reason for this is the electronic clouds circulating around the nucleus which "shield" or protect the nucleus, to an extent, from the influence of the applied magnetic field. The circulation of these electrons results in the generation of a small but finite induced magnetic field which is oriented so as to oppose the external field (Figure 1.7). Thus the nucleus experiences a magnetic field which is slightly less than the applied field. Equation (1.2) can therefore be modified as

$$v = \frac{\gamma B_0 (1 - \sigma)}{2\pi} \tag{1.4}$$

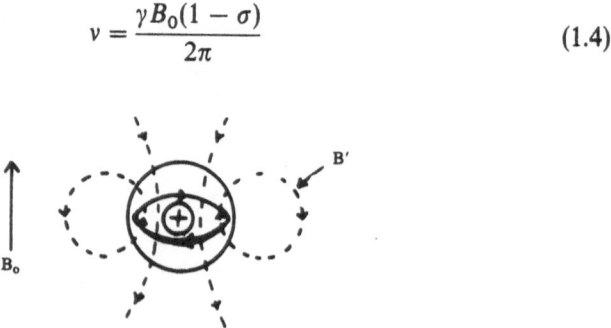

Figure 1.7. The electron circulation around the nucleus produces a resulting magnetic field B'.

where σ is a field-independent shielding factor. If an electron-attracting group is attached to a proton, it will tend to pull electrons away from it and decrease the shielding effect of the electronic clouds around it, thus allowing the proton to come to resonance at a lower value of the applied magnetic field. It therefore follows that the greater the electronegativity of the attached atom the greater will be its "pull" and resonance will occur at a correspondingly lower value of the applied magnetic field. Thus the shielding of methyl group protons decreases in the order: $SiCH_3$, CCH_3, NCH_3, OCH_3. As a result of oxygen being the most electronegative atom in the group, the $-OCH_3$ protons give a signal at a lower value of the applied field than, say, $-CCH_3$ protons; on the other hand, silicon is the least electronegative of the group and the signal of the methyl group attached to it therefore appears at the highest value of the applied magnetic field. Tetramethylsilane [TMS, $(CH_3)_4Si$] affords a sharp signal for the methyl protons at a high value of the magnetic field and is, therefore, conveniently used as an internal standard when recording NMR spectra, and the "chemical shifts" or resonance positions of various protons in a molecule are reported in the literature as distances from the TMS signal. The NMR spectrum is normally calibrated in cycles per second (cps or Hz, frequency units). However, it follows from Equation (1.4) that the frequency at which a resonance signal is observed for a particular proton is dependent on the applied magnetic field, B_0. Thus if the resonance of the methyl protons of an acetyl group appears at 120 Hz on a 60-MHz NMR spectrometer, it would appear at 200 Hz on a 100-MHz NMR spectrometer. Since there are several types of NMR spectrometers with different magnetic field strengths, it is necessary that the position of a resonance signal be given in units which are independent of the field of the magnet. A parameter δ is used for this purpose which is defined as

$$\delta = \frac{v_s - v_{TMS}}{Z} \tag{1.5}$$

where v_s is the resonance frequency, in Hz, of the protons, v_{TMS} is the resonance frequency, in cycles per second, of TMS, and Z is the operating frequency of the instrument in megacycles per second or 10^6 cycles per second. As the numerator in the above equation is in cycles per second while the denominator is in megacycles per second, the units of δ are *parts per million* or ppm. An alternative parameter which was formerly employed is τ (tau), which is equal to $10 - \delta$. Thus, on a 200-MHz instrument the chemical shift will be expressed as

$$\delta = \frac{\text{Distance (in cycles per second) from TMS}}{200} \text{ ppm} \tag{1.6}$$

Protons attached to differently hybridized carbon atoms or elements of different electronegativity, or having different environments exhibit different

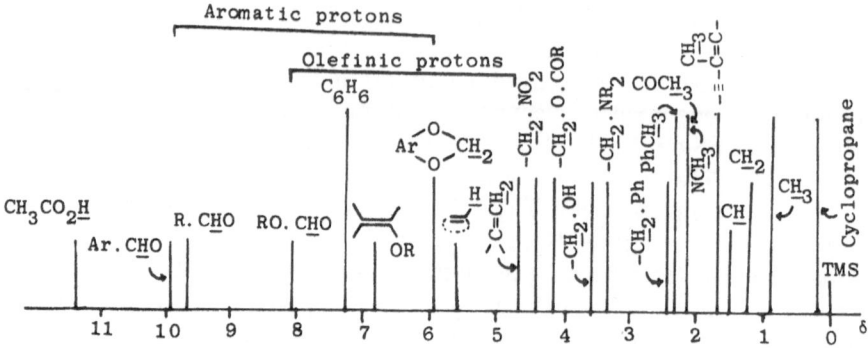

Figure 1.8. Chemical shifts of protons in various organic functional groups.

chemical shifts. This proves very useful in their identification. Some typical chemical shift values of different protons are presented in Figure 1.8.

A useful piece of information which may be derived from an NMR spectrum relates to the relative intensity of the signals. Thus in ethyl alcohol it is found that the intensities of the peaks (as measured from the peak areas under the signals) are not the same but are directly proportional to the number of protons from which the signals arise. Thus the intensity of the CH_3 signal is greater than that of the CH_2 signal, the two peak areas being in a ratio of 3:2. The peak area can be "integrated" by means of a sloping line which runs across the NMR spectrum and which rises in steps whenever a signal occurs underneath it, the extent of the "rise" being directly proportional to the number of underlying protons. Thus in the NMR spectrum of highly purified ethanol (Figure 1.9) the vertical "rise" in the integration line over the OH, CH_2, and CH_3 protons, as measured in millimeters, is given by the ratio 1:2:3 approximately. Integration is often carried out in separate sections of the spectrum in order to prevent noise, impurities, or solvent peaks from being incorporated into the rise of the integration line.

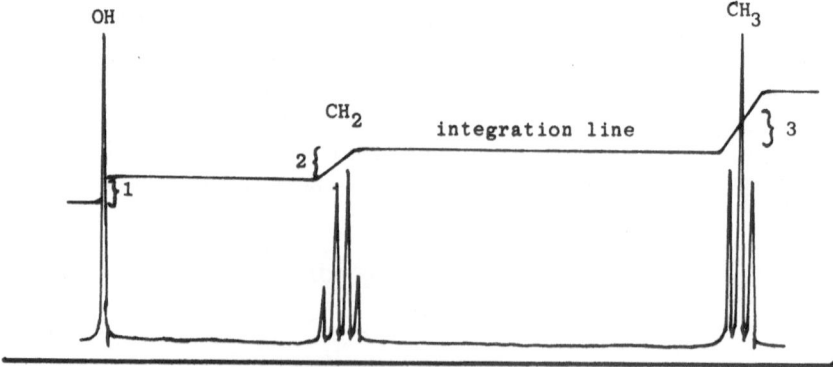

Figure 1.9. 100-MHz ^{1}H-NMR spectrum of ethanol.

1.3.1   Some Factors Affecting Chemical Shift

There are a number of factors which influence the chemical shift at which
proton resonances occur. These may be categorized as (a) local diamagnetic
shielding effects and (b) the effect of neighboring atoms or groups in the
molecule. Each proton in a molecule may be considered to share two bonding
electrons, which are induced to circulate by the external magnetic field in a
manner so as to produce local secondary fields which oppose the applied
field. These local diamagnetic shielding effects are thus dependent on the
electron density around the atom and on the extent to which the C—H
bond may be polarized.

The major factors affecting chemical shifts are discussed in the following
sections.

1.3.1.1   Inductive Effects

As has been mentioned earlier, when a nucleus is placed in a uniform magnetic
field, the electrons surrounding it circulate in such a manner as to produce a
secondary field which reduces the influence of the external magnetic field on
the nucleus. This shielding influence of the surrounding electrons may
decrease if another electronegative atom is attached to the nucleus because
the protective electronic clouds are pulled away towards the neighboring
electronegative atom, causing it to come to resonance at a lower value of the
external magnetic field. On the other hand, groups which donate electrons will
tend to increase the electron density around the nucleus and hence a higher
value of the external applied field will be required for resonance to occur since
the nucleus is more shielded from the influence of the external magnetic field.
The influence of electronegativity on the chemical shift can be seen from
Tables 1.1, 1.2, and 1.3.

The majority of alkanes show resonances between 0.9 and 1.6 ppm. Methyl
groups attached to saturated carbon atoms are the most shielded of aliphatic
protons and absorb in the region 0.6–1.2 ppm. Methylene protons are less
shielded and resonate around 1.3 ppm in saturated hydrocarbons while the
methine protons resonate even further downfield (around 1.5 ppm). This is
true only in the absence of electron-withdrawing groups which change the
chemical shifts drastically. In long-chain alkanes, the methylene protons
usually resonate very close to each other, resulting in a large unresolvable peak
around 1.2–1.3 ppm. These overlapping peaks can often be resolved if the
spectra are recorded with higher-field instruments such as at 400 or 500 MHz.
Methyl groups attached to carbonyls resonate at about 1.9–2.2 ppm and are
usually easily recognized. When methyl groups are attached to oxygen atoms
(methoxyl groups), the methyl protons resonate at about 3.3 ppm but in
methyl groups attached to nitrogen they experience a relatively higher
shielding and resonate at around 2.3 ppm. Methyl groups attached to olefinic
carbon atoms resonate around 1.6–1.8 ppm but if attached to aromatic rings
they are deshielded and resonate around 2.4 ppm. Methylenes and methines

Table 1.1. Chemical shifts of CH$_3$, CH$_2$, and CH protons of groups attached to different functional groups

Methyl protons	δ	Methylene protons	δ	Methine protons	δ
CH$_3$—C	0.9	C—CH$_2$—C	1.4	C—CH—C	1.5
CH$_3$—C—O	1.4	C—CH$_2$—C—O	1.9	C—CH—C—O	2.0
CH$_3$—C=C	1.1	C—CH$_2$—C=C	1.7		
CH$_3$—C=C	1.6	C—CH$_2$—C=C	2.3		
CH$_3$—Ph	2.3	C—CH$_2$—Ph	2.7	C—CH—Ph	3.0
CH$_3$—CO—R	2.2	C—CH$_2$—CO—R	2.4	C—CH—CO—R	2.7

$\mathrm{CH_3-CO-Ph}$	2.6			$\mathrm{-\overset{\vert}{\underset{\vert}{C}}-\overset{-\overset{\vert}{C}-}{\overset{\vert}{CH}}-CO-Ph}$	3.3
$\mathrm{CH_3-CO-OR}$	2.0	$\mathrm{-\overset{\vert}{\underset{\vert}{C}}-CH_2-CO-OR}$	2.2		
$\mathrm{CH_3-CO-OPh}$	2.4				
$\mathrm{CH_3-CO-NR}$	2.0	$\mathrm{-\overset{\vert}{\underset{\vert}{C}}-CH_2-CO-NR}$	2.2		
$\mathrm{CH_3-OR}$	3.3	$\mathrm{-\overset{\vert}{\underset{\vert}{C}}-CH_2-OR}$	3.4	$\mathrm{-\overset{\vert}{\underset{\vert}{C}}-\overset{-\overset{\vert}{C}-}{\overset{\vert}{CH}}-OR}$	3.7
		$\mathrm{-\overset{\vert}{\underset{\vert}{C}}-CH_2-OH}$	3.6	$\mathrm{-\overset{\vert}{\underset{\vert}{C}}-\overset{-\overset{\vert}{C}-}{\overset{\vert}{CH}}-OH}$	3.9
$\mathrm{CH_3-OPh}$	3.8	$\mathrm{-\overset{\vert}{\underset{\vert}{C}}-CH_2-OPh}$	4.3		
$\mathrm{CH_3-O-\underset{\vert}{C}=\underset{\vert}{C}-}$	3.8				
$\mathrm{CH_3-O-COR}$	3.7	$\mathrm{-\overset{\vert}{\underset{\vert}{C}}-CH_2-O-COR}$	4.1	$\mathrm{-\overset{\vert}{\underset{\vert}{C}}-\overset{-\overset{\vert}{C}-}{\overset{\vert}{CH}}-O-COR}$	4.8

Table 1.1. (*Continued*)

Methyl protons	δ	Methylene protons	δ	Methine protons	δ
$CH_3-N\!\!<$	2.3	$>\!\!C-CH_2-N\!\!<$	2.5	$>\!\!C-CH-N\!\!<$	2.8
$CH_3-N-COR$	2.9			$>\!\!C-CH-N-COR$	4.1
CH_3-N-Ph	3.0				
$CH_3-\overset{+}{N}\!\!<$	3.3				
CH_3-S-	2.1	$>\!\!C-CH_2-S-$	2.4	$>\!\!C-CH-S-$	3.2
		$>\!\!C-CH_2-NO_2$	4.4	$>\!\!C-CH-NO_2$	4.7
CH_3-C-NO_2	1.6	$>\!\!C-CH_2-C-NO_2$	2.1		
$CH_3-C=C-CO$	2.0	$>\!\!C-CH_2-C=C-CO$	2.4		

Structure	Shift
C=C–CO–CH₃	1.8
C=C–CO–CH₂–	2.4
–C–CH₂–Cl	3.6
–C–CH–Br	4.3
–C–CH₂–Br	3.5
–C–CH–I	4.3
–C–CH₂–I	3.1
–C–CH–C≡N	2.7
–C–CH₂–C≡N	2.3
cyclopropane CH	0.7
cyclopropane CH₂	0.3
epoxide O–CH	3.1
epoxide O–CH₂	2.6

Table 1.2. Chemical shifts of protons attached to olefinic or acetylenic carbons

Proton	δ	Proton	δ
H—C≡C—	1.8[a]	$\underset{\text{H}}{\overset{}{\diagdown}}$C=C$\diagup$C=O	6.2
H—C≡C—C—O—	2.4[a]	Ph—C=C(—H)—C=O	6.4
H—C≡C—C≡C—	2.8[a]	$\underset{\text{R}}{\overset{\text{H}}{\diagup}}$C=C—OR	5.0
H—C≡C—Ph	2.93[a]		
H—C≡C—CO	3.17[a]	C=C=CH	5.2
$H_2C=C\diagdown\overset{R}{\underset{R'}{}}$	4.65	Ph$\diagdown$C=C$\diagup$H	5.05
$H_2C=C\!-\!C\!-$	4.9	$\underset{\text{Ph}}{\diagup}$C=C$\diagdown$H	5.35
—C=C(H)—C=O	4.8	(cyclic) =H	5.6
$\underset{}{\overset{\text{H}}{}}$C=C—C=O	6.0	N—C(H)=O	7.85
(ring)—H	5.7	R—O—C$\overset{O}{\underset{H}{}}$	8.03
C=C$\diagup$H	5.8	R—C$\overset{O}{\underset{H}{}}$	9.65
—OR			
C=C$\diagdown$H$\diagup$Ph	7.0	Ph—C$\overset{H}{\underset{O}{}}$	9.9
$\underset{\text{Ph}}{\overset{\text{H}}{}}$C=C—C=O	7.8		
(furan ring) H_b, H_a	H_a, 6.23 H_b, 4.86	(pyran ring) H_b, H_a	H_a, 6.24 H_b, 4.54

Table 1.3. Parameters for calculation by modified Shoolery's rule[a] of approximate chemical shifts for methylene and methine protons

X	a	X	a
—Ph	1.3	—C=C—	0.75
—Br	1.9	—C≡C—	0.90
—Cl	2.0	—CO_2H	0.7
—OH	1.7	—CO_2R	0.7
—OR	1.7	—CN	1.2
—OCOR	2.7	—COR	1.2
—OPh	2.3	—SR	1.0
—NH_2	1.0	—I	1.4
—NR_2	1.0		

[a] Modified Shoolery's rule:

$$\text{For } CH_2 \overset{X_1}{\underset{X_2}{\diagdown}} \quad \delta CH_2 = 1.25 + \Sigma a$$

Thus the approximate chemical shift for the methylene protons in cyanoacetic acid $CN—CH_2—CO_2H$) would be calculated thus:

$$\delta CH_2 = 1.25 + (1.2 + 0.7) = 3.15$$

are deshielded by 0.2 and 0.1 ppm, respectively, relative to the positions given above for the methyl group.

The chemical shifts of methyl, methylene, and methine protons are presented in Table 1.1. The methylene protons of cycloalkanes absorb at similar positions to those of alkanes except in cyclopropanes in which the methylenes are significantly shielded (chemical shift: 0.1–0.5 ppm) because of the anisotropy of the cyclopropane C—C bonds. The dependence of the chemical shifts of methyl groups in methyl halides on the electronegativities of halogens can be seen from the fact that with decreasing electronegativities from F to I, the chemical shifts of the methyl groups also decrease from 4.13 ppm in CH_3F to 1.98 ppm in CH_3I (Table 1.4). An example of how the electron-withdrawing effect is transmitted through the carbon chain can be

Table 1.4. Dependence of chemical shifts of methyl protons in methyl halides on electronegativities (E) of halogens.

	CH_3F	CH_3Cl	CH_3Br	CH_3I	CH_3H
$\delta(CH_3)$	4.13	2.84	2.45	1.98	0.13
E(Pauling)	4.0	3.0	2.8	2.5	2.1

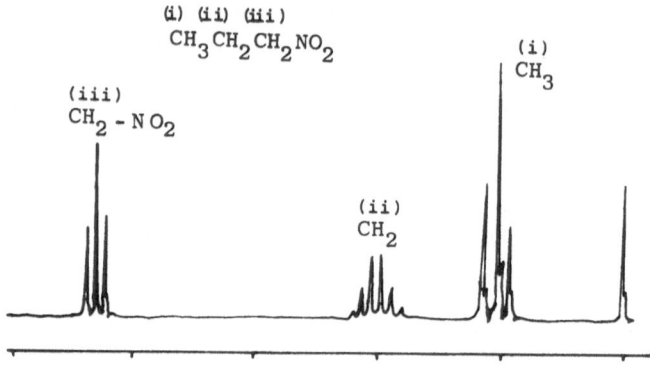

Figure 1.10. ^{1}H-NMR spectrum of nitropropane.

seen in the NMR spectrum of nitropropane (Figure 1.10). It is clearly evident that the resonance of the protons on the carbon to which the NO_2 group is attached appears farthest downfield and is the most affected, being shifted to lower fields by ~ 3.45 ppm from the position of the methyl resonance in propane (0.91 ppm) while a sizable downfield shift of δ 0.72 can be seen for the protons at the adjacent carbon atom (δ_{CH_2} 1.33 in propane) and a small downfield shift of 0.12 ppm occurs in the position of the methyl signal.

1.3.1.2 Hybridization

The state of hybridization of the carbon atom to which a proton is attached significantly influences the chemical shift of the proton. With increasing "s" character of the carbon, the bonding electrons will be drawn closer towards the carbon and away from the hydrogen, thus decreasing the electron density around the proton. This will result in the "deshielding" of the proton and it will therefore come to resonance at greater distances from the TMS signal. It has been calculated that change from sp^3 hybridization in ethane (25% "s" character) to sp hybridization in acetylene (50% "s" character) would cause a downfield shift of 5–8 ppm.

1.3.1.3 van der Waals Effects

If protons are forced into positions which are spatially very close to one other, van der Waals repulsive forces distort the electronic clouds which shield the protons. The reduced spherical symmetry of the electron cloud around the protons gives rise to a paramagnetic contribution to the shielding constants, thus causing deshielding or "downfield" shift in the positions of the proton resonances.

1.3.1.4 Diamagnetic Anisotropy

1.3.1.4.1 Single Bonds. The electronic clouds around protons attached to carbon atoms do not have spherical symmetry, but are asymmetrically distributed. The resulting secondary fields from protons are, therefore,

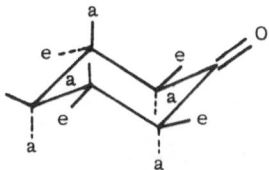

Figure 1.11. In rigid cyclohexanone derivatives, axial protons are more shielded than equatorial protons.

"anisotropic." A random tumbling of molecules will accordingly not result in their averaging to zero but will reinforce or diminish the applied field, depending on the orientation of the bonds. Thus in a rigid cyclohexane ring or in cyclohexanone (Figure 1.11), the axial protons are more shielded than the equatorial protons because of the anisotropy of the carbon framework so that they appear about 0.5 ppm further upfield than their equatorial counterparts.

1.3.1.4.2 Double Bonds. Anisotropic effects are even more marked in the case of double bonds such as C=C, C=O, and C=N which, in contrast to single bonds, do not have axial symmetry. Thus, in a double bond, the applied magnetic field causes an induced circulation of electrons which tends to oppose the applied magnetic field in the center of the double bond but reinforces it at the ends. Figure 1.12 shows the induced circulation of electrons when the double bond is aligned at right angles to the applied field. In solution the tumbling molecules would, however, adopt any one of a large number of orientations, and the average of these various orientations would afford two shielding cones with their apexes meeting at the center of the double bond. Any protons which fall in the regions of these cones (+zone) would be shielded whereas protons falling in the regions beyond the ends of the double bonds (–zone) would experience a deshielding influence (Figure 1.13). Similar situations would hold for C=S, C=N, and NO_2 groups. A striking example of such effects is provided by the aldehyde proton which lies

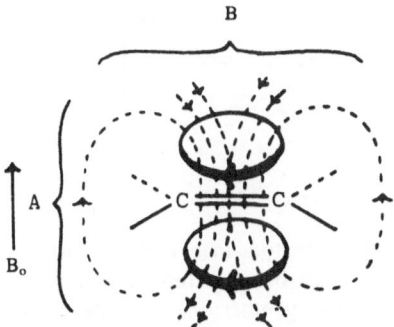

Figure 1.12. Induced diamagnetic circulation of π-electrons in a double bond. Nuclei lying in region A will be deshielded whiie those lying in region B will be shielded.

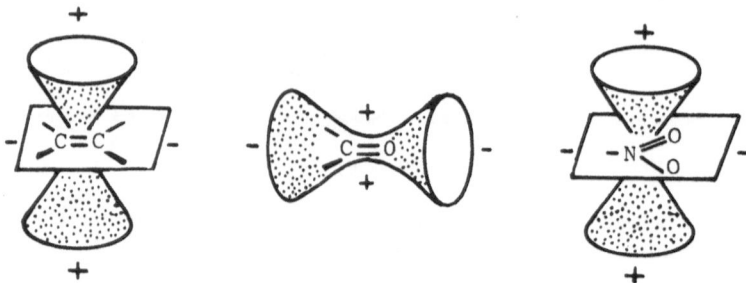

Figure 1.13. The magnetic anisotropic effect of the carbon–carbon double bond, the carbonyl group, and the nitro group, and the resulting shielding (+) and deshielding (−) regions.

in the deshielding zone of the carbonyl group and is therefore shifted considerably downfield. When substituents are attached to double bonds, they can significantly affect chemical shifts depending on their nature. In ethylene [Figure 1.14(a)] the olefinic protons resonate at δ 5.29. In methyl vinyl ketone [Figure 1.14(b)] the proton α to the carbonyl group is strongly shielded as it falls in the shielding region of the carbonyl group. The olefinic protons β to the carbonyl group are deshielded on account of the electropositive character of the β-carbon because of the contribution of the resonance structure shown in Figure 1.14(c). When a methoxy substituent is present, as in Figure 1.14(d), the electron donation to the β-carbon atom by the methoxyl group significantly shields these protons. The effects of substitution of different groups on the chemical shifts of olefinic bonds are summarized in Table 1.2.

1.3.1.4.3 Triple Bonds. Triple bonds, such as those encountered in acetylenes or nitriles, have axial symmetry. As is shown in Figure 1.15, the induced circulation of electrons in acetylenes or nitriles results in the generation of a secondary magnetic field which tends to oppose the applied field at the ends but reinforces it at a certain distance along the axis. Since

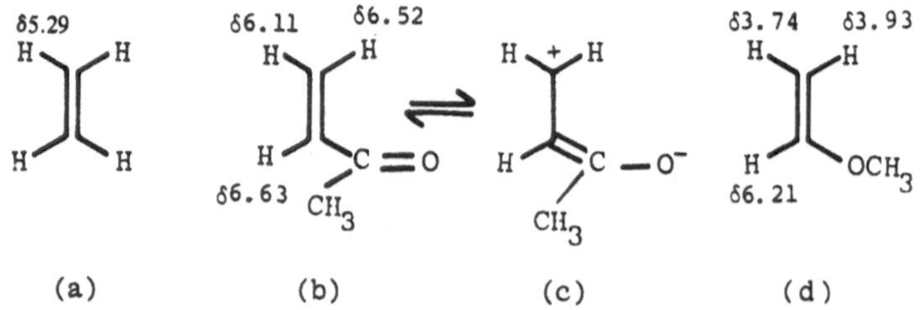

Figure 1.14. Chemical shifts of protons in (a) ethylene; (b)/(c) methyl vinyl ketone; the protons on the carbon β to the carbonyl group are deshielded because of its electron-withdrawing effect; (d) methyl vinyl ether; the opposite effect is now seen, the protons on the carbon β to the ethereal oxygen being strongly shielded due to its electron-donating effect.

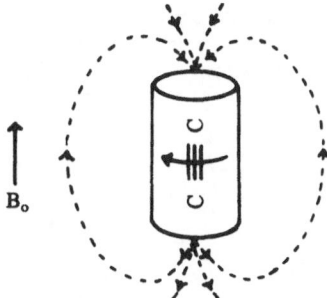

Figure 1.15. Diamagnetic circulation of π-electrons about the axis of a triple bond.

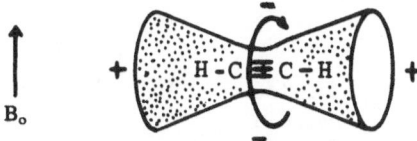

Figure 1.16. The magnetic anisotropic effect of the triple bond.

acetylenic protons fall in the shielding zone of the cone (see Figure 1.16), they are shifted significantly upfield and resonate at a value of δ 2.4, in contrast to olefinic protons which resonate at δ 4.6–6.0. If, however, another proton is brought over the center of the triple bond, it will be deshielded. Thus H-5 is shifted downfield by δ 1.71 ppm in 4-ethynylphenanthrene as compared to the position of the corresponding proton in phenanthrene (Figure 1.17).

1.3.1.4.4   Aromatic Rings.   In aromatic rings the π-electrons delocalized over the ring atoms are induced to circulate above and below the plane of the ring (in two loops on the two sides of the σ-bond framework). This circulation is such as to oppose the applied magnetic field at the center of the ring but reinforce it in the region where the aromatic protons are located. This results in a strong deshielding of the aromatic protons as shown in Table 1.5 (Figure 1.18). On the other hand, if a nucleus is held above the center of the

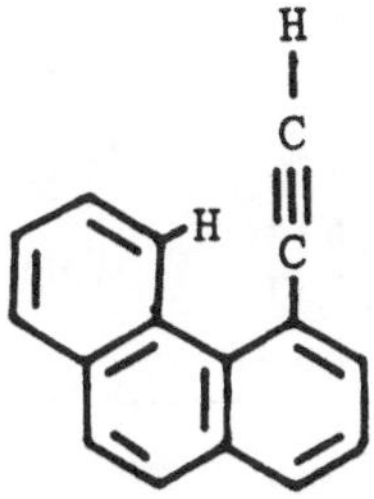

Figure 1.17. 4-Ethynylphenanthrene; the proton shown on the benzene ring lies in the deshielding region of the triple bond.

Table 1.5. Chemical shifts for aromatic protons

Protons	δ values	Protons	δ values
Benzene	7.27	Imidazole (H_b, H_b, N, N, H, H_a)	H_a, 7.70 H_b, 7.14
Naphthalene (H_a, H_b)	H_a, 7.78 H_b, 7.42	(H_c, H_b, N, N, CH_3, H_a)	H_a, 7.40 H_b, 6.86 H_c, 7.05
Pyridine (H_c, H_b, N, H_a)	H_a, 8.50 H_b, 6.99 H_c, 7.36	Thiazole (H_b, S, H_c, N, H_a)	H_a, 8.88 H_b, 7.41 H_c, 7.98
Pyrazine (N, N, H)	8.50	Benzofuran (H_c, H_d, H_b, H_e, O, H_a, H_f)	H_a, 7.52 H_b, 6.66 H_c, 7.49 H_d, 7.13 H_e, 7.19 H_f, 7.42
Pyrimidine (H_b, H_c, N, N, H_a)	H_a, 9.26 H_b, 8.77 H_c, 7.37	Quinoline N-oxide (H_c, H_b, H_a, H_d, N^+, O^-)	H_a, 8.57 H_b, 7.27 H_c, 7.74 H_d, 8.75
Pyridazine (H_b, H_a, N, N)	H_a, 9.24 H_b, 7.50	Isoquinoline (H_d, H_c, H_e, H_b, H_f, N, H_g, H_a)	H_a, 9.13 H_b, 8.45 H_c, 7.47 H_d, 7.70 H_e, 7.57 H_f, 7.49 H_g, 7.86
N-methylpyrazole (H_c, H_b, N, N, H_a, CH_3)	H_a, 7.36 H_b, 6.14 H_c, 7.30	Thiophene (H_b, S, H_a)	H_a, 7.17 H_b, 7.06

Table 1.5. (*Continued*)

Protons	δ values	Protons	δ values
Furan	H$_a$, 7.36 H$_b$, 6.28		H$_a$, 9.15 H$_b$, 8.60 H$_c$, 7.09
Pyrrole	H$_a$, 6.53 H$_b$, 6.06		H$_a$, 8.98 H$_b$, 8.43 H$_c$, 7.34 H$_d$, 8.24
	H$_a$, 7.0 H$_b$, 6.54 H$_c$, 6.34		H$_a$, 8.77 H$_b$, 8.14
	H$_a$, 13.7 H$_b$, 7.55 H$_c$, 6.25 H$_d$, 7.55		H$_a$, 8.73 H$_b$, 8.04 H$_c$, 7.66
	H$_a$, 7.29 H$_b$, 7.26 H$_c$, 7.71 H$_d$, 7.27 H$_e$, 7.30 H$_f$, 7.77		H$_a$, 9.44 H$_b$, 7.93 H$_c$, 7.85
			H$_a$, 9.18
	H$_a$, 6.57 H$_b$, 7.26 H$_c$, 6.15 H$_d$, 7.31		H$_a$, 9.22 H$_b$, 9.28 H$_c$, 7.84 H$_d$, 7.57 H$_e$, 7.83 H$_f$, 8.00
	H$_a$, 8.81 H$_b$, 7.26 H$_c$, 8.00 H$_d$, 7.68 H$_e$, 7.43 H$_f$, 7.68 H$_g$, 8.05		

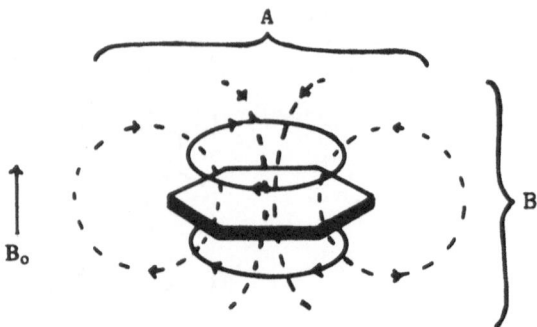

Figure 1.18. The circulation of π-electrons around a benzene ring produces a magnetic field: shielding (A) and deshielding (B) zones around benzene ring.

aromatic ring, it will be strongly shielded. Thus in paracyclophane the central methylene protons are located in the shielding region of the aromatic ring and are therefore shifted upfield by 0.45 ppm as compared to the methylene protons in a normal aliphatic chain (Figure 1.19).

A particularly illustrative example of the effect of ring currents can be seen in cyclic conjugated systems known as annulenes. Those annulenes containing $4n + 2$ π-electrons exhibit aromatic character. Thus (18)-annulene (Figure 1.20) exhibits a similar ring current as found in benzene, the circulation

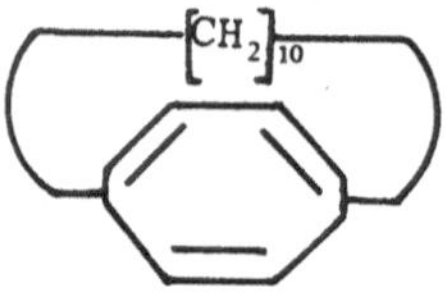

Figure 1.19. Paracyclophane.

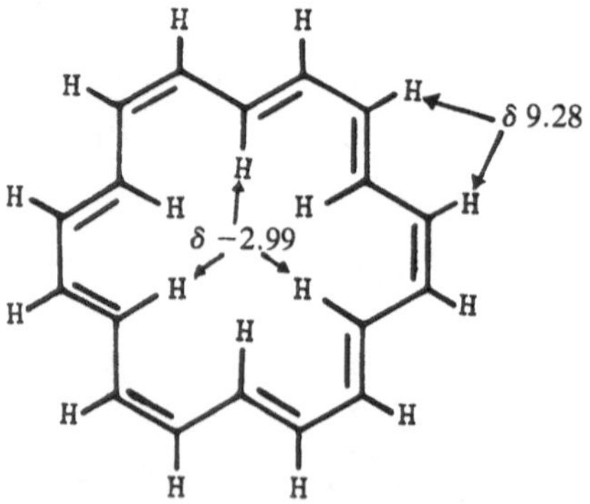

Figure 1.20. (18)-Annulene.

of π-electrons creating an induced magnetic field which opposes the applied magnetic field in the center of the ring and reinforces it around the periphery. As a result, the protons lying inside the ring of (18)-annulene are strongly shielded (δ −2.99) while those on the outer side are strongly deshielded (δ 9.28). In the case of the non-aromatic annulenes containing $4n$ π-electrons, e.g., (16)-annulene, quantum mechanics predicts a *paramagnetic* ring current effect (paramagnetic anisotropy) instead of the diamagnetic effect discussed above, which results in the opposite situation, i.e., inner protons are deshielded and outer ones are shielded.

It may be noted that ring current effects are only discernible in planar conjugated cyclic molecules. In nonplanar molecules, e.g., cyclooctatetraene (Figure 1.21), which is tub-shaped, the delocalization of π-electrons is severely restricted, and the protons of cyclooctatetraene therefore resonate in the normal olefinic region (δ 5.80, cf. 5.85 in cyclohexadiene).

The effect of substituents on the chemical shifts of protons in aromatic rings is shown in Table 1.6. Strongly electron-withdrawing groups, such as nitro or aldehyde, result in significant deshielding of the aromatic protons, the *ortho* and *para* protons being more affected than the *meta* protons on account of the greater transmission of electropositive character to the *ortho* and *para* carbons by mesomerism. Conversely, the presence of amino groups on benzene rings causes a shielding of the *ortho* and *para* carbons because of the donation of electrons from the nitrogen lone pair to these carbon atoms, resulting in an increase in the electron density at these positions [Figure 1.22(a), (b)].

1.3.1.4.5 Cyclopropanes. Protons which fall in a region perpendicular to the plane of the cyclopropane ring experience a significant shielding effect. This is illustrated in Figure 1.23. The resonance of axial protons is normally about 0.5 ppm downfield from that of equatorial protons in cyclohexanes, but this order is reversed in $3\alpha,5\alpha$-cyclosteriods. Thus in Figure 1.23(a) the equatorial hydrogen atom is forced into the shielding zone of the cyclopropane ring and therefore resonates at δ 4.47, which is 0.60 ppm upfield from the resonance of the corresponding axial proton [Figure 1.23(b)].

1.3.1.5 Solvent Effects

Substances dissolved in aromatic solvents generally give signals at higher fields than when dissolved in aliphatic solvents. These effects are attributed to the diamagnetic anisotropy of aromatic rings, and are more significant when

Figure 1.21. Cyclooctatetraene.

Table 1.6. Chemical shifts of *ortho*, *meta*, and *para* protons in monosubstituted benzenes

	δ, in ppm from benzene		
Substituent	*Ortho*	*Meta*	*Para*
NO_2	−0.97	−0.30	−0.42
CHO	−0.73	−0.23	−0.42
CO_2H	−0.63	−0.10	−0.17
CO_2R	−0.80	−0.15	−0.20
COR	−0.6	−0.3	−0.3
$CONH_2$	−0.5	−0.2	−0.2
NH_3^+	−0.4	−0.2	−0.2
CN	−0.3	−0.3	−0.3
NHCOR	−0.4	0.2	0.3
I	−0.3	0.2	0.1
⤷	−0.2	−0.2	0.2
OCOR	−0.2	0.1	0.2
SR	−0.1	0.1	0.2
Cl	0.0	0.0	0.0
Br	0.0	0.0	0.0
CH_2Cl	0.0	0.0	0.0
CH_3	0.15	0.1	0.1
OCH_3	0.23	0.23	0.23
OH	0.37	0.37	0.37
$N(CH_3)_2$	0.5	0.2	0.5
NH_2	0.77	0.13	0.40

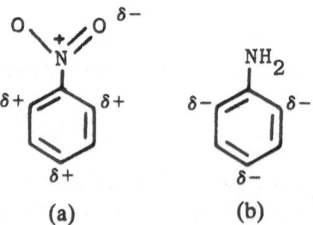

       (a)   (b)

Figure 1.22. (a) Nitrobenzene, showing the deshielding influence of the electron-withdrawing nitro group on the *ortho* and *para* protons. (b) Aniline, showing the shielding influence of the electron-donating amino group on the *ortho* and *para* protons.

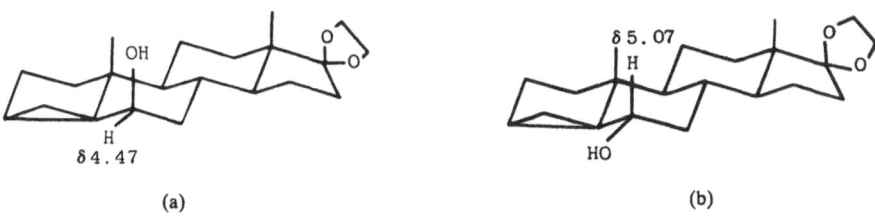

(a)   (b)

Figure 1.23. The equatorial proton geminal to the hydroxyl group lies in the shielding region of the cyclopropane ring in compound (a) and its resonance is thus 0.60 ppm upfield in comparison to that of (b) in which it occupies an axial position which is away from the shielding zone.

intermolecular solute–solvent associations lead to complex formation. As a result of dipole–dipole or Van der Waals interactions between the solute and solvent, some orientations of the solute molecules may be more favored than others, giving rise to shifts in the resonance frequencies of the protons of the solute molecules.

1.3.1.6 Concentration and Temperature

As is evident from the chemical shifts given in Table 1.7, resonances of exchangeable protons such as those of OH, NH, or SH groups are markedly dependent on concentration, solvent, and temperature. Thus if we examine the NMR spectrum of acetic acid we find that the position of the carboxyl proton resonance is several ppm upfield from the position of the hydroxyl proton resonance in the NMR spectrum of water. However, if we examine an NMR spectrum of a mixture of acetic acid and water, we find that only a single resonance line occurs, lying between the chemical shifts of COOH and OH protons. The position of this resonance is in fact a linear function of the percentage of acetic acid in water. This is so because the OH protons in acetic acid and water are exchanging rapidly so that we "see" a time-averaged situation in NMR spectroscopy. In general, when exchange of protons is possible, two distinct peaks will be observed in a mixture where the rate of exchange is slow. In cases where the exchange is rapid, as in the example of acetic acid–water quoted above, only one sharp resonance line will be observed. If the exchange rate is intermediate, peak broadening will be observed. (This is discussed in more detail in Section 3.2.) The broadening of NH signals is, however, generally due to the fact that ^{14}N has a nuclear spin and behaves like an electric quadrupole. A proton attached to a nitrogen therefore "sees" it in several magnetic quantum states, and the resulting peak broadening is attributed to *quadrupole relaxation* associated with nitrogen. Resonances which are suspected to be due to NH, OH, or SH groups may be made to disappear by shaking with D_2O or, in the case of amides, with a $D_2O/NaOD$ mixture. The rate of exchange of OH, NH, or SH protons can be substantially increased by addition of acid or by increase of temperature.

Table 1.7. Chemical shifts and characteristics of —OH, NH, and SH protons

Protons	δ values	Special features
R—OH (pure)	0.5–4.0	Can appear as a multiplet because of coupling with adjacent protons, but usually occurs as a broadened singlet; position is temperature, solvent, and concentration dependent. H-bonded enols may occur from δ 11.0 to 16.0.
R—OH (with a trace of acid)	4–8 (variable)	Occurs as a sharp singlet; position of line is concentration dependent.
Ar—OH	4.5 (without H-bonding) 9.0 (with H-bonding)	Position of signals dependent on solvent, concentration, and temperature.
R—CO$_2$H	10–13	Position largely independent of concentration.
R—NH$_2$, RNHR′	2–5	Usually sharp single line but may be broadened; position is solvent dependent; addition of a trace of acid causes it to shift to an intermediate position between that for RNH$_2$ and H$_2$O; addition of excess concentrated acid results in complete suppression of exchange caused by

		protonation and the disappearance of signals because of large coupling (triplet $J \sim 50$ Hz) with nitrogen ($I = 1$).
ArNH$_2$, Ar.NHR	3.5–6.5	Lines often broadened.
R.CONH$_2$	5.0–8.5	Very broad, often unobservable lines sharpened by addition of alkali; position is solvent dependent.
R.CO.NH.CO.R'	9.0–12.0	Often broadened lines, but sharper than those of primary amides.
R.SH	1.0–2.0	Often broadened.
S—H	5.0	
Ar.SH	3.5	
=NOH	10.0–12.0	Can be broad.

Note: OH and SH signals are removable by addition of D$_2$O, signals of amines (NH or NH$_2$) by addition of D$_2$O/DCl (trace), while those of amide and imide NH protons are removed by addition of D$_2$O/NaOD.

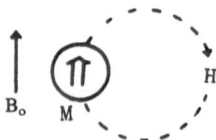

Figure 1.24. Paramagnetic moment caused by anisotropic distribution of charge deshields metal nucleus but shields protons.

1.3.1.7 Hydrogen Bonding

Hydrogen bonding results in downfield shifts of proton resonances from their positions in the unbonded state. The low-field values of phenols or carboxylic acids are attributed to hydrogen bonding. Increase of temperature or dilution with carbon tetrachloride can often break up hydrogen bonds and cause an upfield shift (in the hydroxyl resonance of ethanol, for instance).

1.3.1.8 Organometallic Compounds

Protons directly bonded to metals, as in transition metal hydrides, are strongly shielded and their resonances may appear as far upfield as -30 ppm. The shielding is due to the asymmetrical distribution of charge in the valence orbitals, which results in a strong diamagnetic shift of the proton resonances (Figure 1.24).

1.3.1.9 Contact Shifts

Organometallic compounds in which the metal is paramagnetic exhibit chemical shifts covering a range of 200 ppm for protons, and an even greater range for other nuclei. These shifts can occur by a *contact* or a *pseudocontact* interaction. In the contact interaction, some unpaired electron density is transferred from the metal to the organic ligand, which can cause shielding or deshielding effects, depending on electron spin correlation effects and electron distribution. Thus in the nickel complex depicted in Figure 1.25 the α-, β-, and

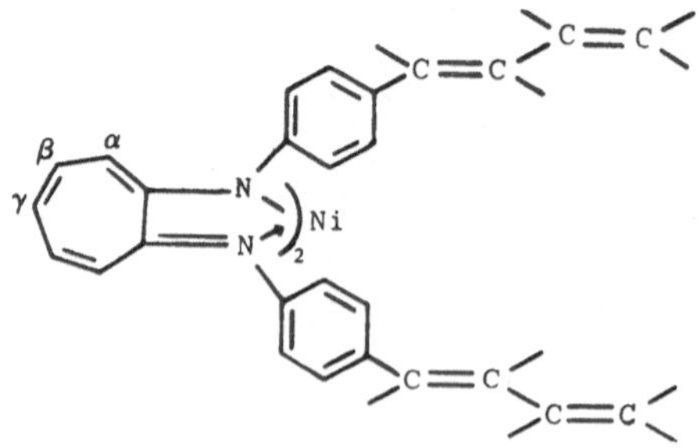

Figure 1.25. Complexation with the paramagnetic nickel ion results in the α, β, and γ protons indicated having chemical shifts of -42, $+47$, and -64 ppm, respectively.

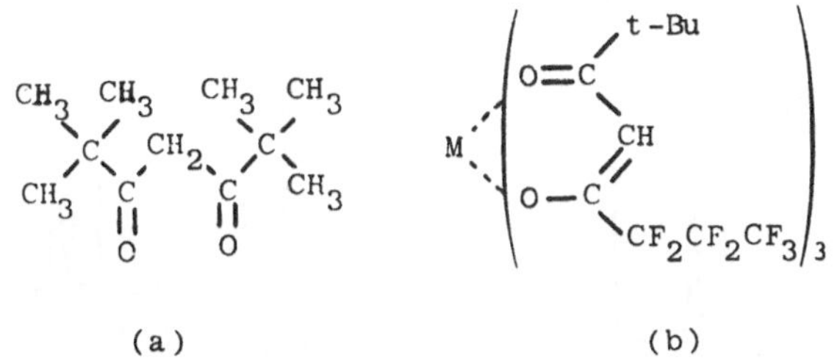

(a)   (b)

Figure 1.26. Complexation of diketones such as (a) with europium or praseodymium afford lanthanide shift reagents; (b) shows a heptafluorodimethyloctanedionato complex, M representing the complexing metal ion.

γ-protons resonate at $\delta -42$, $\delta +47$, and $\delta -64$, respectively. These large shifts are thought to arise because of donation of unpaired electrons from the paramagnetic ion (nickel in the complex cited) through the bonding atom (nitrogen) to the rest of the molecule. In order for such a shift to occur, a finite electron density at the nucleus has to be postulated. This type of electron–nucleus "contact" interaction was first described by Fermi.*

1.3.1.10 Lanthanide Shift Reagents

Another related interaction, known as the *pseudocontact interaction*, is observed when alcohols or amines, for example, are allowed to interact with certain rare earth metal complexes possessing unpaired electrons in their valence orbitals. The direction of the shift caused by interaction with the rare earth metal complex depends on the anisotropy of the magnetic susceptibility, X, and on the angle between the main axis of susceptibility and the vector R to the nucleus. If the proton is at a distance R from the center of the anisotropic group, and if the direction in which the proton lies makes an angle θ with the direction of the magnetic susceptibility, then the change in shielding, $\Delta\sigma$, is given by equation (1.7):

$$\Delta\sigma = \frac{1}{3R^3}(1 - 3\cos^2\theta)(X^{\parallel} - X^{\perp}) \qquad (1.7)$$

(the signs $\parallel$ and $\perp$ indicate the direction of the proton–metal axis with respect to the applied field B_0, i.e., parallel or perpendicular).

The complex formed from the diketone shown in Figure 1.26(a) and europium(III), i.e., the tris(dipivaloylmethanato)europium(III) complex [Eu(DPM)$_3$], is one such reagent which interacts with alcohols or amines to

---

* E. Fermi, *Z. Physik* **60**, 320 (1930); D.R. Eaton and W.D. Phillips, *Advan. Magn. Resonance* **1**, 103 (1965).

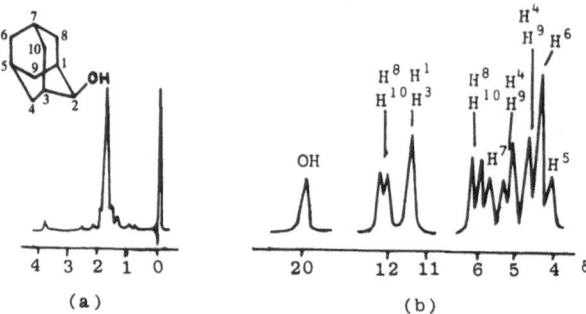

Figure 1.27. ^{1}H-NMR spectrum of 2-adamantanol: (a) normal spectrum; (b) spectrum in the presence of tris(dipivaloylmethanato)europium(III).

afford large downfield shifts resulting in great simplification in the interpretation of overlapping proton resonances. The ^{1}H-NMR spectra of 2-adamantanol before addition of the Eu(DPM)$_3$ complex [Figure 1.27(a)] and after addition of the complex [Figure 1.27(b)] can serve as an example.[†] The heptafluorodimethyloctanedionato complexes (FOD) of europium or praseodymium [Figure 1.26(b)] are also often used.

In general, the closer the proton is to the OH or NR$_2$ group interacting with the europium complex, the greater will be the downfield shift. With praseodymium complexes, shifts in the opposite direction are observed. Occasionally the nuclei may lie at an angle $\theta > 55°$ so that the sign of the factor $1 - 3\cos^2\theta$ changes, and shifts in the opposite direction are observed.

1.3.1.11 Charged Species

As chemical shifts are significantly affected by electron density, the presence of negative and positive charges in compounds causes shielding and deshielding, respectively, of protons attached to carbon atoms on which such charges are located. This is illustrated by comparison of chemical shift values of the cyclopentadienyl anion [Figure 1.28(a)] and the cycloheptatrienyl cation [Figure 1.28(b)]. Since all the protons in both species are equivalent, only one peak is observed in each case; in the spectrum of the cyclopentadienyl anion this singlet appears at δ 5.42 (about δ 1.85 *upfield* from benzene protons which resonate at δ 7.27) while in that of the cycloheptatrienyl cation the singlet appears at δ 9.17 (about δ 1.90 *downfield* from the benzene resonance), as shown in Figure 1.28. Similarly, the methine proton in the isopropyl cation [Figure 1.29(a)] is shifted downfield to δ 13.50 while the methyl protons on the adjacent carbon are also affected (δ 5.06). The effect of charges is dramatically illustrated when the NMR spectra of allyl cations and anions are compared [Figure 1.29(b), (c)]. In the case of the allyl cation the terminal methylenes are

---

[†] *Technical Bulletin*, No. 4, Varian Associates, Palo Alto, California.

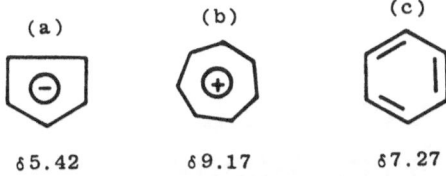

Figure 1.28. Upfield shift of aromatic proton resonances in anionic compounds and downfield shift of aromatic proton resonances in cationic compounds.

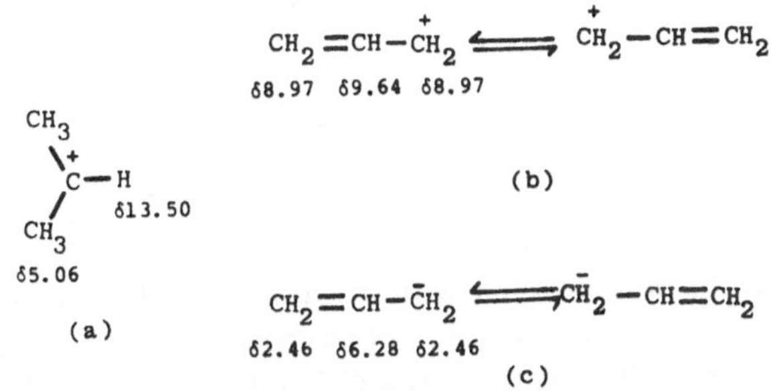

$$CH_2 = CH - \overset{+}{CH_2} \rightleftharpoons \overset{+}{CH_2} - CH = CH_2$$
$$\delta 8.97 \quad \delta 9.64 \quad \delta 8.97$$

$$\begin{array}{c} CH_3 \\ \diagdown \overset{+}{C} - H \\ \diagup \quad \delta 13.50 \\ CH_3 \\ \delta 5.06 \end{array}$$

(b)

(a)

$$CH_2 = CH - \overset{-}{CH_2} \rightleftharpoons \overset{-}{CH_2} - CH = CH_2$$
$$\delta 2.46 \quad \delta 6.28 \quad \delta 2.46$$

(c)

Figure 1.29. Effect of charges on the chemical shifts of some cations and anions.

deshielded by the positive charge and shifted downfield to δ 8.97 while in the anion they are shielded and therefore shifted upfield to δ 2.46. Both terminal methylenes are identical on account of the delocalization of charges.

RECOMMENDED READING

1. H. Günther, *NMR Spectroscopy*, John Wiley and Sons, New York (1980).
2. L.M. Jackman and S. Sternhell, *Applications of Nuclear Magnetic Resonance Spectroscopy in Organic Chemistry*, International Series in Organic Chemistry, Vol. 10, Pergamon Press, Oxford (1969).
3. F.A. Bovey, *Nuclear Magnetic Resonance Spectroscopy*, Academic Press, New York (1969).
4. E.D. Becker, *High Resolution NMR*, Academic Press, New York (1980).
5. M.L. Martin and G.J. Martin, *Practical NMR Spectroscopy*, Heyden and Son Ltd., London (1980).
6. T. Clerc and E. Pretsch, *Kernresonanz Spektroskopie*, Akademische Verlagsgesellschaft, Frankfurt (1973).
7. A. Carrington and A.D. McLachlan, *Introduction to Magnetic Resonance*, Chapman and Hall, London (1979).

Chapter 2

Spin–Spin Coupling in
¹H-NMR Spectroscopy

2.1 INTRODUCTION

So far we have been concerned with positions at which different protons resonate. These signals, however, do not necessarily occur as single peaks but may be split into doublets, triplets, quartets, or multiplets. Thus ethyl chloride (Figure 2.1) would afford a triplet for the methyl group and a quartet for the methylene protons. This splitting occurs because nonequivalent protons attached to the same carbon or to adjacent carbon atoms exhibit spin–spin coupling. The splitting of the signals does not affect the overall integrated intensities of the multiplets which still correspond to the number of underlying protons. The magnitude of splitting is expressed as the "coupling constant," J, in cycles per seconds or Hz. It is independent of the strength of the applied magnetic field but depends on the molecular stereochemistry and diminishes with, among other factors, an increasing number of bonds between the coupled protons. Coupling constants may be either positive or negative in sign, though by just looking at an NMR spectrum it is not possible to determine the sign of the J value.

Let us take the simple case of 2-bromo-5-nitropyrrole (Figure 2.2). The C-3 and C-4 protons appear not as single peaks but as two doublets (Figure 2.3), the low-field doublet arising from the C-4 proton and the higher-field doublet from the C-3 proton. The magnitude of splitting of the C-4 proton resonance (i.e., the coupling constant) is found to be identical to the magnitude of splitting of that of the C-3 proton.

$$\overset{\text{triplet}}{\text{CH}_3}\!-\!\overset{\text{quartet}}{\text{CH}_2}\!-\!\text{Cl}$$

Figure 2.1. Ethyl chloride; the methyl group resonates as a triplet while the methylene group resonates as a quartet.

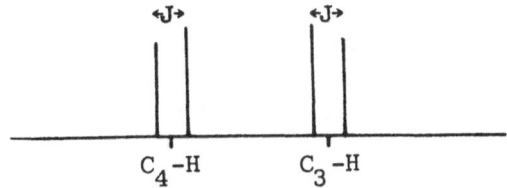

Figure 2.2. 2-Bromo-5-nitropyrrole.

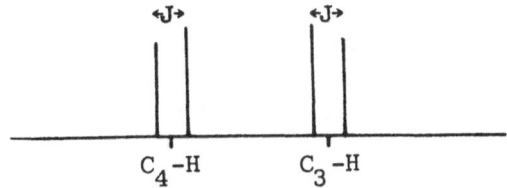

Figure 2.3. Splitting of resonances of C-3 and C-4 protons of 2-bromo-5-nitropyrrole due to spin–spin coupling.

Why does splitting occur? Let us consider the two vicinal C-3 and C-4 protons. The C-3 proton would give rise to a singlet if C-4 was not occupied by a proton but by another group. As has been mentioned earlier, protons have a spin quantum number of $\frac{1}{2}$, which means that they can exist in one of two possible magnetic states. The C-3 proton will therefore "see" the C-4 proton in one of these two possible states. In one of these states, the C-4 proton will be aligned with the applied magnetic field while in the other state it will be aligned against the applied field. The magnetic field experienced by the C-3 proton will be influenced by these two opposing magnetic states of the C-4 proton. When the C-3 proton sees the C-4 proton in the spin state in which the latter is aligned with the external field, it will experience a magnetic field slightly greater than the external field, and will therefore require a slightly lower value of the external applied field to come to resonance. Conversely, when the C-3 proton sees the C-4 proton in a spin state in which the latter is opposed to the applied field, the C-3 proton will experience a field which is slightly lower than the external field, and will therefore require a slightly higher value of the external field to bring it to a state of resonance. The C-3 proton therefore comes to resonance not at one value of the applied field (which would have resulted in the generation of a singlet) but at two different values, resulting in the recording of a doublet. Similarly, the magnetic field experienced by the C-4 proton will be influenced by the two magnetic states of the C-3 proton and the C-4 proton will also appear as a doublet. The extent to which the C-3 proton interacts with the C-4 proton is exactly equal to the extent to which the C-4 proton interacts with the C-3 proton, with the result that the two J values will be identical. The chemical shifts of the C-3 and C-4 protons would be the respective midpoints of the two doublets.

How is the influence of the magnetic state of one proton transmitted to a neighboring proton? As the local fields are averaged to zero by rapid tumbling of molecules, the magnetic effects cannot be transmitted through the interconnecting bonds. A number of mechanisms have been advanced to explain how protons existing in the two possible spin states can modify the magnetic fields experienced by adjacent protons. Let us consider the case of two protons A and B attached to two vicinal carbon atoms C and D. We have to proceed on the basis that (a) nuclear and electronic spins tend to be paired, i.e., have opposite spin, (b) that the two bonding electrons in a covalent bond also have opposite spins, (c) that if there are two bonding electrons between atoms A and B, then at any one time there is a high probability of finding one electron close to A and the other close to B, and (d) that the two electrons nearest to the carbon in separate bonding orbitals do *not* tend to be paired, but will have parallel spins.

Figure 2.4 shows two protons H_A and H_B attached to two adjacent carbon atoms. If B_0 represents the direction of the applied magnetic field, then in the lower-energy magnetic state proton *A* will have α spin, i.e., it will be aligned with the external field. Because of pairing of nuclear and electronic spins, electron c nearest to A will have the opposite β spin, and this will cause the second bonding electron d to have a spin (α) opposite to that of c. Electron e will tend to be parallel to electron d (i.e., have α spin) and, by the Pauli principle, be antiparallel to f. Similarly, electron g will be parallel to f but antiparallel to h. This will result in proton B having the opposite spin to that of A. One can therefore see immediately how the spin state of one proton influences the spin state of the proton on an adjacent carbon atom. Thus when proton A is aligned with the external field, proton B will be aligned against it, and vice versa. Because of this, the transition energies between the two spin states are altered, resulting in the formation of a doublet.

In the above example we have considered two protons attached to two "vicinal" (adjacent) carbon atoms. A similar picture can be drawn if both

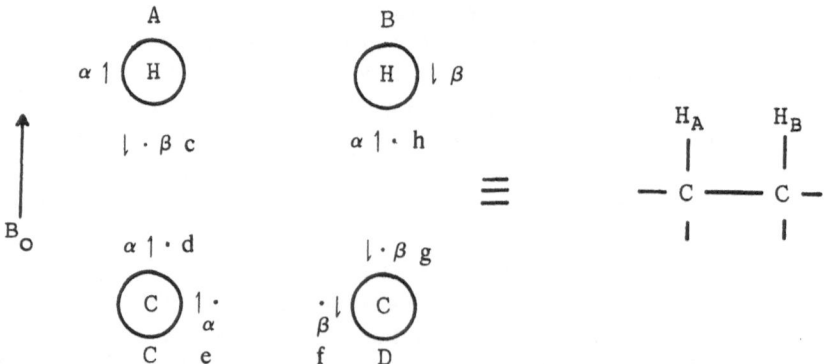

Figure 2.4. Relative spin orientations of bonding electrons in H—C—C—H system.

protons are attached to the same carbon, that is, for "geminal" protons. If H_A is in the lower-energy spin state, i.e., aligned parallel to the applied field B_0, then the bonding electron nearest to H_A will have the opposite spin due to the pairing of nuclear and electronic spins. The other bonding electron (nearest to the carbon) would therefore have the opposite spin (i.e., parallel to the applied field B_0). The electron in the second bonding orbital of the carbon atom would also have a parallel spin, this being the more stable arrangement. This would result in the bonding electron nearest to H_B having an antiparallel orientation, and H_B thereby having a parallel orientation with respect to the applied field (Figure 2.5). Thus H_A will end up having a parallel spin to H_B when both protons are attached to the same carbon atom (Figure 2.5) but an antiparallel orientation when they are attached to adjacent carbon atoms. The transmission of magnetic polarization described above is governed by the Pauli principle and Hund's rule.

An alternative way of looking at the phenomenon of spin–spin coupling emerges from a consideration of the energy level diagram of a two-spin system. If we have a nucleus A coupled to a nucleus X, then one can see four different orientations of the nuclear spins of A and X with respect to the applied field B_0: (a) both A and X can be parallel to B_0, (b) A can be parallel to B_0 while X can be antiparallel, (c) A can be antiparallel to B_0 while X can be parallel, and (d) both A and X can be antiparallel with respect to B_0 (Figure 2.6).

In the absence of spin–spin coupling, two transitions, A_1 and A_2, will occur for the nucleus A (Figure 2.7), and as both transitions are of equal energy, only a single resonance line will be observed for A. In the presence of spin–spin coupling, however, the eigenvalues of the spin system are altered as shown in Figure 2.7. The states (ii) and (iii) in which the spins of A and X are antiparallel to each other are by convention the stable arrangements, and as a result of spin–spin coupling the eigenvalues of both (ii) and (iii) are lowered (stabilized) while those of (i) and (iv) are raised (destabilized). This means that the transitions (A_1') and (A_2') are no longer of the same energy, as was the case previously, but instead A_1' becomes greater than A_2'. There will therefore be two different values of the applied magnetic field at which the nucleus A will come to the state of resonance, resulting in the formation of a doublet. An identical argument applies for X, which will also appear as a doublet.

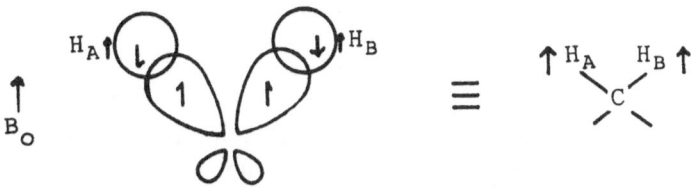

Figure 2.5. Relative spin orientations of bonding electrons in H—C—H system.

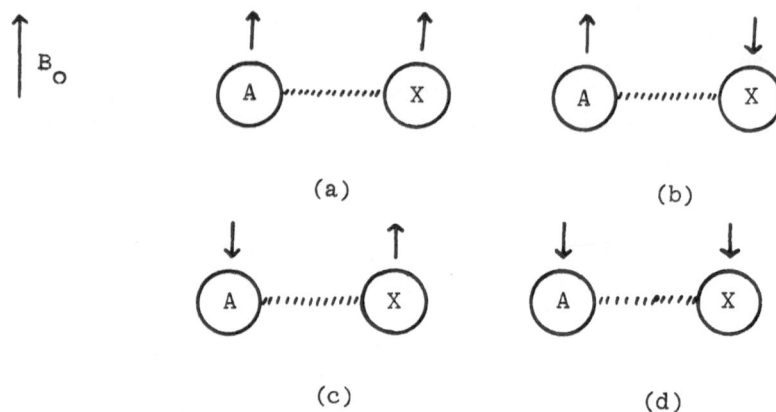

Figure 2.6. Relative orientations of the spins of nuclei A and X with respect to applied field B_0.

It was shown earlier that protons attached to adjacent carbon atoms tend to have antiparallel nuclear moments with respect to one another in the lower-energy state (Figure 2.4). By definition, the coupling constants between such protons are generally positive. On the other hand, protons attached to the

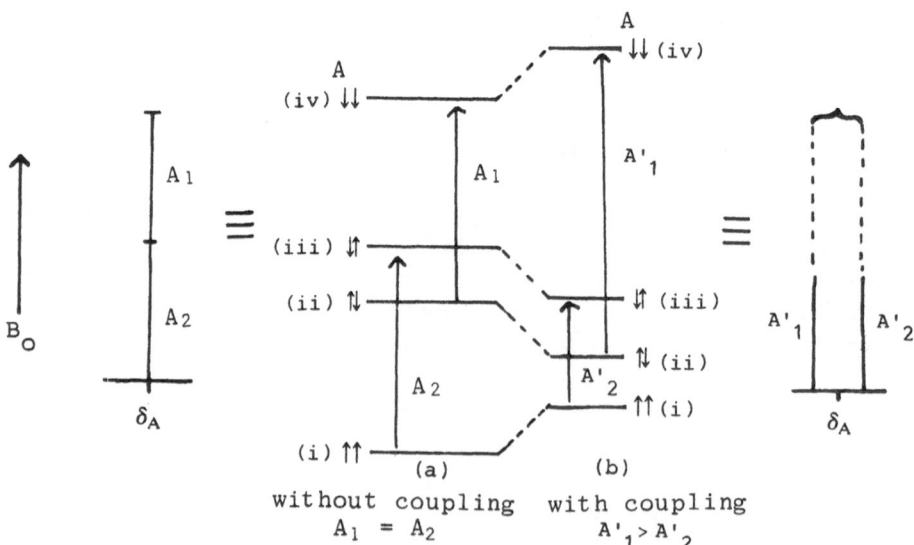

Figure 2.7. Effect of spin–spin coupling on the eigenvalues of coupled nuclei. (a) Without spin–spin coupling, the two transitions A_1 and A_2 are of the same energy and only one line is observed. (b) With spin–spin coupling the transitions A_1' and A_2' are of different energies and thus two lines appear in the NMR spectrum.

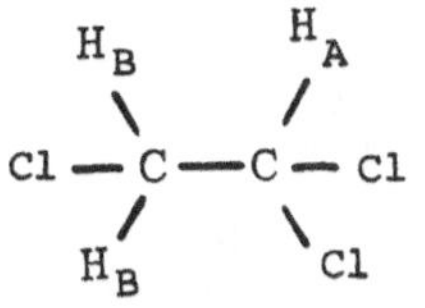

$$H_B \qquad H_A$$
$$\backslash \qquad /$$
$$Cl - C - C - Cl$$
$$/ \qquad \backslash$$
$$H_B \qquad Cl$$

Figure 2.8. 1,1,2-Trichloroethane.

same carbon atom tend to have their nuclear moments parallel to one another in the lower-energy state (Figure 2.5). The coupling constants between such geminal methylene protons are usually negative.

So far we have been concerned with cases in which there were only two protons coupled to one another. What happens if there are more than two protons coupled together? If we look at the NMR spectrum of 1,1,2-trichloroethane (Figure 2.8) we find that there is a doublet centered at δ 3.96 which integrates for two underlying protons while a three-line pattern appears centered at δ 5.78 in which the central line is twice as intense as the outer two lines. This "triplet" integrates for one proton (Figure 2.9).

The chemical shifts as well as the integration suggest that the downfield triplet is due to the methine proton A whereas it is the methylene protons B and B' which give rise to the doublet. The explanation for the CH_2 protons appearing as a doublet is identical to that provided earlier for the two-spin system, i.e., the vicinal methine proton (A) exists in two different spin states which influence the eigenvalues of the spin states of the two methylene protons (B and B'), resulting in the appearance of transitions between these states as a doublet. As both the CH_2 protons are identical to each other, only one doublet is actually seen, but its intensity is doubled so that it integrates for two protons. To understand why proton A appears as a triplet, one has to consider the orientations of the magnetic moments of the two adjacent protons B and B'. The nuclear spins of these methylene protons (see Figure 2.10) can exist in the following states: (a) both B and B' can be parallel, (b) B can be parallel while B'

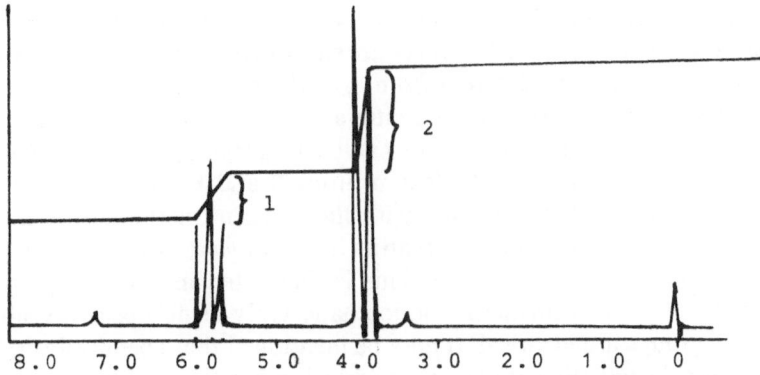

Figure 2.9. ^{1}H-NMR spectrum of 1,1,2-trichloroethane.

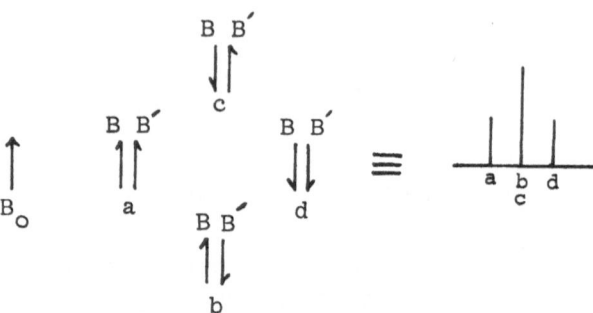

Figure 2.10. The relative orientations of the nuclear spins of the methylene protons in 1,1,2-trichloroethane which cause the methine signal to be split into a triplet with an intensity of 1:2:1.

can be antiparallel, (c) B can be antiparallel while B′ can be parallel, and (d) both B and B′ can be antiparallel. Since protons B and B′ are equivalent, the situations (b) and (c) are indistinguishable. The field experienced by the proton A is (a) slightly diminished when both the B and B′ protons are parallel but aligned against the applied field, (b) unmodified when B and B′ are antiparallel to one another, and (c) enhanced when B and B′ are parallel but aligned with the applied field. Thus the various orientations of the CH_2 protons will result in proton A experiencing three (and not four) different fields, resulting in the appearance of its resonance as a triplet. Moreover since states (b) and (c) of the B and B′ protons are equivalent, the central peak in the triplet of the A proton will be twice as intense as the outer peaks of the triplet.

Another way of looking at the splitting phenomenon is to consider proton A being first split into a doublet by proton B. Each of the two peaks of the doublet will then be further split into two more doublets by proton B′. However, since the two coupling constants are equal, i.e., $J_{AB} = J_{AB′}$, the two doublets will overlap in the middle to afford a triplet with a relative intensity of 1:2:1 (Figure 2.11).

It may be noted that the methine proton A will only appear as a triplet as long as the methylene protons B and B′ are equivalent, in which case J_{AB} will be equal to $J_{AB′}$. If J_{AB} was different from $J_{AB′}$, the resulting peaks (doublet of doublets) would not overlap in the center and a total of four peaks would be observed from which the two different J values (J_{AB} and $J_{AB′}$) could be deduced. It is also notable that the methylene protons B and B′ do not exhibit any coupling with each other but only with the adjacent methine proton. This brings us to the important observation that *coupling between chemically equivalent protons is not observed*. This is illustrated in Figure 2.12 which shows the NMR spectrum of 1,2-diiodoethane. Only a single peak is visible as both CH_2 groups are identical. It may be mentioned that chemically equivalent protons *do* actually couple with each other but the coupling cannot be observed.

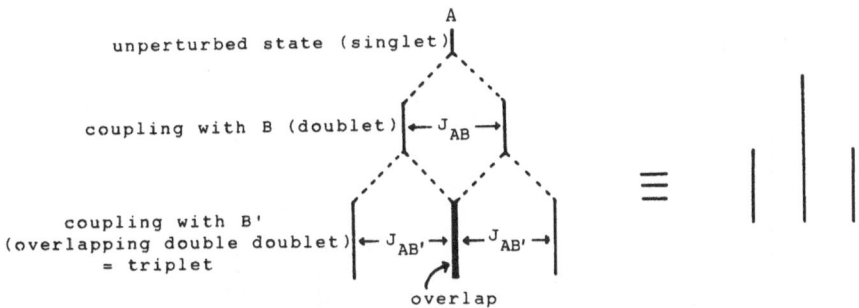

Figure 2.11. A "coupling tree" showing the coupling of A first with B and then with B′ protons to form a triplet. The central peak of the triplet is double the intensity of the outer peaks due to overlap.

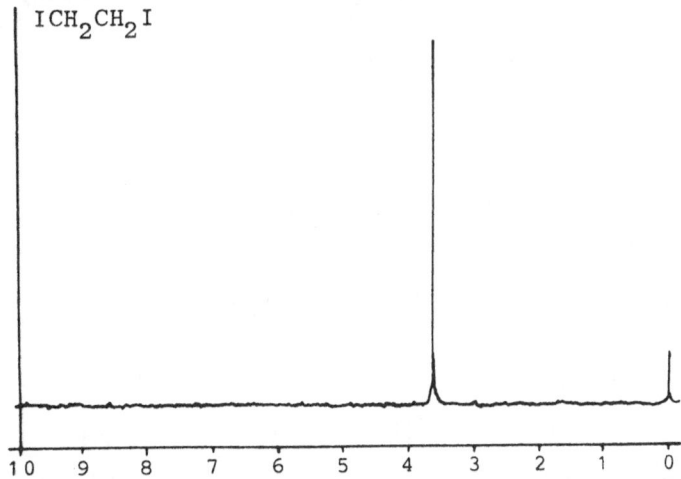

Figure 2.12. The methylene protons of 1,2-diiodoethane do not show any spin–spin coupling as they are chemically equivalent. All four protons therefore give only one peak.

In an ethyl group (CH_2—CH_3), as in iodoethane (Figure 2.13), the methyl protons appear as a triplet for reasons identical to those described above for the appearance of the H_A proton in 1,1,1-trichloroethane as a triplet. The only difference is that the triplet is now more intense, integrating for three underlying protons. As all three protons of the methyl group are equivalent, on account of the rapid rotation about the C—C bond, only one triplet is seen. The methylene protons on the other hand experience *four* different fields caused by the four possible combinations of the spins of the three protons of the methyl group. In Figure 2.14 all three methyl protons are aligned in (a) so as to reinforce the external magnetic field while in (d) all are aligned against the field to reduce its effect. In (b) two of the methyl protons are aligned with

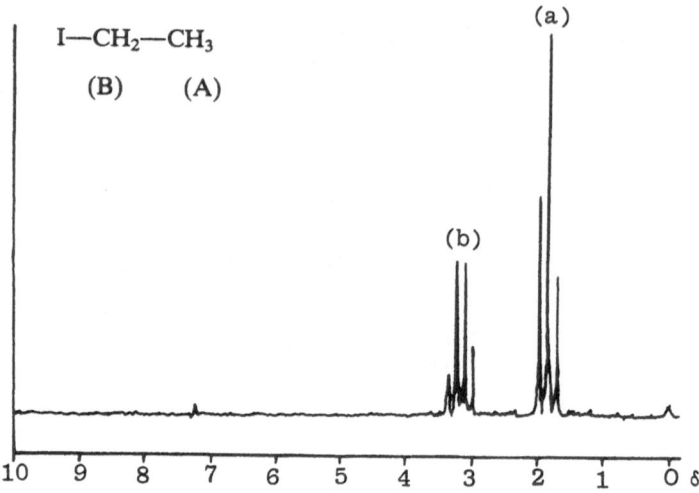

Figure 2.13. ¹H-NMR spectrum of iodoethane.

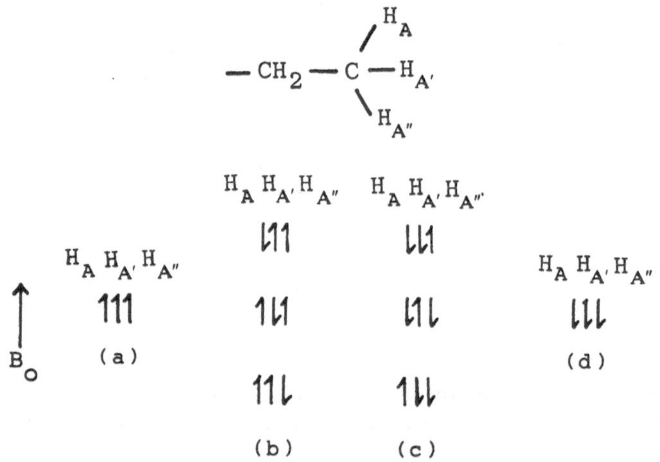

Figure 2.14. Four different orientations (a)–(d) of the methyl protons in an ethyl group which cause the methylene proton signal to be split into a 1:3:3:1 quartet.

the field while the third is aligned against it. In (c) the opposite situation to (b) is prevailing. The methylene protons therefore give rise to a quartet as they will resonate at four different modifications of the applied field. Since the number of combinations of orientations contributing to (b) and (c) is three times greater than the contributions to (a) or (d), the probability of encountering situations (b) and (c) is therefore three times as high as the probability of encountering situations (a) or (d). The quartet therefore appears with a relative intensity ratio of 1:3:3:1, the outer peaks being only one-third the intensity of the inner ones.

As a generalization one can say that if a proton (or protons) is coupled with n other *equivalent* protons, then it will give rise to $(n + 1)$ peaks. Thus in an ethyl group, the methylene protons are coupled to three adjacent equivalent protons and hence their resonance is split into $(3 + 1) = 4$ peaks. The methyl protons are coupled to the two equivalent methylene protons and their resonance is therefore split into $(2 + 1) = 3$ peaks. This generalization is correct only when the coupled nuclei have a spin of $\frac{1}{2}$. A wider generalization would be

$$S = 2nI + 1$$

where S is the number of lines formed by coupling, n is the number of equivalent nuclei which are causing the coupling, and I is the spin quantum number of the nuclei causing the coupling.

One should bear in mind that n and I refer to the nuclei with which the multiplet is coupled. Thus in considering the multiplicity of the CH_2 protons in the $-CH_2-CH_3$ group, n will refer to the number of protons with which the CH_2 protons are coupled, which in this case is three, whereas I will refer to the spin quantum number of the nuclei with which the CH_2 protons are coupled, which in this case is $\frac{1}{2}$.

A more general prediction for the intensities of the multiplets can also be made. They are related to the coefficients of the binomial expansion $(a + b)^n$. These can be easily deduced from the Pascal triangle, where each coefficient is the sum of the two terms above it.

Thus in isopropyl alcohol the methine proton is coupled with six protons. Since $n = 6$, these interactions will give rise to $6 + 1 = 7$ peaks which will have a relative intensity of 1:6:15:20:15:6:1 (follows from (g) in the Pascal triangle, Figure 2.15).

These rules are only valid when the chemical shift differences between the coupled protons are much greater than the coupling constants. The spectra are then said to obey the *first-order* approximation. When the chemical shift differences become smaller, more complex spectra are obtained and as some of the simplifying assumptions are no longer valid, the spectra are said to be *second order*.

In isopropyl alcohol the methine proton appears as seven peaks because the protons of both methyl groups are coupled to the $-CH$ proton with the same coupling constant. However, one can often encounter situations when a proton is coupled to neighboring protons with different coupling constants. In such cases, it may be stated that if a proton is coupled to n_a, n_b, n_c ... sets of chemically equivalent protons, it will afford a resonance with a multiplicity of $[(n_a + 1)(n_b + 1)(n_c + 1)...]$. Let us consider, for instance, a proton A which is coupled to one neighboring proton M ($n_a = 1$) with a coupling constant of 10 Hz and to two other neighboring protons, X and X' ($n_b = 2$) with a coupling constant of 7 Hz (Figure 2.16). According to the generalization given above, one would expect a multiplicity of $(1 + 1)(2 + 1) = 6$. This can be shown in the form of a "tree" (Figure 2.16). H_A is first split into a doublet by proton M. Each

													n

				1						(a)	0
		1	:			1				(b)	1
	1	:	2	:	1					(c)	2
1	:	3	:	3	:	1				(d)	3
1	:	4	:	6	:	4	:	1		(e)	4

1 : 5 : 10 : 10 : 5 : 1   (f) 5

1 : 6 : 15 : 20 : 15 : 6 : 1 (g) 6

Figure 2.15. Pascal triangle, showing the relative intensities of peaks of a multiplet arising from coupling with equivalent protons; n represents the number of equivalent protons with which coupling is taking place.

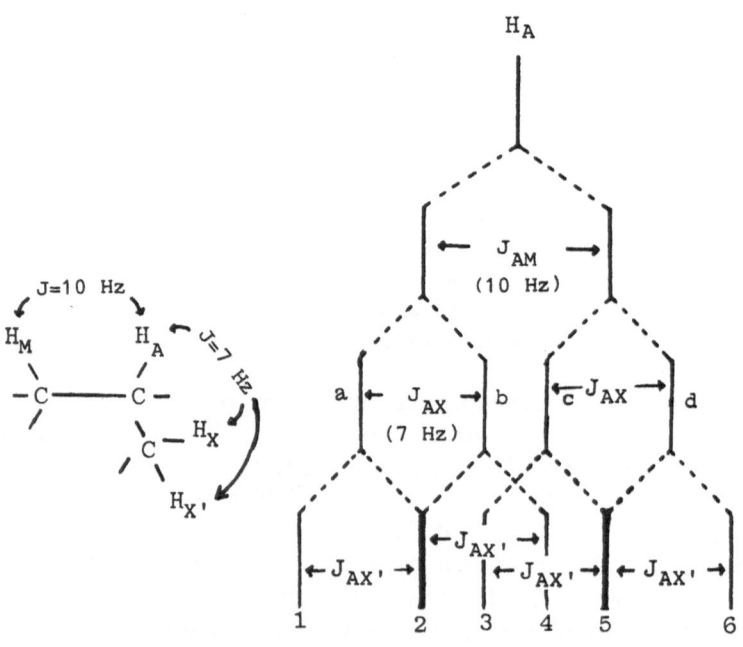

Figure 2.16. A coupling tree showing coupling of H_A with one proton H_M ($J_{AM} =$ 10 Hz) and two protons H_X and $H_{X'}$ ($J_{AX} = J_{AX'} = 7$ Hz). Note that overlap occurs at peaks 2 and 5 which results in their appearing with double the intensity of the other peaks.

of the two peaks are further split into two more peaks by coupling with H_X. Since J_{AM} is greater than J_{AX}, there is no overlapping of the central peaks. These four peaks are then again split by $H_{X'}$ into further doublets, but since $J_{AX} = J_{AX'}$, peaks 2 and 5 are formed by the overlapping of two peaks with the result that six peaks are observed with peaks 2 and 5 having twice the intensity of peaks 1, 3, 4, or 6. It is clear from this that if the coupling constant J_{AX} had been different from $J_{AX'}$, then no overlapping at peaks 2 and 5 would have occurred, resulting in the formation of eight peaks. In other words, H_A would have been regarded as having three different sets of neighbors, and n_a, n_b, n_c being equal to 1, the multiplicity would have been obtained from the product of $(1 + 1)(1 + 1)(1 + 1) = 8$. The chemical shift of A is the center point between lines 3 and 4 (Figure 2.16). The coupling constant $J_{AX} = J_{AX'}$ can be read from the distance in Hz between lines 1 and 2, while the distance between lines 1 and 3 provides J_{AM}. It may be noted that if the coupling tree shown in Figure 2.16 was drawn with the coupling to one of the other protons, say H_X, being shown first, it would have made no difference with regard to the end picture (Figure 2.17). The distance between lines 1 and 2 (or 5 and 6) in Hz gives the coupling constant of H_A with H_X and $H_{X'}$, respectively (J_{AX} and $J_{AX'}$). The distance between lines 1 and 3 or between 4 and 6 gives the coupling constant of H_A with H_M (J_{AM}).

If H_A was coupled to only two protons, say H_M and H_X, and not to three protons, and if the two coupling constants were different, then a four-line pattern (corresponding to lines a, b, c, and d in Figures 2.16 and 2.17) would

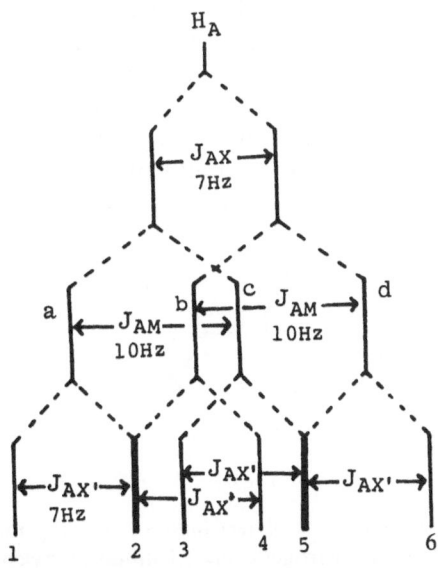

Figure 2.17. Same as Figure 2.16 but with J_{AX} shown before J_{AM} in the coupling tree. Note that the line positions and intensities remain unchanged.

have appeared. The two coupling constants J_{AM} and J_{AX} would have been deduced from the distance in Hz between lines a and c (J_{AM}), and between lines a and b (J_{AX}) (see Figure 2.17).

2.2 CHEMICAL AND MAGNETIC EQUIVALENCE OF NUCLEI

In the NMR spectrum of *tert*-butanol [Figure 2.18(a)], the protons present on each methyl group do not show any coupling with one another. This is because of the rapid rotation about the C—C bonds; if this rotation could be slowed down, by lowering the temperature, then the methyl protons would begin to show coupling among themselves.

Two or more nuclei are said to be *chemically equivalent* when they are in identical chemical environments. They will then experience identical shielding effects and, therefore, have the same chemical shift. *Magnetically equivalent* nuclei on the other hand will not only be chemically equivalent but also *have the same coupling constant with every other nucleus in the molecule*. Let us illustrate this by a number of examples. The methyl protons in *tert*-butanol, mentioned above, are chemically and magnetically equivalent. The two hydrogens H_a and H_b of 1,1-difluoroethylene [Figure 2.18(b)], however, are chemically equivalent but they are magnetically nonequivalent. Thus H_a is coupled to the fluorine nearest to it with one coupling constant (J_{cis}) and to the other fluorine with a different coupling constant (J_{trans}). The same situation prevails with H_b. Thus H_a and H_b do not satisfy our conditions for magnetic equivalence which is that in order to be magnetically equivalent they should be coupled to *all* other nuclei in the molecule with the *same* coupling constants. Similarly, in *p*-dibromobenzene the four hydrogen atoms are chemically equivalent but magnetically nonequivalent. Thus H_a is not magnetically equivalent to H_b in Figure 2.18(c) since the coupling constant between H_a and H_c is different from the coupling constant between H_b and H_c.

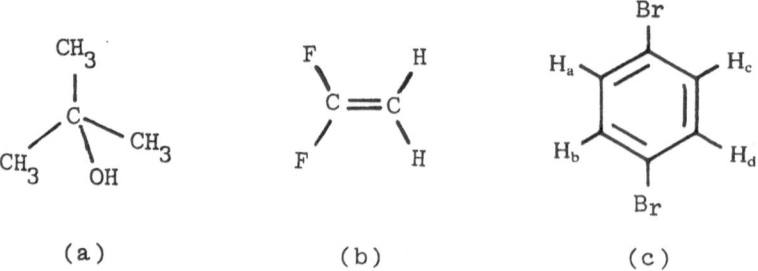

(a)   (b)   (c)

Figure 2.18. (a) Methyl hydrogens of *tert*-butanol showing magnetic and chemical equivalence. (b) Methylene hydrogens of 1,1-difluoroethylene showing chemical equivalence but magnetic nonequivalence. (c) Aromatic protons of *p*-dibromobenzene showing chemical equivalence but magnetic nonequivalence.

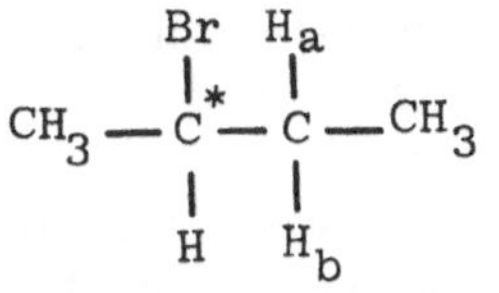

$$\begin{array}{c} \text{Br} \quad \text{H}_a \\ | \quad\; | \\ \text{CH}_3 - \overset{*}{\text{C}} - \text{C} - \text{CH}_3 \\ | \quad\; | \\ \text{H} \quad \text{H}_b \end{array}$$

Figure 2.19. 2-Bromobutane; H_a and H_b are diastereotopic protons α to an asymmetric center.

Two protons attached to a carbon atom which is α to an asymmetric center often exhibit nonequivalence. Let us consider protons H_a and H_b in 2-bromobutane (Figure 2.19). These protons are α to a chiral center (marked with an asterisk). In order to understand the reason for the nonequivalence, one has to examine the steric disposition of groups in the various possible conformers, as demonstrated by the Newman projections (Figure 2.20). As is usual in these projections, one looks down the C—C bond. The asymmetric carbon atom is drawn as the large circle at the rear and the three groups (Br, CH_3, and H) are seen emerging from it. The carbon atom to which H_a, H_b, and a CH_3 group are attached is represented by the point in the center. By keeping the disposition of the rear carbon atom constant and by rotating the carbon atom in front, one obtains the three staggered conformations shown in Figure 2.20. It is evident that H_a and H_b are never in the same environment. Thus in Figure 2.20(a) H_b lies between Br and CH_3, while H_a is between H and Br. If we now rotate the front carbon atom in a clockwise direction to bring H_a to the position that H_b occupies in Figure 2.20(a), we find that H_b now comes to lie between CH_3 and H, and not between Br and H [see Figure 2.20(b)]. Thus by rotation about the C—C bond, we find that one cannot arrive at any conformation in which the environments of H_a and H_b are identical.

Two protons attached to a carbon α to a chiral center therefore often appear at different chemical shifts and exhibit coupling with one another. Such protons are said to be *diastereotopic*. Moreover, when two protons are

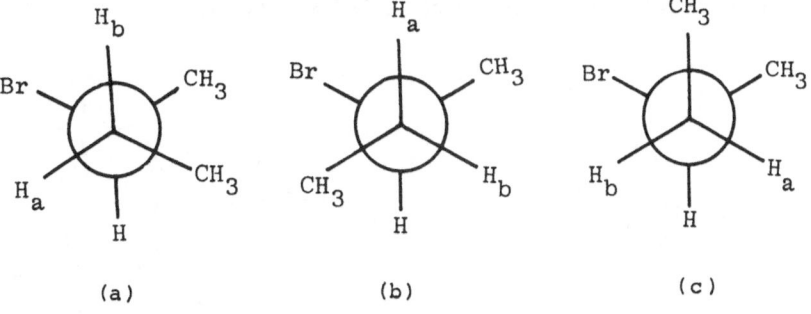

(a)   (b)   (c)

Figure 2.20. Newman projections of 2-bromobutane. Note that the front carbon atom (central point) is rotated while the rear carbon atom (circle) is kept stationary. The three staggered conformations show that H_a and H_b will continue to be in different spatial environments whichever conformation is adopted.

attached to a carbon atom to which two different groups are attached, they are said to be *enantiotopic*. Replacement of one of these protons with a third different group would create a chiral center, while replacement of the other proton would result in the formation of a mirror image (enantiomer) of the first chiral compound. The chemical shifts of such protons are identical in achiral solvents but may differ in chiral solvents. In other words, when two methylene protons are near a center of asymmetry they are said to be "diastereotopic" as their environments become nonequivalent. However, when the environment of one of the methylene hydrogen atoms is a mirror image of the environment of the other hydrogen atom, they are said to be "enantiotopic." Protons can experience an asymmetric environment, even if there is no chiral center in the molecule. Thus, in bromoacetal [Figure 2.21(a)] the methylene protons H_a and H_b of the ethoxy group exhibit nonequivalence. The carbon atom to which the ethoxy group is attached (marked with an asterisk) has three different groups attached to it (H, CH_2Br, and the other OC_2H_5) with the consequence that protons H_a and H_b are in an asymmetric environment. Such carbon atoms in which replacement of one of the two identical groups (OC_2H_5 in the case of bromoacetal) would lead to the formation of a chiral center are said to be *prochiral*. On account of their nonequivalence, the protons H_a and H_b in bromoacetal couple with each other. They can also resonate at different chemical shifts, resulting in the formation of two doublets (one for $H_{a'}$ and the other for $H_{b'}$). Each of these four signals would be further split by the adjacent methyl protons into quartets, so that a sixteen-line pattern should be observed. Actually only eight lines are observed due to overlapping in bromoacetal but all sixteen lines for the methylene protons can be seen in propynal acetal [Figure 2.21(b)].

Nonequivalence can also arise by restricted rotation imposed by steric or electronic factors. In such cases the affected protons will appear as separate peaks at lower temperatures, but as the temperature is raised the lines move closer together, and finally merge into a single peak. The methyl groups of dimethylformamide are nonequivalent at room temperature because of restriction of rotation about the N—CO bond due to the partial double

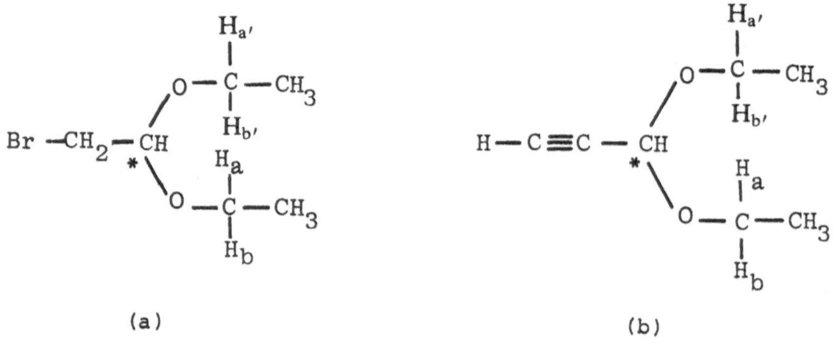

(a)   (b)

Figure 2.21. Bromoacetal (a) and propynal acetal (b). Prochiral carbon atom giving rise to nonequivalence of the methylene protons is marked with an asterisk.

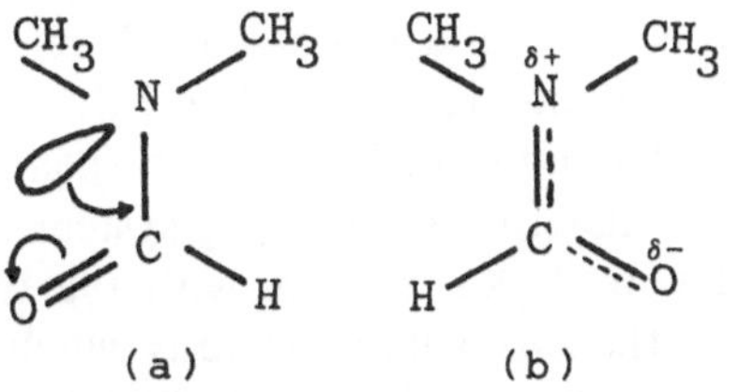

Figure 2.22. Partial double bond character of the bond connecting the nitrogen with the C—O group gives rise to nonequivalence of the methyl groups in dimethylformamide.

bond character in amides [Figure 2.22(a), (b)], and two separate signals for the two methyl groups are observed. At higher temperatures, however, the two peaks fuse into one six-proton singlet because of more rapid rotation at higher temperature.

2.3 DIFFERENT SPIN SYSTEMS

It is often convenient to refer to the type and number of nuclei of interest using the letters of the alphabet. Thus if three protons with close chemical shifts are coupled to each other, they are referred to as an ABC system. Two protons coupled to each other with very different chemical shifts constitute an AX system. If three protons are coupled to one another and one falls at low field, another at an intermediate value, and the third at a high-field position, they are termed as an AMX system. A system in which two identical protons are coupled to a third one which is close to the first two in its chemical shift is classified as A_2B. The A_2B_2 system similarly represents two identical protons (A_2) which are close in chemical shift and coupled to two other identical protons (B_2) [e.g., Figure 2.23(a)]. When two protons are chemically equivalent but magnetically nonequivalent, they are often differentiated by maintaining the same letter of the alphabet but placing a prime symbol on it (e.g., A'). Thus in p-chlorotoluene [Figure 2.23(b)], the two protons *ortho* to CH_3 (as well as those *meta* to it) are chemically equivalent but magnetically nonequivalent

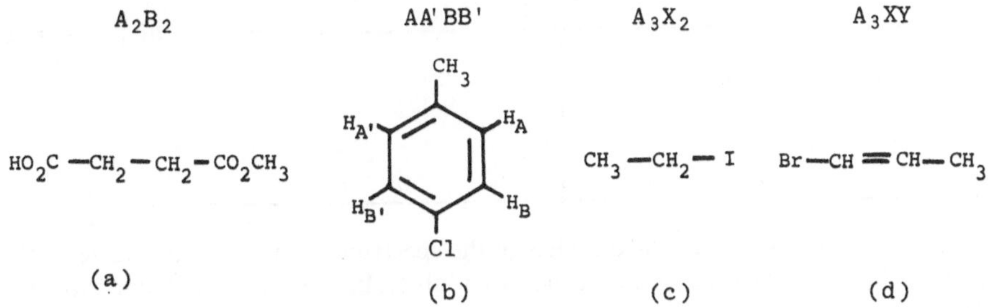

Figure 2.23. Typical A_2B_2, AA'BB', A_3X_2, and A_3XY systems.

(for reasons explained earlier), and this would therefore be regarded as an AA′ BB′ system. In ethyl iodide [Figure 2.23(c)] the three hydrogens of the methyl group are equivalent and well removed in chemical shift from the two methylene protons; this therefore is an A_3X_2 system. Bromopropene [Figure 2.23(d)] constitutes an A_3XY system, with the three identical methyl protons (A_3) coupled to the two olefinic protons which are close in chemical shifts to one another (hence X and Y) but well away from the methyl protons.

2.3.1 Two Coupled Nuclei (AX, AB, A_2 Systems)

2.3.1.1 AX and AB Systems

When the chemical shifts of two protons coupled to each other are very different, or more precisely when their difference in chemical shifts (in Hz) divided by the coupling constant in Hz ($\Delta\delta/J$) is greater than 7–10, then essentially first-order AX spectra are observed; the resonance of each proton is split into a doublet and the intensity of the lines is about the same. The

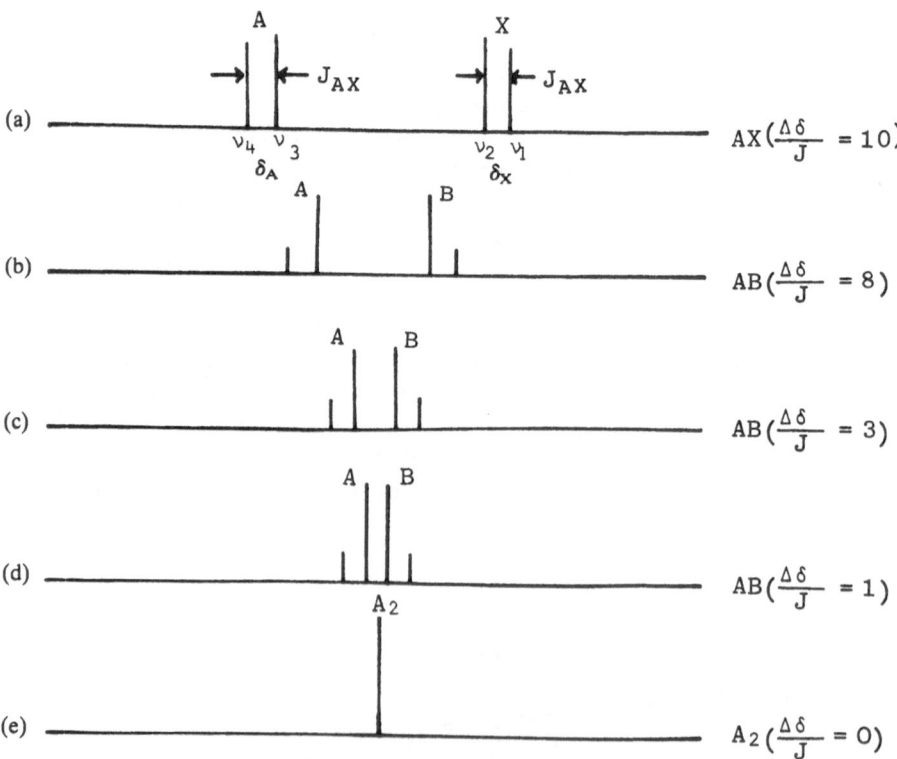

Figure 2.24. Distortions of the doublets in the spectrum of two coupled nuclei with decrease in $\Delta\delta/J$. The outer peaks of the two doublets become smaller as compared to the inner ones.

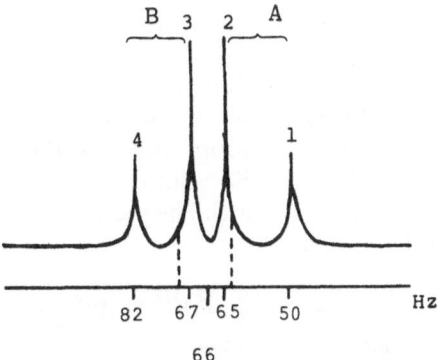

Figure 2.25. Chemical shifts of a distorted AB double doublet lie nearer the larger peaks (see dotted lines). Spectrum recorded at 60 MHz.

coupling constants can be read directly by measuring the separation of the lines in the doublets. The chemical shifts of A and X protons in the *AX system* are given by the midpoint between the two lines of each doublet [Figure 2.24(a)].

However, as $\Delta\delta/J$ becomes smaller, the inner lines of the two doublets become bigger and the outer lines become smaller. In such *AB spectra* [Figure 2.24(b)–(d)] the chemical shifts of the A and B protons are no longer at the midpoints of the doublets but are moved nearer to the inner larger peaks and away from the smaller outer peaks. The greater the distortion of the doublet, i.e., the more intense the inner peak as compared to the outer one, the closer is the chemical shift of the proton concerned to the inner peak. The chemical shift can be calculated in such cases from the relationship

$$\delta_A - \delta_B = [(v_4 - v_1)(v_3 - v_2)]^{1/2}$$

where v_1, v_2, v_3, and v_4 are the frequencies (in Hz) from TMS at which lines 1, 2, 3, and 4 appear.* Let us consider a typical AB pattern recorded on a 60-MHz instrument, shown in Figure 2.25. Suppose peak 1 appears at 50 Hz, peak 2 at 65 Hz, peak 3 at 67 Hz, and peak 4 at 82 Hz from the TMS signal. As mentioned earlier, the coupling constant J_{AB} can be immediately derived by measuring the distance (in Hz) between peaks 1 and 2 or between peak 3 and 4, since the distortion caused by lowering of the $\Delta\delta/J$ ratio affects the resonance frequencies but not the coupling constants. In order to deduce the chemical shifts of A and B (which are now *not* the center points of lines 3 and 4, or of lines 1 and 2) we apply the formula given above:

$$\delta_A - \delta_B = [(82 - 50)(67 - 65)]^{1/2} = [(32)(2)]^{1/2} = 8 \text{ Hz}$$

---

* A formal derivation of this equation is available in several standard texts, and it is therefore not repeated here.

As the center point of the AB pattern is at 66 Hz, and as δ_A and δ_B are 8 Hz apart and equal distances from the center, δ_A can be calculated to be 66 Hz − 4 Hz = 62 Hz, while δ_B will be 66 Hz + 4 Hz = 70 Hz from TMS. Now if we record the same spectrum at 400 MHz, since the chemical shift (but not the coupling constants) is proportional to the operating frequency, proton B will appear at (70 × 400)/60 = 466.6 Hz while proton A will appear at (62 × 400)/60 = 413.3 Hz from TMS. The precise positions of the four lines can be derived from the formula

$$\nu_4 - \nu_2 = \nu_3 - \nu_1 = [(\delta_A - \delta_B)^2 + (J_{AB})^2]^{1/2} = [(466.6 - 413.3)^2 + (15)^2]^{1/2}$$

$$= (3066)^{1/2} = 55.3 \text{ Hz}$$

Since the center of the pattern occurs at (466.6 + 413.3) Hz/2 = 439.95 Hz, and since the coupling constant remains unaltered at 15 Hz, one can work out the line positions at 400 MHz as follows. Since the distance between peaks 2 and 4 (or between peaks 1 and 3) is 55.3 Hz, and the distance between peaks 3 and 4 (or between peaks 1 and 2) is 15 Hz, the distance between peaks 2 and 3 must be 55.3 − 15 = 40.3 Hz. Since the center of the symmetrically disposed pattern falls at 439.95, peaks 2 and 3 must be 40.3/2 = 20.15 Hz away from the center. Thus peak 2 appears at 439.95 − 20.15 = 419.80 Hz, peak 1 appears at 419.8 − 15 = 404.8 Hz, peak 3 appears at 439.95 + 20.15 = 460.10 Hz, and peak 4 appears at 460.10 + 15 = 475.10 Hz (Figure 2.26). Moreover, relative intensities of the lines can be derived from the equation

$$\frac{I_2}{I_1} = \frac{I_3}{I_4} = \frac{\nu_4 - \nu_1}{\nu_3 - \nu_2} = \frac{475.1 - 404.8}{460.10 - 419.8} = \frac{70.3}{40.3} = 1.74$$

where I_1, I_2, I_3, and I_4 are the intensities of lines 1, 2, 3, and 4. The ratio of the intensities of the peaks in the 60-MHz spectrum were, on the other hand, given by

$$\frac{I_2}{I_1} = \frac{I_3}{I_4} = \frac{82 - 50}{67 - 65} = \frac{32}{2} = 16$$

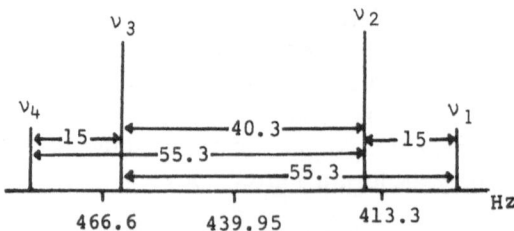

Figure 2.26. Second-order distortions decrease on more powerful instruments. The spectrum corresponds to that shown in Figure 2.25 but was recorded on a 400-MHz instrument. Note that the distance between the inner two peaks is now greater (40.3 Hz) and the distortion between the sizes of the outer and inner peaks is smaller.

One can immediately see that peak 2 (and 3) was 16 times as intense as peak 1 (and 4) on the 60-MHz instrument but on the 400-MHz instrument it is only 1.74 times as intense. Thus the distortion resulting in the inner peaks (2 and 3) becoming more intense than the outer peaks (1 and 4) is much more severe on the 60-MHz instrument than on the 400-MHz instrument. In other words, the "second-order" character of a spectrum decreases and the "first-order" character increases as one goes to more powerful instruments. Moreover, the separation of the inner pair of lines increases from 2 Hz to 40.3 Hz. Further, since the signal strength is dependent on the square of the frequency, the size of peak 3, for instance, at 400 MHz would be related to the size of the same peak at 60 MHz by the ratio $(460.1)^2/(67)^2 = 47.18$, approximately. The advantages of recording NMR spectra at higher MHz are thus clear from the above example: they become simpler with lower distortions due to reduction of second-order effects, and the signal-to-noise ratio improves significantly, thus reducing the recording time. This factor is of considerable importance, particularly in ^{13}C-NMR where the sample quantities can often be a serious limitation when recording spectra.

It may be noted that a "double doublet" obtained in an AB system should not be confused with a "quartet" obtained as a part of an AX_3 system. As mentioned earlier, the distances between the lines in a quartet will all be identical, and the relative intensities of the peaks will be in a ratio of $1:3:3:1$, though distortions in the intensities may occur as $\Delta\delta/J$ becomes smaller. In an AB system, on the other hand, the distance between the two central peaks is different from the distance between the outer and inner peaks.

2.3.1.2 A_2 System

In an A_2 *system* both nuclei are equivalent and therefore do not show any spin–spin coupling.

2.3.2 Three Coupled Nuclei (AX_2, AMX, ABX, ABC, AB_2 Systems)

2.3.2.1 AX_2 System

In the AX_2 system, the two X nuclei will afford a two-proton doublet while the A proton will afford a $1:2:1$ triplet. The chemical shifts and coupling constants are measured directly.

2.3.2.2 AMX System

In an AMX system, three nuclei with fairly different chemical shifts are coupled with each other with their three respective coupling constants. Since the $\Delta\delta/J$ ratio is large, the spectrum is essentially first order and each of the three nuclei will give a four-line pattern. Thus A will first be split by M into a doublet, and each of the peaks of the doublet will then be split by X into a double doublet.

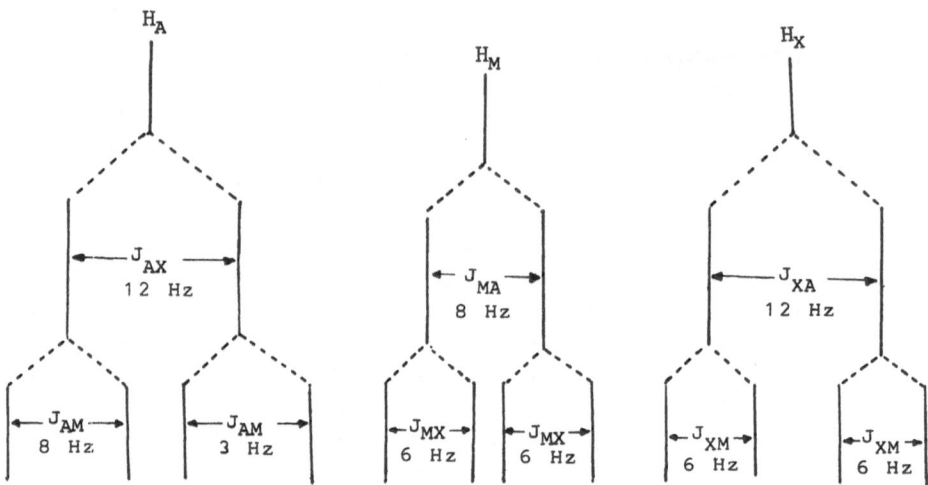

Figure 2.27. AMX system.

Similarly, M and X will give their corresponding double doublets by interaction with the other two nuclei so that twelve peaks will be seen all together. Thus if $J_{AM} = 8$ Hz, $J_{MX} = 6$ Hz, and $J_{AX} = 12$ Hz, then the splitting pattern shown in Figure 2.27 can be drawn from the corresponding tree.

All twelve lines will be of equal intensity and will occur in groups of four. In order to find out the two coupling constants in each double doublet, a simple procedure is to measure the distance between the first and third peaks to afford the larger coupling constant, and the distance between the first and second peaks to give the smaller coupling constant. Each coupling constant occurs twice. H_A is first split into a doublet by H_X ($J_{AX} = 12$ Hz) and then by H_M ($J_{AM} = 8$ Hz). Similarly, H_M is split by H_A and then by H_X ($J_{MA} = 8$ Hz, $J_{MX} = 6$ Hz), and H_X is split by H_A and H_M ($J_{XA} = 12$ Hz, $J_{XM} = 6$ Hz). If two of the coupling constants are accidentally equal, then one of the protons appears as a 1:2:1 triplet and only eleven lines are observable. If both coupling constants happen to be equal then the protons appear as three separate triplets and a nine-line pattern is visible.

2.3.2.3 ABX System

If the chemical shift differences between two of the three nuclei (A and B) are small as compared to their coupling constants (i.e., $\Delta\delta/J$ is small for H_A and H_B), while the difference between the chemical shifts of these nuclei and the third nucleus (X) is large as compared to the coupling constants between nucleus X and nuclei A and B, then one gets an ABX spectrum in which the AB portion of the spectrum can contain up to eight lines while the X portion can contain six lines. Figure 2.28 shows that the ABX system is an intermediate

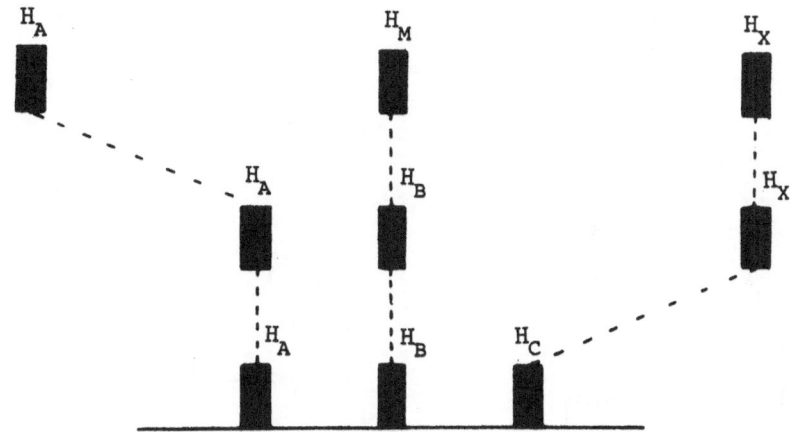

Figure 2.28. Chemical shift relationships between protons of an AMX system (*top*), ABX system (*middle*), and ABC system (*bottom*).

case between the AMX and ABC systems. Examples of ABX systems may be found in vinyl compounds such as that in Figure 2.29(a) in which the two geminal protons have close values of chemical shifts but the one attached to the same carbon as R has a distinctly different shift. Other examples of ABX systems are often encountered in substituted pyrroles, furans, or thiophenes [Figure 2.29(b)], trisubstituted benzenes [Figure 2.29(c)], or trisubstituted ethanes [Figure 2.29(d)]. As comprehensive discussions on the calculation of ABX and other more complex spectra have been published in other texts (see recommended reading list at the end of this chapter) only the salient features will be presented here.

On the basis of first-order arguments, one can draw the "coupling tree" for the ABX system shown in Figure 2.30. This would afford four lines for the X part which would be of equal intensity, and eight lines for the A and B protons. Note that the line separation is not necessarily equal to the coupling constant—hence S (separation) has been used to indicate the distances between the lines, rather than J values.

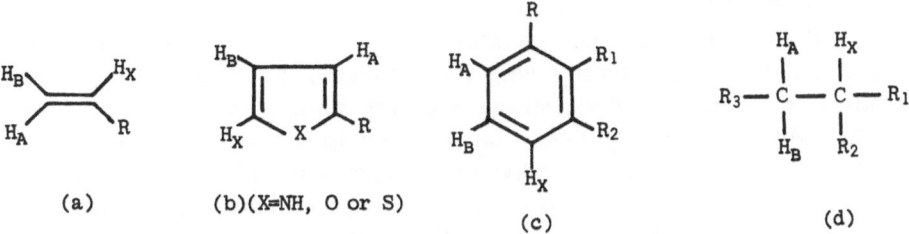

(a)   (b)(X=NH, O or S)   (c)   (d)

Figure 2.29. Some examples of ABX systems.

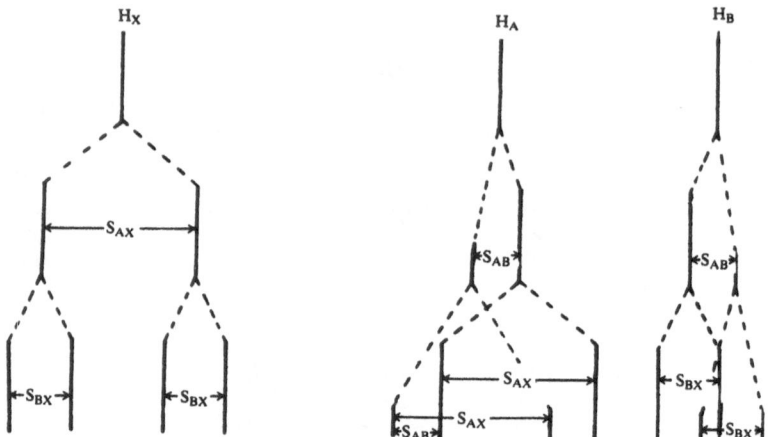

Figure 2.30. ABX system. Note the relative intensities of the two groups of peaks (four in each) for the H_A and H_B protons; they follow the order weak–strong–weak–strong for H_A and strong–weak–strong–weak for H_B (going from left to right).

In practice, due to second-order perturbation effects, the X proton may give rise to six lines, and the pseudoquartets from the A and B protons may partly overlap. The ABX patterns obtained are dependent on the signs of the coupling constants, J_{AX} and J_{BX}, but are unaffected by the sign of J_{AB}. The AB component of the ABX system may give rise to any of the three patterns shown in Figure 2.31.

Conventionally, one pseudoquartet is numbered 1, 3, 5, and 7 while the other pseudoquartet is numbered 2, 4, 6, and 8. The recognition of the lines in each pseudoquartet is facilitated by noting that the following relations apply in the peak positions: $v_3 - v_1 = v_4 - v_2 = v_7 - v_5 = v_8 - v_6 = J_{AB}$, so that J_{AB} can be read directly from the spectrum.

A useful procedure for analyzing ABX spectra and other more complex systems is that of *subspectral analysis*. This is based on the fact that many complex spectra can be seen to contain one or more simpler subspectra which readily lend themselves to separate analysis. Thus the AB part of the ABX spectrum can be divided into two AB-type subspectra.

The spectrum of vinyl acetate $CH{=}CH{-}OCOCH_3$ (Figure 2.32) corresponds fairly closely to the spectrum deduced in Figure 2.30 by application of first-order considerations. The X region shows four lines of approximately equal intensity while the AB region shows two AB quartets with a weak–strong–weak–strong intensity pattern just visible. More commonly, however, one finds that more complex spectra are obtained, depending on the difference between the chemical shifts of H_A and H_B ($v_A - v_B$) and the relative signs of the coupling constants, J_{AX} and J_{BX} (see Figure 2.31). The dependence of the ABX spectra on $v_A - v_B$ is illustrated in Figure 2.33.

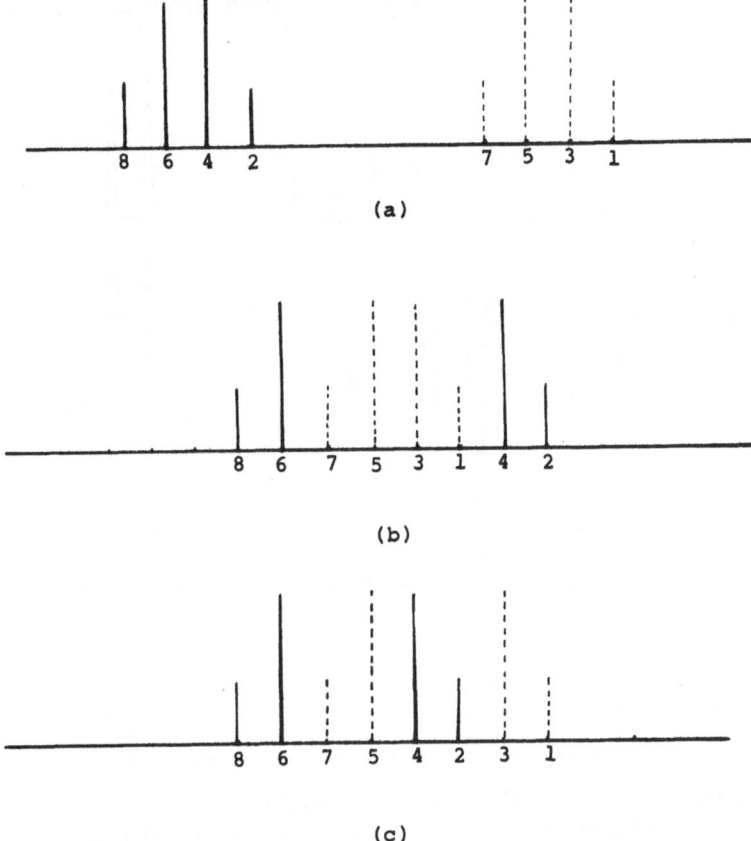

Figure 2.31. AB portion of an ABX system. (a) Relative signs of coupling constants, J_{AX} and J_{BX} are the same. (b) Relative signs of coupling constants, J_{AX} and J_{BX} are different. (c) Relative signs of J_{AX} and J_{BX} may be same or different.

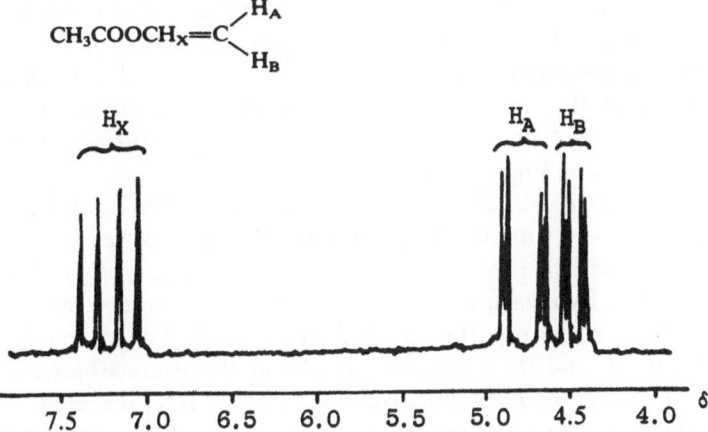

Figure 2.32. Olefinic region of the ^{1}H-NMR spectrum of vinyl acetate.

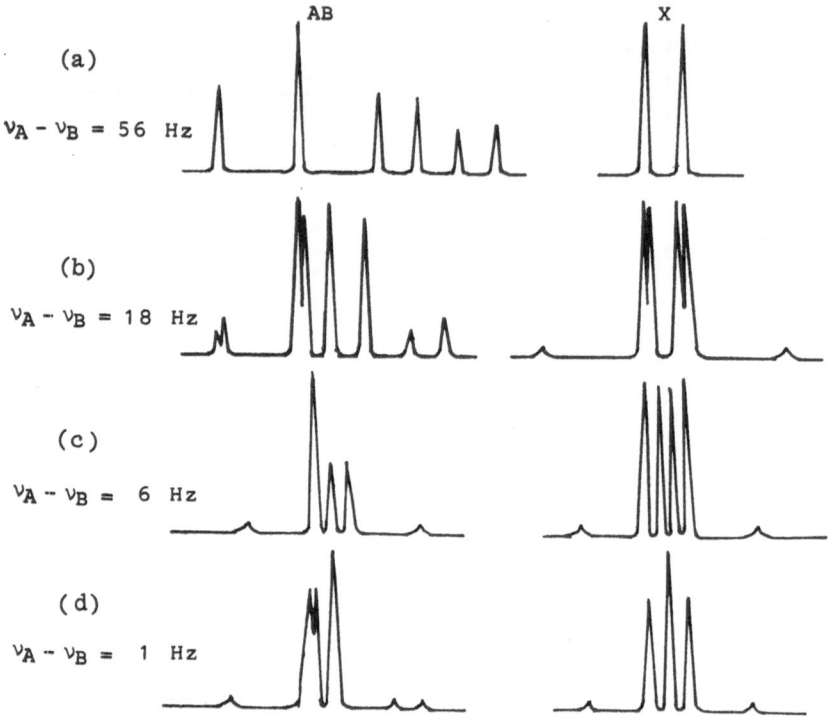

Figure 2.33. Illustration of the effect of decreasing difference between chemical shifts of H_A and H_B on the pattern of an ABX system ($J_{AB} = 16$ Hz, $J_{AX} = 0.2$ Hz, $J_{BX} = 8$ Hz).

The X region of the spectrum can afford a four-line pattern in which all the four lines may be of equal intensity and v_X is the center point of the pattern. If, however, a six-line pattern is obtained, v_X can still be directly obtained as the pattern is symmetrical about the midpoint. The frequencies and intensities of ABX spectra are shown in Table 2.1. It is notable that lines 9 and 12 are the two most intense of the six lines in the X part of the spectrum and that the separation between them is $J_{AX} + J_{BX}$. The six-line patterns obtained in the X region are shown in Figure 2.34.

The values that can be directly *measured* from the AB and X portions of the ABX spectrum are summarized in Figure 2.35. The distance between the midpoints of the two pseudoquartets is equal to $\frac{1}{2}(J_{AX} + J_{BX})$. This can be counterchecked by measuring the distance between the two strongest lines (lines 9 and 12) in the X part of the spectrum, which is equal to ($J_{AX} + J_{BX}$). The value of $2D_+$ and $2D_-$ can be measured from the distance between the first and third lines in each pseudoquartet (the larger is $2D_+$, the smaller is $2D_-$). This can also be checked from the separations of the lines in the X region (Figure 2.35). The measured values are then introduced into the following

Table 2.1. Frequencies and intensities of ABX spectra

Line no.*	Assignment	Energy (i.e., frequency)	Relative intensity
1	B	$\nu_{AB} + \tfrac{1}{4}(-2J_{AB} - J_{AX} - J_{BX}) - D_-$	$1 - \sin 2\theta_-$
2	B	$\nu_{AB} + \tfrac{1}{4}(-2J_{AB} + J_{AX} + J_{BX}) - D_+$	$1 - \sin 2\theta_+$
3	B	$\nu_{AB} + \tfrac{1}{4}(+2J_{AB} - J_{AX} - J_{BX}) - D_-$	$1 + \sin 2\theta_-$
4	B	$\nu_{AB} + \tfrac{1}{4}(+2J_{AB} + J_{AX} + J_{BX}) - D_+$	$1 + \sin 2\theta_+$
5	A	$\nu_{AB} + \tfrac{1}{4}(-2J_{AB} - J_{AX} - J_{BX}) + D_-$	$1 + \sin 2\theta_-$
6	A	$\nu_{AB} + \tfrac{1}{4}(-2J_{AB} + J_{AX} + J_{BX}) + D_+$	$1 + \sin 2\theta_+$
7	A	$\nu_{AB} + \tfrac{1}{4}(+2J_{AB} - J_{AX} - J_{BX}) + D_-$	$1 - \sin 2\theta_-$
8	A	$\nu_{AB} + \tfrac{1}{4}(+2J_{AB} + J_{AX} + J_{BX}) + D_+$	$1 - \sin 2\theta_+$
9	X	$\nu_X - \tfrac{1}{2}(J_{AX} + J_{BX})$	1
10	X	$\nu_X + D_+ - D_-$	$\cos^2(\theta_+ - \theta_-)$
11	X	$\nu_X - D_+ + D_-$	$\cos^2(\theta_+ - \theta_-)$
12	X	$\nu_X + \tfrac{1}{2}(J_{AX} + J_{BX})$	1
14	Comb. (X)	$\nu_X - (D_+ + D_-)$	$\sin^2(\theta_+ - \theta_-)$
15	Comb. (X)	$\nu_X + (D_+ + D_-)$	$\sin^2(\theta_+ - \theta_-)$

* Line 13 is of zero intensity, and is therefore not included.

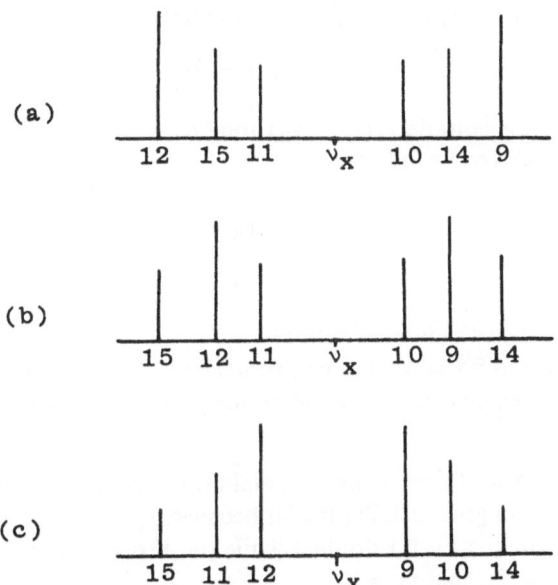

Figure 2.34. The X portion of an ABX system. The six-line pattern can vary in relative intensities as shown in (a), (b), and (c) but it is symmetrical about the midpoint, which corresponds to the chemical shift of X.

AB part of spectrum   **X part of spectrum**

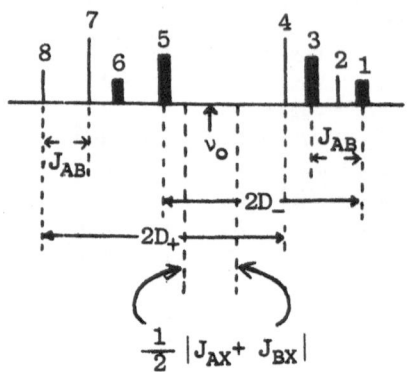
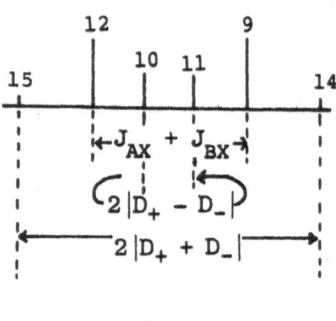

Figure 2.35. Measurable information in an ABX system. *Left*: AB part of spectrum; *right*: X part of spectrum. Line 13 is forbidden in the ABX case.

equations:

$$(v_A - v_B) + \tfrac{1}{2}(J_{AX} - J_{BX}) = (4D_+^2 - J_{AB}^2)^{1/2}$$

$$(v_A - v_B) - \tfrac{1}{2}(J_{AX} - J_{BX}) = (4D_-^2 - J_{AB}^2)^{1/2}$$

By solving these equations $(v_A - v_B)$ and $(J_{AX} - J_{BX})$ are obtained. v_A and v_B can then be derived as follows:

$$v_A = v_0 + \tfrac{1}{2}(v_A - v_B)$$

$$v_B = v_0 + \tfrac{1}{2}(v_A - v_B)$$

J_{AX} and J_{BX} can be calculated similarly.

Sometimes the ABX spectrum can be *deceptively simple* (i.e., one or more subspectra can degenerate into a single line), giving rise to one of the two following situations:

a. The AB region contains the normal eight lines for the two pseudoquartets but the X region gives a 1:2:1 triplet because $J_{AX} = J_{BX}$ *or* $v_A = v_B$.
b. The AB region gives a 1:1 doublet while the X region gives a 1:2:1 triplet because $J_{AX} = J_{BX}$ *and* $v_A = v_B$.

2.3.2.4   ABC System

Here the chemical shifts of the coupling protons are very close, and $\Delta\delta/J$ is hence small. ABC spectra are exemplified by the NMR spectra of trisubstituted benzenes, e.g., 1,2,4-trichlorobenzene, disubstituted pyridines, vinyl compounds, and substituted thiophenes, pyrroles, and epoxides. A total of

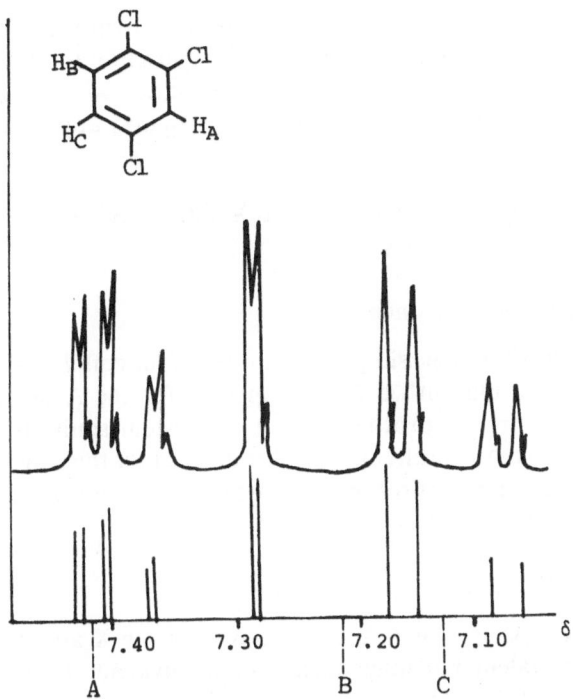

Figure 2.36. An ABC system illustrated by the ^{1}H-NMR spectrum of 1,2,4-trichloro-benzene. *Above*: observed spectrum; *below*: calculated spectrum.

fifteen lines may be obtained, and the relationships between these lines described above for the ABX system are also valid for the ABC system. The NMR spectrum of 1,2,4-trichlorobenzene serves to illustrate an ABC system (Figure 2.36). The analysis of an ABC system is difficult, and its discussion is beyond the scope of this book.

2.3.2.5   AB$_2$ System

When two magnetically equivalent protons are coupled to a third proton which lies close to them, and $\Delta\delta/J$ is small, the group of three protons is said to constitute an AB$_2$ system. Symmetrically substituted 1,2,3-benzene derivatives and monosubstituted allene derivatives provide good examples, and up to 9 lines can be obtained.

2.3.2.6   Virtual Coupling

Occasionally one comes across a situation where proton A is coupled to another proton, B, which in turn is coupled to a third proton, X. Since H_X is not coupled to H_X, H_X should on the basis of first-order arguments, afford only a doublet by coupling with H_B. However, if $\Delta\delta/J$ is small for H_A and H_B

(i.e., H_A and H_B are *strongly* coupled), then they may "mix their spins" and the first-order arguments no longer apply, resulting in a more complex splitting of H_X. This is known as *"virtual coupling."* In cases where H_A and H_B are chemically (but not magnetically) equivalent, H_X may appear as a triplet.

2.3.3   Four Coupled Nuclei (AX_3, A_2X_2, $AA'XX'$, A_2B_2, AB_3, $AA'BB'$, ABCD Systems)

2.3.3.1 AX_3 and A_2X_2 Systems

The AX_3 system affords the simplest example of four coupled nuclei, affording a first-order spectrum with a 1:3:3:1 quartet for the A proton and a 1:1 doublet for the three X protons. The A_2X_2 system similarly affords a 1:1 doublet for the two A protons and another doublet for the two X protons. The doublets for the A and X protons are mirror images of one another and are therefore easily recognized.

2.3.3.2   AA'XX' System

In the AA'XX' system, the A and A' (as well as the X and X') protons are chemically equivalent but magnetically nonequivalent. There are therefore only two chemical shifts (one for A and A', and the other for X and X') but four coupling constants ($J_{AA'}$, J_{AX}, $J_{AX'}$, and $J_{XX'}$). The entire spectrum can contain up to twenty-four lines but as four of the transitions are degenerate, only twenty lines can be observed at the most, ten for the AA' portion and the other ten for the XX' portion. The AA' section is symmetric about v_A while the XX' portion is symmetric about v_X and hence the chemical shifts are easily obtained. The AA' portion of an AA'XX' system is given in Figure 2.37. The following parameters can be measured from the spectrum:

(i) The separation of lines 1, 2 and 3, 4 $= J_{AX} + J_{AX'}$ (this is the most intense doublet).
(ii) The separation of lines 5 and 6 (or of 7 and 8) $= J_{AA'} + J_{XX'}$.
(iii) The separation of lines 9 and 10 (or of lines 11 and 12) $= J_{AA'} - J_{XX'}$.
(iv) Lines 5, 6, 7, and 8, and lines 9, 10, 11, and 12 form two quartets, each centered on v_A.
(v) The spacing of lines 5 and 7 (or of 6 and 8) gives $[(J_{AA'} + J_{XX'})^2 + (J_{AX} - J_{AX'})^2]^{1/2}$.
(vi) The spacing of lines 9 and 11 (or of 10 and 12) gives $[(J_{AA'} - J_{XX'})^2 + (J_{AX} - J_{AX'})^2]^{1/2}$.

It may be noted that the relative signs of $J_{AA'}$ and $J_{XX'}$ cannot be directly determined, nor can one decide from the spectrum alone which value obtained is $J_{AA'}$ and which is $J_{XX'}$. Similarly, one cannot distinguish between J_{AX} and $J_{AX'}$. The signs of J_{AX} and $J_{AX'}$ can, however, be determined by checking whether ($J_{AX} + J_{AX'}$) is larger or smaller than ($J_{AX} - J_{AX'}$).

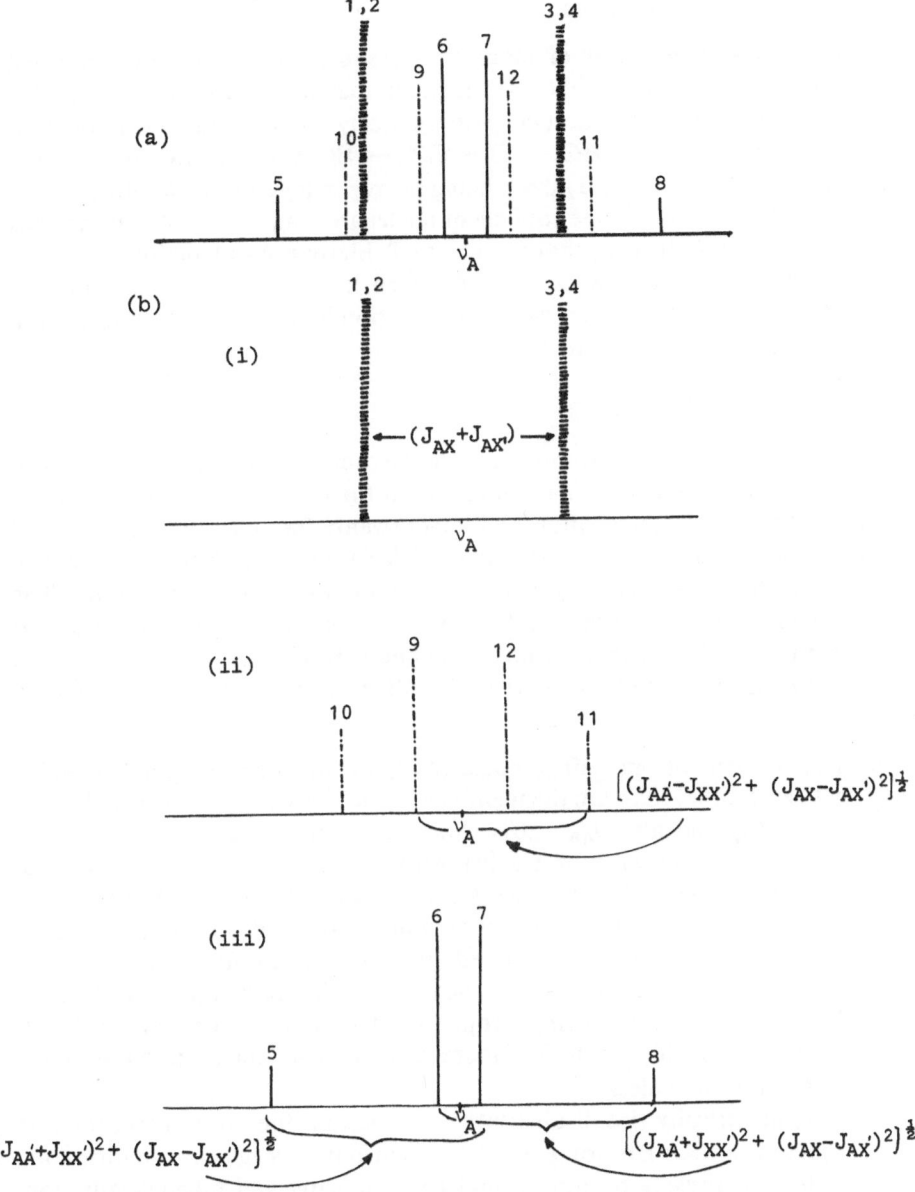

Figure 2.37. (a) AA'XX' system (AA' portion), with the following parameters: $J_{AA'}$ = 8.5 Hz, $J_{XX'}$ = 2.8 Hz, J_{AX} = 12.5 Hz, $J_{AX'}$ = 2.9 Hz. The XX' portion would be identical to the AA' portion. (b) The same but broken down into three subspectra (i), (ii), and (iii).

2.3.3.3 A_2B_2 and AB_3 Systems

When two different pairs of magnetically equivalent nuclei have chemical shifts which are close so that $\Delta\delta/J$ is small, they form an A_2B_2 system. This may be represented by the two pairs of freely rotating methylene protons in a molecule such as $HO—CH_2—CH_2—OR$. The spectra obtained are symmetrical, the A_2 and B_2 parts being mirror images of one another.

In the AB_3 system, three equivalent nuclei are coupled to a fourth nucleus which lies close to them so that $\Delta\delta/J$ is small. Methanol and methylmercaptan (CH_3SH) serve as examples. Up to sixteen lines can be obtained in AB_3 spectra, but two of these are weak combination lines and it is therefore more common to encounter up to 14 lines.

2.3.3.4   AA′BB′ System

The AA′BB′ system is encountered when the chemical shift difference between H_A and H_B is of comparable magnitude to the coupling constants between them. The AA′BB′ system affords a total of twenty-four lines (lines 1, 2 and 3, 4 are not degenerate as was the case in the AA′XX′ system). There are again two chemical shifts (since $v_A = v_{A'}$ and $v_B = v_{B'}$) but four coupling constants. The AA′ and BB′ portions of the spectrum are not independently symmetrical, but the entire AA′BB′ system is symmetrical about $\frac{1}{2}(v_A + v_B)$.

There are three main categories of AA′BB′ spectra which are commonly encountered:

a. The members of one pair of nuclei (AA′) are much more strongly coupled with each other than are the members of the other pair of nuclei (BB′), i.e., $J_{AA'} \gg J_{BB'}$ and also $J_{AB} \gg J_{AB'}$. This situation is usually seen in *ortho*-disubstituted benzenes or 1,4-disubstituted butadienes. The spectrum of *o*-dichlorobenzene (Figure 2.38) serves to illustrate this situation.

b. J_{AB} is much greater than all other coupling constants and $\Delta v_{AB} \gg J_{AA'}$ or $J_{BB'}$. This is usually encountered in *para*-disubstituted benzenes, particularly if the two groups attached to the ring have different electron-donating and -withdrawing properties. This is illustrated in the NMR spectrum of *p*-chloroanisole (Figure 2.39), in which the pattern appears as four distorted triplets.

c. Unsymmetrically 1,2-disubstituted ethanes in the *trans* conformation represent an AA′BB′ (or AA′XX′) system but in a *gauche* conformation all four methylene protons would be chemically and magnetically non-equivalent and would therefore represent an ABCD system. If M and N denote the two groups attached, then structures (a) and (b) in Figure 2.40 represent the two situations. The vicinal couplings J_{AB} and $J_{AB'}$ may vary considerably with the solvent, depending on the population of the various conformers.

The AA′BB′ system may be exemplified by the spectrum of a substance shown in Figure 2.41. The structure may be identified by the following reasoning: Microanalysis and accurate mass measurement on the molecular ion es-

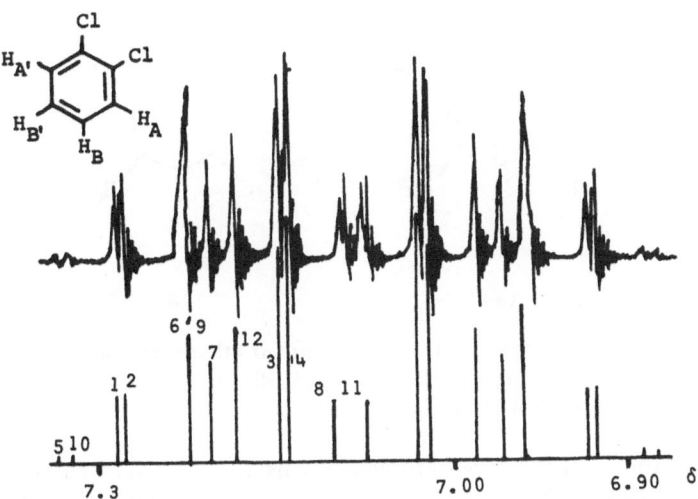

Figure 2.38. Observed (*above*) and calculated (*below*) ^{1}H-NMR spectra of *o*-dichloro-benzene.

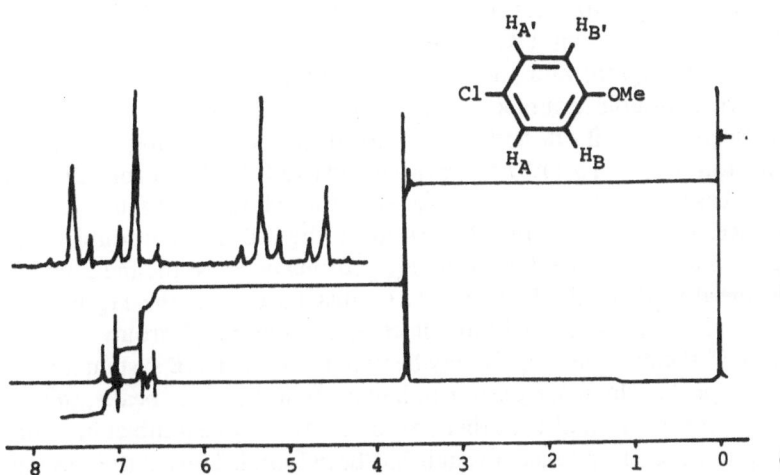

Figure 2.39. ^{1}H-NMR spectrum of *p*-chloroanisole. The aromatic region has been expanded for clarity. The four distorted triplets can be recognized.

tablished the formula of the compound to be $C_9H_{10}O_3$. Measurement of the height of the "steps" in the integration line shows that the relative intensity of signals at δ 3.57, δ 3.77, δ 6.94 (half of AB double doublet), δ 7.26 (other half of AB double doublet), and δ 11.83 is 2:3:2:2:1, respectively. The number of double bond equivalents (D.B.E) as calculated from the formula

$$\text{D.B.E.} = (C + 1) - \frac{H}{2} + \frac{N}{2} = (9 + 1) - \frac{10}{2} = 5$$

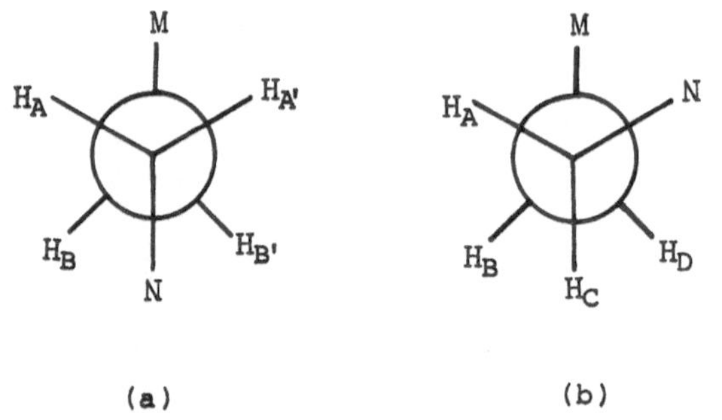

(a)   (b)

Figure 2.40. Conformations of a 1,2-disubstituted ethane drawn as Newman projections: (a) *trans* conformation affords an AA'BB' (or AA'XX') system; (b) *gauche* conformation affords an ABCD system.

comes out to be 5 (the oxygen atoms are considered when calculating double bond equivalents). Our knowledge of chemical shifts tells us that the two-proton singlet at δ 3.57 is consistent with a CH_2 group flanked by a benzene ring and a carbonyl or ester group (Shoolery's rules), while the three-proton singlet at δ 3.77 is characteristic of methoxyl protons. The symmetrically disposed group comprising the two double doublets at δ 6.94 and δ 7.26 are readily recognized as an AB system (actually AA'BB'). These peaks appear in the aromatic region, and the only substitution pattern that will afford a symmetrical double doublet corresponding to an AA'BB' system for the aromatic protons is if one had a *para*-substituted benzene ring. The downfield one-proton singlet at δ 11.87 can be assigned to a COOH proton. On the basis of this reasoning, we can assign the structure shown in Figure 2.42 to the substance. Further, the upfield portion of the AB double doublet, i.e., the signal at δ 6.94, must be due to the H_A protons *ortho* to the methoxyl group while the downfield doublet at δ 7.21 must be due to the H_B protons on account of the greater shielding effect of the methoxyl group at the *ortho* position of the benzene ring. It may be noted that in an AB system it is always the inner peaks which are greater in height than the outer peaks; this can be helpful in the location of the other half of an AB double doublet in a complex overlapping spectrum, once one half has been found. Thus in the spectrum of ω-nitrostyrene (Figure 2.43), the two olefinic protons are part of an AB system. The greater intensity of line A_2 relative to line A_1 indicates immediately that the other half of the AB double doublet is located to the *right* and not to the left of A_1; this is verified by measuring the distance between B_1 and B_2 and noting that it is identical to the distance between A_1 and A_2 (J_{AB}). This also shows that it is line B_1 and not line E which constitutes the outer line of the double doublet (since the B_2–E distance would be smaller than the A_1–A_2 distance).

2.3.3.5 ABCD System

When four interacting nuclei are magnetically and chemically nonequivalent, then ABCD spectra are obtained. These can be highly complex and are usually

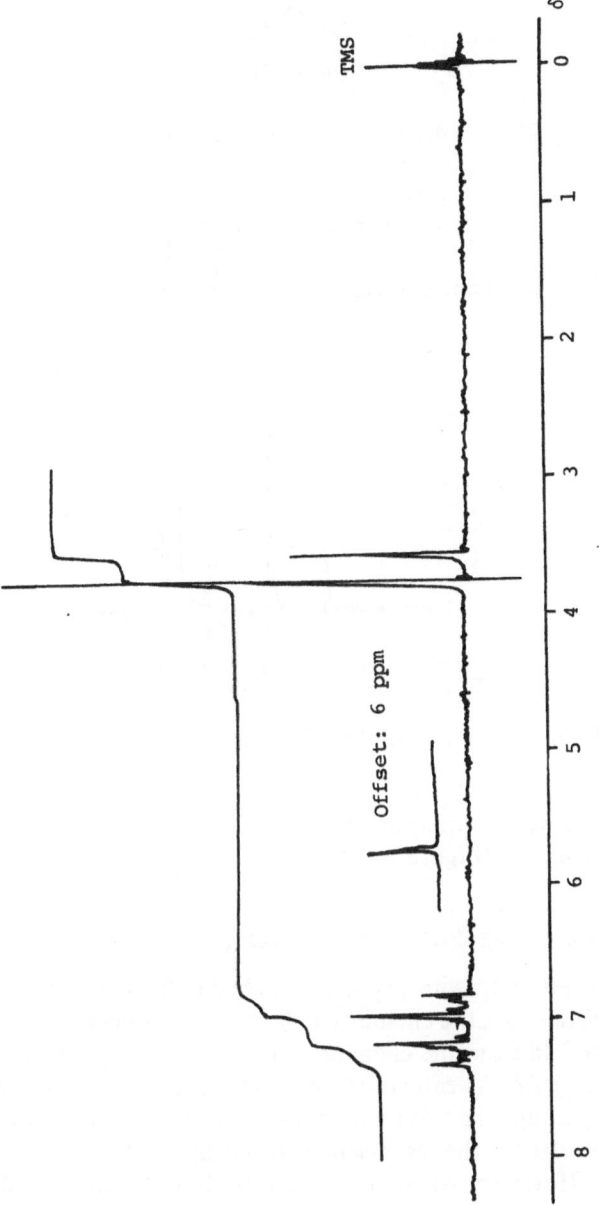

Figure 2.41. AA'BB' system (δ 6.7–7.4 region) illustrated by the ^{1}H-NMR spectrum of a *para*-disubstituted aromatic compound.

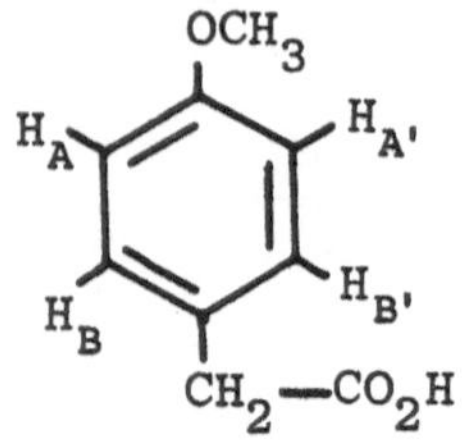

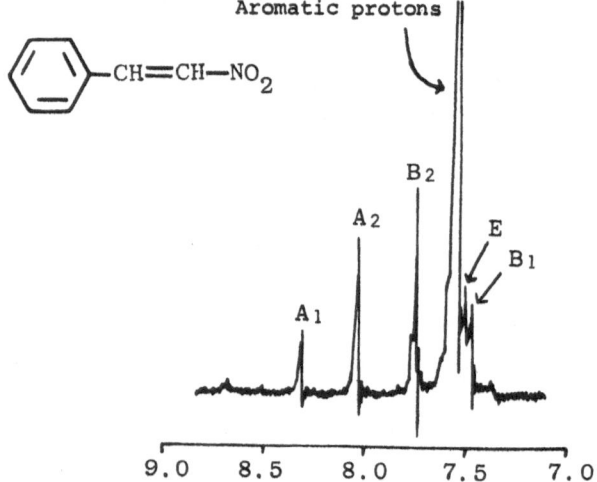

Figure 2.42. p-Methoxyphenylacetic acid.

Figure 2.43. Low-field region of the ^{1}H-NMR spectrum of ω-nitrostyrene.

not amenable to ready analysis. These are found in *ortho*-disubstituted benzenes, epoxides, etc. (Figure 2.44).

2.3.4   Five Interacting Nuclei (A_2X_3, A_2B_3, AA′BB′C)

When an electron-withdrawing group is attached to the methylene of an ethyl group (CH_3—CH_2—X), the chemical shift of the methylene protons is sufficiently far downfield from the chemical shift of the methyl protons so as to afford a fairly large $\Delta\delta/J_{AB}$ ratio (10 to 20) and a largely first-order A_2X_3 pattern results, i.e., an upfield 1:2:1 triplet for the methyl protons and a downfield 1:3:3:1 quartet for the methylene protons (e.g., CH_3—CH_2—OH discussed earlier). However, when the chemical shifts of the coupled nuclei are fairly close to one another, such as when another carbon atom is attached to the methylene of the ethyl group (CH_3—CH_2—CH_2—R), additional lines and distortions result in the A_2B_3 system. Theoretically 240 lines can result from five interacting nuclei with low $\Delta\delta/J$ ratios but usually most of

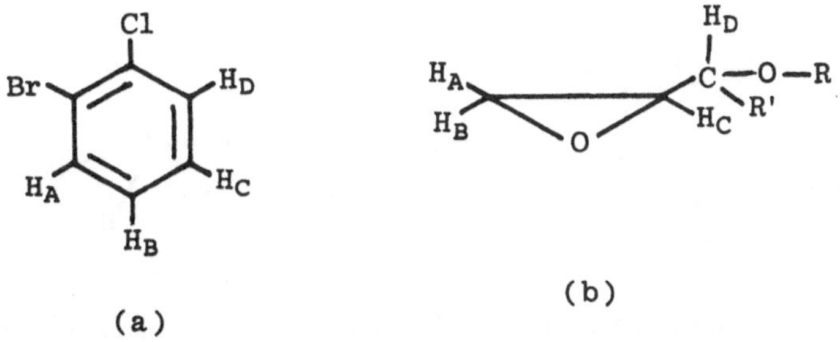

Figure 2.44. Some typical ABCD systems found in (a) disubstituted benzenes and (b) epoxides.

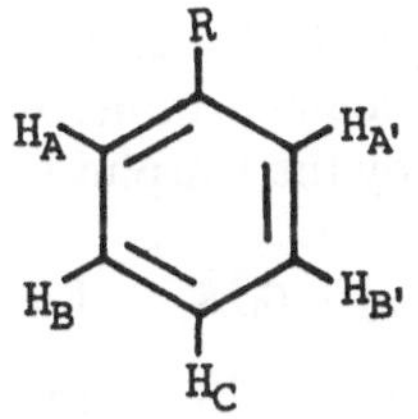

Figure 2.45. An AA'BB'C system exemplified by a monosubstituted benzene.

these are not observed. Ethyl alcohol shows some second-order hyperfine splitting of the lines of the quartet.

Monosubstituted benzenes (Figure 2.45) represent the AA'BB'C system, which again is highly complex and can afford up to 110 lines. These may occur in one or two groups of peaks.

2.4 FACTORS AFFECTING COUPLING CONSTANTS

Coupling constants between coupled protons can vary both in sign and in magnitude. Some of the more important factors affecting coupling constants are (i) dihedral angle between the C—H protons, (ii) electronegativity of substituents R on the H—C—C—H moiety, (iii) the H—C—C valence angles ϕ and ϕ', and (iv) the bond length between the carbons bearing the protons.

2.4.1 Dependence on Dihedral Angle

2.4.1.1 Vicinal Protons

If one looks down the C—C bond between the carbon atoms to which two coupled vicinal protons are attached, the angle which the protons make with

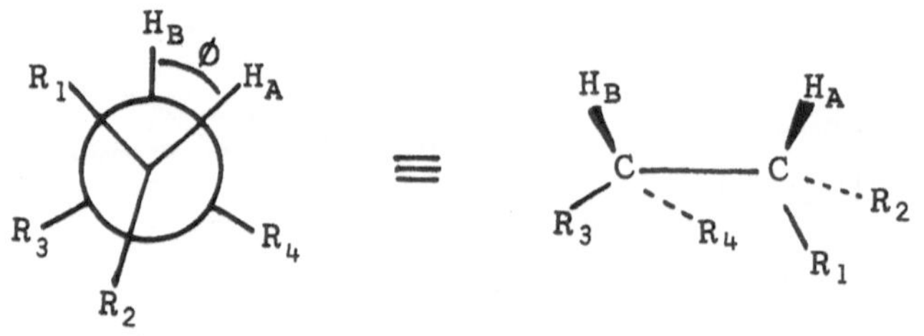

Figure 2.46. Angle θ between H_A and H_B represents the vicinal dihedral angle.

one another is known as the dihedral angle. A Newman projection shown in Figure 2.46 illustrates this angle. The near carbon atom is represented by a point to which H_A, R_1, and R_2 are attached while the rear carbon atom is represented by a circle to which H_B, R_3, and R_4 are attached. The dependence of the coupling constants of vicinal protons on the dihedral angle ϕ between them is given approximately by the Karplus equation:

$$J = (8.5 \cos^2 \phi) - 0.28 \qquad \text{for } \phi = 0°\text{--}90°$$

$$J = (9.5 \cos^2 \phi) - 0.28 \qquad \text{for } \phi = 90°\text{--}180°$$

Coupling constants between vicinal protons are usually positive in sign. A rough idea of how the dihedral angle affects coupling constants may be obtained from Figure 2.47, which shows that the coupling constants are largest

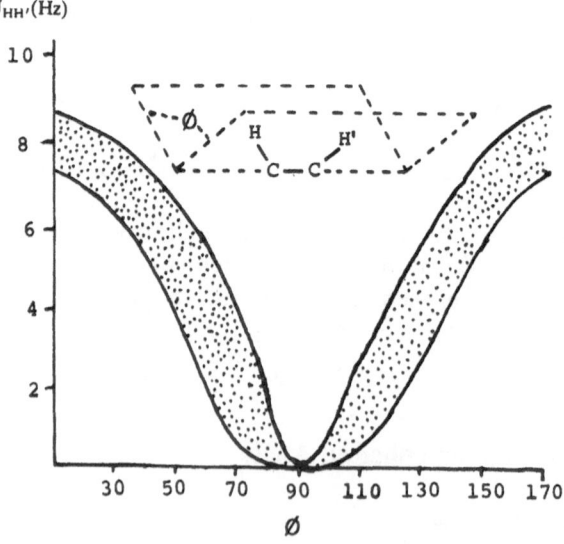

Figure 2.47. Dependence of coupling constant J of vicinal protons H and H' on the vicinal dihedral angle ϕ.

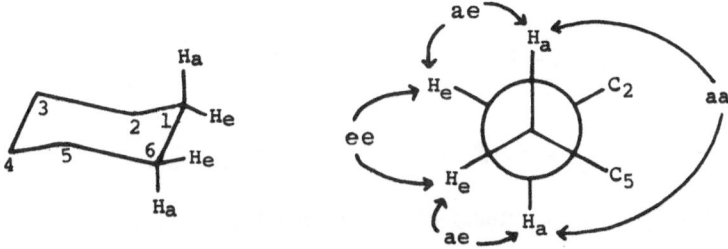

Figure 2.48. Newman projection of C_6 on C_1 in a cyclohexane ring showing axial–axial (aa), axial–equatorial (ae), and equatorial–equatorial (ee) interactions.

when the vicinal protons are *trans*-coplanar (i.e., $\phi = 180°$), zero when the respective H—C—C planes are at $90°$ to one another, and large when the protons are *cis*-coplanar (i.e., $\phi = 0°$).

This is of particular importance in assigning the stereochemistry of groups attached to cyclic systems, e.g., triterpenoids, steroids, etc. Thus in a cyclohexane system, the methylene protons can be either axial or equatorial. Since the dihedral angle is largest ($\sim 180°$) when the vicinal protons are diaxial to one another (i.e., angle between H_a on C-1 and H_a on C-6 is $\sim 180°$; Figure 2.48) the coupling constants are accordingly larger. When, however, the vicinal protons are in an axial–equatorial or equatorial–equatorial relationship, the dihedral angle is about $60°$ or less and the coupling constants are correspondingly smaller (Table 2.2).

A modified Karplus equation applied to vicinal coupling constants in olefins predicts a larger *trans* coupling than *cis* coupling. This is observed to be true; vicinal protons on olefinic bonds in a *trans* disposition ($\phi = 180°$) afford J values of 12–18 Hz while *cis* protons ($\phi = 0°$) afford J values of 7–10 Hz in acyclic systems (Figure 2.49).

2.4.1.2 Geminal Protons

The coupling constants between nonequivalent *geminal* protons, i.e., protons attached to the *same* carbon, are also dependent on the angle between them.

Table 2.2. Relationship between dihedral angle and vicinal coupling constants

Dihedral relationship	J (calculated using Karplus equation) (Hz)	J (observed) (Hz)
Axial–axial ($\phi_{aa} = 180°$)	9	8–14 (usually 8–12)
Axial–equatorial ($\phi_{ae} = 60°$)	1.8	1–7 (usually 2–3)
Equatorial–equatorial ($\phi_{ee} = 60°$)	1.8	1–7 (usually 2–3)

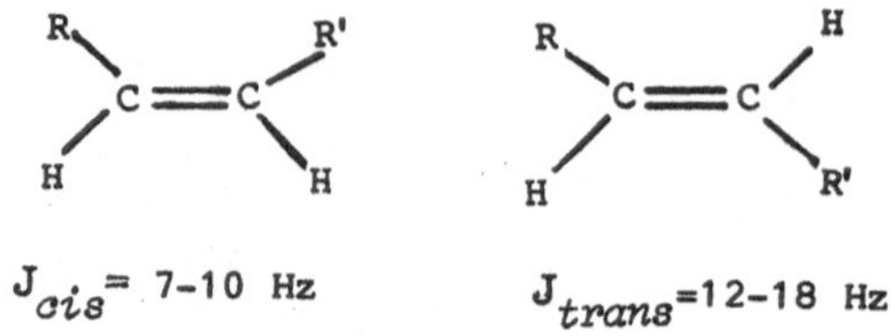

$$J_{cis} = 7\text{-}10 \text{ Hz} \qquad\qquad J_{trans} = 12\text{-}18 \text{ Hz}$$

Figure 2.49. Relative magnitudes of ¹H–¹H coupling constants in *cis* and *trans* olefins.

Thus methylene protons exhibit the following ranges of J values: terminal methylene groups $J = 0$ to 3 Hz; in cyclopropane rings, $J = -4$ to -9 Hz; in cyclobutanes, $J = -7$ to -14 Hz; in cyclopentanes, $J = -10$ to -14 Hz; in cyclohexane, $J = -12$ to -14 Hz (Figure 2.50). Coupling constants between geminal protons are usually negative in sign.

2.4.2 Dependence of Coupling Constants on Electronegativity of Substituents

2.4.2.1 Vicinal Protons

Vicinal coupling constants (^3J) decrease if an electronegative substituent is attached to the H—C—C—H moiety. The relationship between the electronegativity of a substituent X and the coupling constant between vicinal

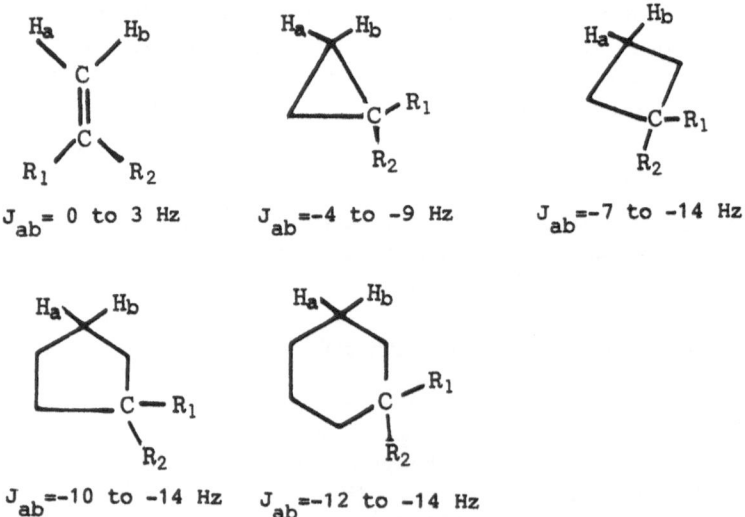

$$J_{ab} = 0 \text{ to } 3 \text{ Hz} \qquad J_{ab} = -4 \text{ to } -9 \text{ Hz} \qquad J_{ab} = -7 \text{ to } -14 \text{ Hz}$$

$$J_{ab} = -10 \text{ to } -14 \text{ Hz} \qquad J_{ab} = -12 \text{ to } -14 \text{ Hz}$$

Figure 2.50. Variation of ¹H–¹H coupling constants with dihedral angle for non-equivalent geminal protons.

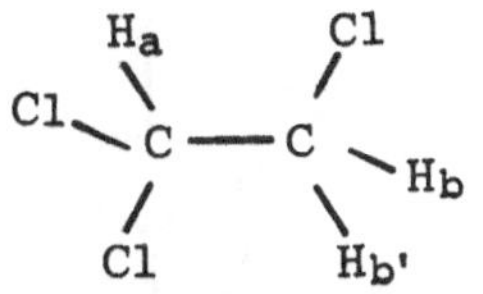

Ha   Cl
  \   /
Cl   C — C
  /   \
Cl   Hb
  \   /
 Cl   Hb'

Figure 2.51. 1,1,2-Trichloroethane.

protons on a freely rotating alkyl chain is given by the equation

$$J = 7.9 - (0.7\Delta E)$$

where ΔE is the difference in Huggins electronegativities of the substituent X and hydrogen ($\Delta E = E_X - E_H$). The effect is additive if more than one electronegative substituent is attached to the vicinal carbon atoms. Thus in 1,1,2-trichloroethane (Figure 2.51) there are three electronegative substituents attached, and the Huggins electronegativities of chlorine and hydrogen are 3.15 and 2.20, respectively; ΔE is therefore 0.95. The calculated value of J_{ab} would therefore be

$$J_{ab} = 7.9 - (0.7 \times 3 \times 0.95) = 5.9 \text{ Hz}$$

In cyclic systems the effect of the electronegative substituent X on the vicinal coupling constant is dependent on the steric disposition of X with respect to the coupled vicinal protons. The maximum effect of the substituent in reducing the vicinal coupling constant is observed when the substituent and a proton on the vicinal carbon are *trans*-coplanar. In Figure 2.52 the substituent X is in an equatorial disposition and is not *trans*-coplanar with respect to any of the coupled protons. However, in Figure 2.53 the substituent X and the axial proton attached to C-2 are *trans*-coplanar. Thus, while the dihedral angles between the protons (connected with thick lines) are about 60° in both cases, J_{ae} is 5.5 Hz (± 1 Hz) when the substituent X is equatorial

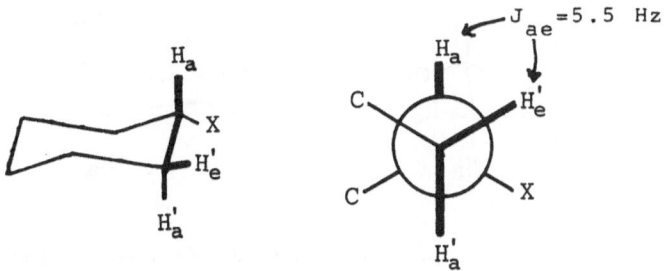

Figure 2.52. The electronegative substituent X is not *trans*-coplanar with respect to H_a' or H_e'.

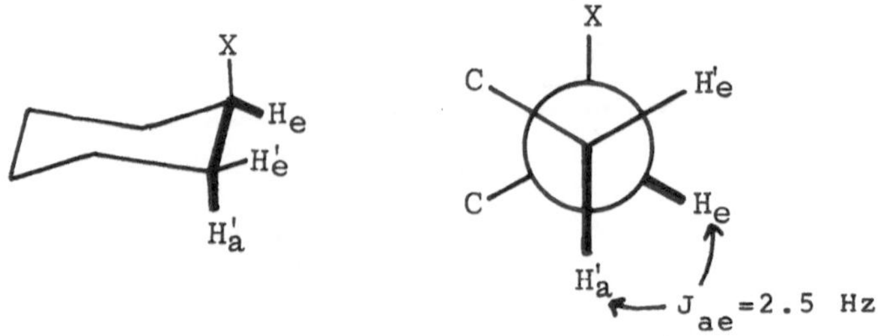

Figure 2.53. Substituent X and H'_a are *trans*-coplanar. Hence J_{ae} is reduced to 2.5 Hz.

(Figure 2.52) (X = OH, Br, OAc) but J_{ae} is only 2.5 Hz (± 1 Hz) when the substituent X is axial (Figure 2.53). It appears that the electron-withdrawing effect of X in reducing the coupling constants is most effectively transmitted through the bonds when X and one of the coupling protons are *trans*-coplanar. This is very useful information as it allows the stereochemistry of the substituent X to be deduced from the magnitude of the axial–equatorial coupling constant. The proton on the same carbon as X is shifted downfield on account of the electron-withdrawing effect of X and can usually be readily recognized. When X is in an equatorial disposition (Figure 2.52), larger coupling interactions J_{aa} and J_{ae} are discernible, but when X is in an axial disposition (Figure 2.53) smaller couplings J_{ae} and J_{ee} are obtained. This is illustrated by the coupling constants obtained between the axial or equatorial protons attached to carbon atoms bearing, say, an acetoxy group in cyclohexanes (Figure 2.54). The double doublet is often distorted on account of second-order perturbations but the width of the multiplet at half-height is still a useful indication of the orientation of X. Thus in steroids which often bear an —OH group at C-3, when the —OH is equatorial, the total width of the multiplet for the C-3 axial proton may be as high as 30 Hz, but if —OH is axial, then the width of the multiplet is much smaller (10–15 Hz).

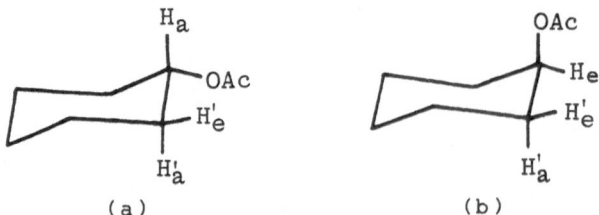

(a)   (b)

Figure 2.54. (a) Cyclohexyl acetate with equatorial —OAc group; larger coupling interactions (axial–axial ~10 Hz, axial–equatorial ~5 Hz) are observed. (b) Cyclohexyl acetate with axial —OAc group; smaller coupling interactions (equatorial–axial ≃ equatorial–equatorial = 2–3 Hz) are observed.

2.4.2.2 Geminal Protons

Geminal coupling constants (^2J) increase algebraically with increasing electronegativity of X in the system RCH_2X. Since they are usually negative in sign, their negative value decreases. However, if a substituent is attached to the carbon adjacent to the methylene group, the opposite effect is observed. This is illustrated in Tables 2.3 and 2.4.

The coupling among the equivalent geminal protons in methane or methyl chloride is not visible but the coupling constants can be calculated from data obtained after isotopic substitution. Replacement of one of the protons by deuterium results in H—D coupling being observed and the proton–proton coupling constant may then be obtained from the equation

$$J_{H,H} = 6.55 \, J_{H,D}$$

Neighboring π bonds make a negative contribution to the coupling constants, i.e., their absolute value increases. If the methylene system (or methyl group) is freely rotating and can adopt any of the possible conformations with respect to the adjacent π bond, then the negative contribution is about 2 Hz for each adjacent π system. Thus, as shown in Table 2.5, the ^2J value for the geminal protons in methane is -12.4 Hz with no adjacent π bonds, -14.5 Hz in toluene with one adjacent bond, -16.9 Hz in acetonitrile

Table 2.3. Effect of α-substitution on J_{gem}

Compound	$J_{gem}(Hz)$	Compound	$J_{gem}(Hz)$
CH_4	-12.4	(benzodioxole) CH_2	± 1.5
CH_3Cl	-10.8	(dioxolane) CH_2	0
CH_3F	-9.6	(aziridine) HN CH_2	$+2.0$
CH_2Cl_2	-7.5	(oxirane) O CH_2	$+5.5$
(dioxane) CH_2	-6.0	$RN{=}CH_2$	$+16.5$
		$O{=}CH_2$	$+42.2$

Table 2.4. Effect of β-substitution on J_{gem}

Compound	J_{gem}(Hz)
	−4.9
	−6.8
	−9.7
	+2.5
	−1.4
	−3.2

Table 2.5. Influence of adjacent π bonds on J_{gem}

Compound	J_{gem}(Hz)	Number of adjacent π bonds
CH_4	−12.4	0
	−14.5	1
CH_3—C≡N	−16.9	2
	−20.4	4

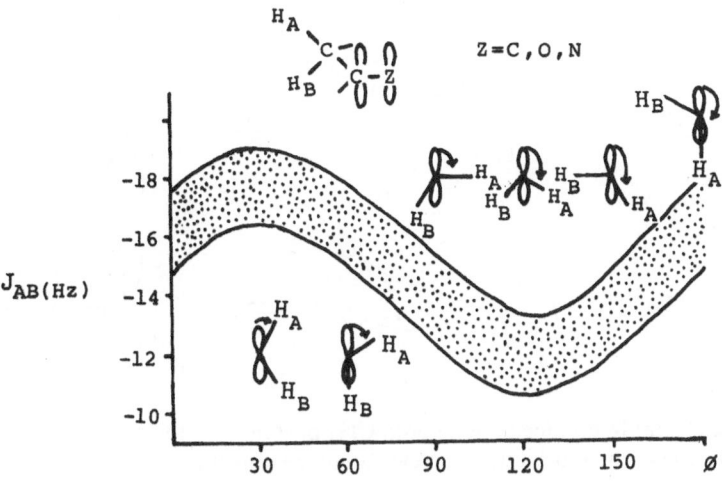

Figure 2.55. The effect of variation of the angle ϕ between the π-orbitals and the neighboring proton H_A on the geminal $H-H$ coupling constant J_{AB}.

with two adjacent π bonds, and -20.4 Hz in malononitrile with four adjacent π bonds.

In ring systems, the methylene protons have a rigid disposition with respect to the adjacent π system and it has been found that the geminal coupling constant is then dependent on the angle ϕ between the C—H bond and the π bond. This is illustrated in Figure 2.55 which shows that the largest effect is observed when both methylene protons lie on one side of the π-orbital, i.e., if the plane P_1 in which the methylene protons lie is parallel to plane P_2 in which the π-orbitals are situated (Figure 2.56). This is encountered in cyclopentanone [Figure 2.57(a)] with $J = -16$ to -17 Hz. The effect is smaller

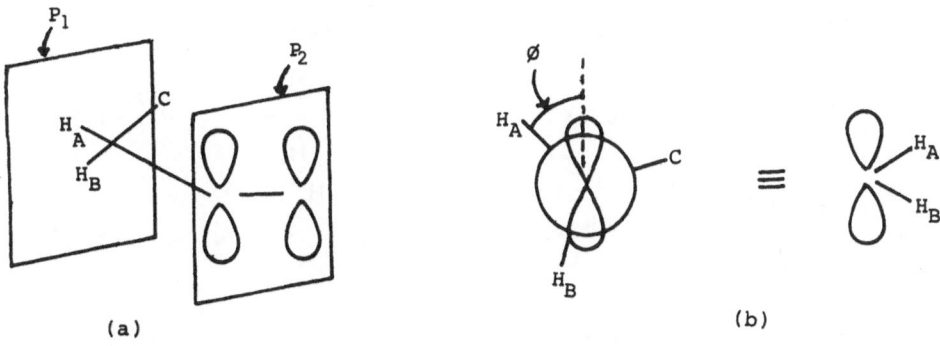

Figure 2.56. (a) Plane P_1 in which the CH_2 protons lie is shown parallel to plane P_2 in which the π-orbitals are situated. (b) The angle ϕ between H_A and the π-orbitals shown in a Newman projection.

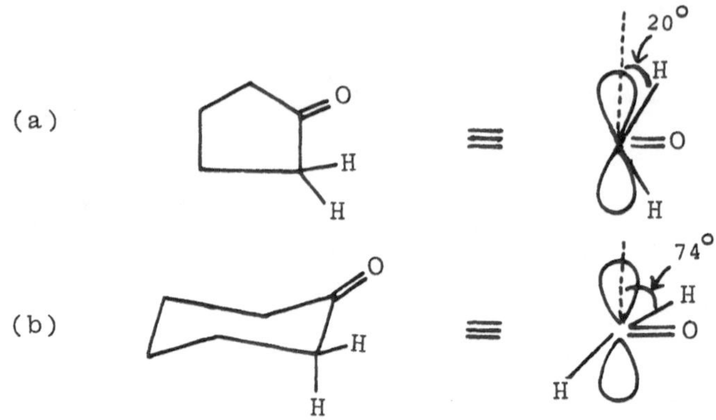

Figure 2.57. (a) In cyclopentanone both CH_2 protons α to the carbonyl group lie on the same side of the π-orbitals in a Newman projection. (b) In cyclohexanone they are found to lie on opposite sides.

when the methylene protons lie on the opposite sides of the orbitals, e.g., in cyclohexanone ($J_{gem} = -12$ to -13 Hz). This is shown in Figure 2.57(b).

2.4.3   Dependence of ^3J on H—C—C—H Valence Angles

Vicinal coupling constants (^3J) are markedly dependent on the angles θ and θ' between the C—H and C—C bonds. This effect is best illustrated in cyclic olefins in which the dihedral angle is zero, and $\theta = \theta'$. With decreasing ring size, the angles θ and θ' increase, and the coupling constants between the vicinal protons decrease. This is illustrated in Table 2.6.

Vicinal coupling constants in cyclic systems are presented in Table 2.7.

Table 2.6. Dependence of vicinal coupling constants ^3J on H—C—C—H angle

Compound	$^3J_{ab}$(Hz)	Compound	$^3J_{ab}$(Hz)
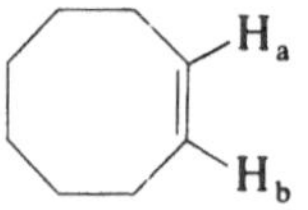 H_a / H_b (cyclooctene)	10–13	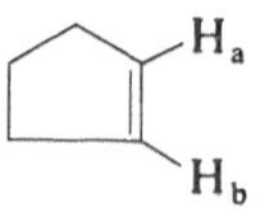 H_a / H_b (cyclopentene)	5.1–7.0
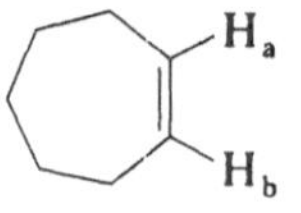 H_a / H_b (cycloheptene)	9–12.6	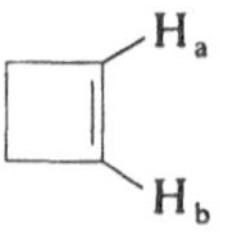 H_a / H_b (cyclobutene)	2.5–3.7
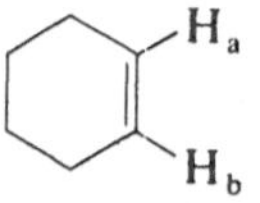 H_a / H_b (cyclohexene)	8.8–11.0	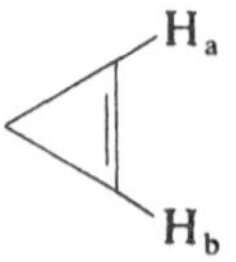 H_a / H_b (cyclopropene)	0.5–1.5

Table 2.7. Vicinal coupling constants in cyclic systems

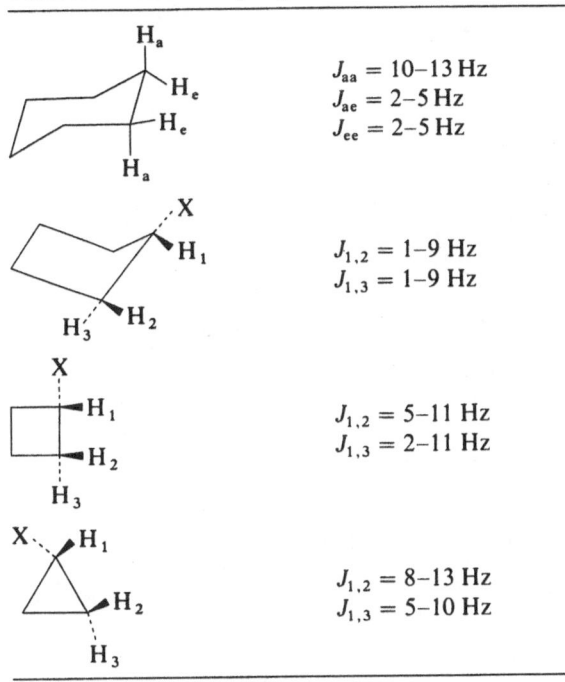

$J_{aa} = 10\text{–}13\,\text{Hz}$
$J_{ae} = 2\text{–}5\,\text{Hz}$
$J_{ee} = 2\text{–}5\,\text{Hz}$

$J_{1,2} = 1\text{–}9\,\text{Hz}$
$J_{1,3} = 1\text{–}9\,\text{Hz}$

$J_{1,2} = 5\text{–}11\,\text{Hz}$
$J_{1,3} = 2\text{–}11\,\text{Hz}$

$J_{1,2} = 8\text{–}13\,\text{Hz}$
$J_{1,3} = 5\text{–}10\,\text{Hz}$

In five-membered rings $J_{1,2}$ and $J_{1,3}$ are of similar magnitude. This is so because $J_{1,2}$ is at a maximum value, since the 1 and 2 protons are eclipsed, while the 1 and 3 protons cannot adopt a coplanar relationship without inducing ring strain. It is therefore difficult to predict the stereochemistry of groups attached to cyclopentane rings on the basis of coupling constants. In four-membered rings $J_{1,2}$ can often be bigger than $J_{1,3}$, and in three-membered rings $J_{1,2}$ is usually bigger than $J_{1,3}$ (see Table 2.7).

2.4.4 Dependence of ^3J on C—C Bond Lengths/π Bond Orders

Vicinal coupling constants are very sensitive to small differences in the length of the C—C bond to which the protons are attached and to the π bond orders. It has been shown that coupling constants decrease with increasing C—C bond lengths or decreasing π bond orders in unsaturated cyclic systems. Thus in ethylene, with a π bond order of 1, $J = 11.5$ Hz, while in benzene, with a π bond order of 0.67, $J = 8.0$ Hz. In naphthalene the 1,2-bond has a higher π bond order than the 2,3-bond, and this is reflected in the higher ^3J value of the 1,2-protons (8.1 Hz) as compared to the ^3J value of the 2,3-protons (6.4 Hz) (Table 2.8).

Spin–spin coupling constants in some common systems are presented in Table 2.9.

Table 2.8. Dependence of ^3J on π bond orders.

Compound	^3J (Hz)	π bond order
	11.5	1
	8.0	0.67
	8.1	0.72
	6.4	0.6

Table 2.9. Some typical coupling constants

System	$J_{HH'}$(Hz) Range	Usual value
	0--25	-10--18
$CH_3-CH_2'-$	6-8	~7
	0-8	~7
$CH_3-CH'-$ CH_3	5-7	6
	0-1	0
	4-10	5-7
	6-13	10-13
	0-3	2
	5-8	7

Table 2.9. (*Continued*)

	$J_{HH'}(Hz)$	
System	Range	Usual value

System	Range	Usual value
$\overset{H}{\diagdown}C{=}C\overset{H'}{\diagup}$ (cis)	0–12	7–10
$C{=}C$ with H', H trans	12–18	14–16
${=}C\overset{H}{\underset{H'}{\diagup}}$	0––3.5	−2
$-CH{=}\overset{\mid}{C}-CH'{<}$	0––3	−0.5––2
${>}CH-\overset{\mid}{C}{=}\overset{\mid}{C}-CH'$	0–2	1
${>}CH-C{\equiv}CH'$	−2––3	−2.5
${>}CH-C{\equiv}C-\overset{\mid}{C}H'$	2–3	2.5
benzene ring (H, H', H'', H''')	$J_{HH'}$, 6–10	8
	$J_{HH''}$, 0–3	2
	$J_{HH'''}$, 0–1	1
furan	$J_{HH'}$, 1.5–2	1.8
	$J_{H'H''}$, 3–4	3.5
	$J_{HH''}$, 0.5–1	0.8
	$J_{HH'''}$, 1–2	1.6
pyrrole	$J_{HH'}$, 2–3	2.6
	$J_{H'H''}$, 3–4	3.5
	$J_{HH''}$, 0–2	1.5
	$J_{HH'''}$, 0–2.5	2.1
thiophene	$J_{HH'}$, 2–6	5.0
	$J_{H'H''}$, 2–4	3.5
	$J_{HH''}$, 0–2	1.3
	$J_{HH'''}$, 0–3	2.7
pyridine	$J_{HH'}$, 4–7	5.5
	$J_{H'H''}$, 6–9	7.5
	$J_{HH''}$, 0–3	2.0
	$J_{HH'''}$, 0–1.5	1.0
	$J_{H'H'''}$, 0–2	1.6
	$J_{HH''''}$, 0–0.5	0.4

2.5 LONG-RANGE SPIN–SPIN COUPLING

In the above discussion we have been concerned with spin–spin coupling between protons attached to the same carbon (^2J, geminal coupling) or between protons attached to adjacent carbon atoms (^3J, vicinal coupling). Couplings are also possible over four or five bonds (designated as ^4J and ^5J, respectively) though they are smaller in magnitude (0–3 Hz). When couplings are transmitted only through σ bonds, they usually decrease by an order of magnitude with each additional intervening bond.

2.5.1 Long-Range Coupling in Unsaturated Compounds

2.5.1.1 Allylic Coupling

When a proton is attached to a methylene group α to an olefinic linkage (Figure 2.58), a discernible coupling may occur between H_A and H_B (*cisoid*-allylic coupling) or H_A and H_C (*transoid*-allylic coupling). These compounds have σ and π bonds, and both contribute to the coupling interaction, the π contribution, $^4J(\pi)$, being negative while the σ contribution, $^4J(\sigma)$, is positive. The magnitude of the coupling is dependent on the torsional angle ϕ made by the C—H bond of the allylic carbon atom and a line parallel to the π bonds (dashed line in Figure 2.58) as well as on the extent of σ or π contributions. When $\phi = 0°$ or $180°$, there is a large π contribution but when it is $90°$ or $270°$, the $^4J(\pi)$ contribution is negligible. As $^4J(\sigma)$ and $^4J(\pi)$ are opposite in sign, they negate one another. With the successive introduction of large alkyl groups at the sp^3-hybridized carbon, the angle ϕ approaches $270°$ and the coupling constant therefore becomes smaller because of the disappearance of $^4J(\pi)$.

In ring systems such as those in Figure 2.59, when $\phi = 0°$ or $180°$ the $^4J(\pi)$ contribution is large and a coupling constant of 4.1 Hz is observed. On the other hand, when $\phi = 90°$ (or $270°$), the $^4J(\pi)$ contribution disappears and the $^4J(\sigma)$ contribution appears, resulting in a coupling constant of $+0.5$ to -1.1 Hz. The conformational dependence of allylic H—H couplings is shown in Figure 2.60.

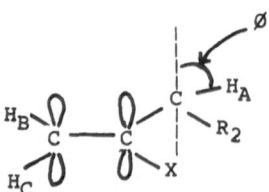

Figure 2.58. $^4J_{AB}$ is *cisoid*-allylic coupling while $^4J_{AC}$ is *transoid*-allylic coupling. Torsional angle ϕ is the angle which C—H_A makes with the dashed line through the allylic carbon atom drawn parallel to the plane in which the π bonds lie.

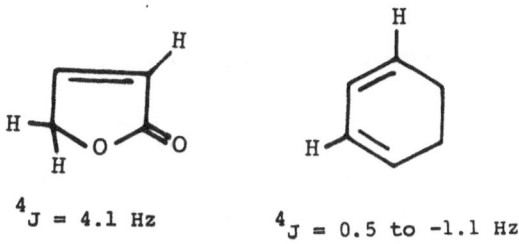

$^4J = 4.1$ Hz $\qquad\qquad$ $^4J = 0.5$ to -1.1 Hz

Figure 2.59. ^4J-Allylic coupling between ring protons.

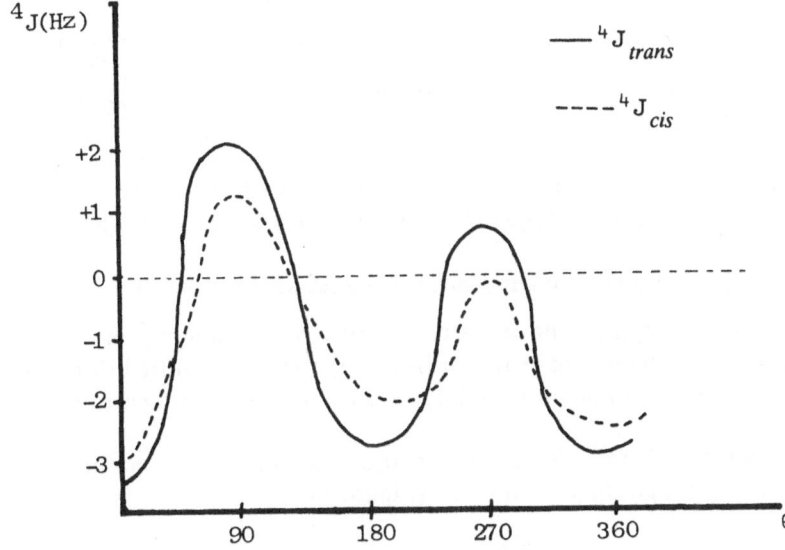

Figure 2.60. Dependence of allylic H—H couplings, $^4J_{cis}$ and $^4J_{trans}$, on the conformation of the coupled allylic protons.

2.5.1.2 Homoallylic Coupling (^5J)

The coupling between H and H′ in Figure 2.61 is known as homoallylic (^5J) coupling. ^5J values are usually in the range of 0–2.0 Hz. As $^5J(\pi)$ is positive in sign, the $^5J(\pi)$ and $^5J(\sigma)$ contributions augment each other. Again, the magnitude of the coupling constant is dependent on the angles θ and θ' made

Figure 2.61. Homoallylic (^5J) coupling is the coupling between H and H′.

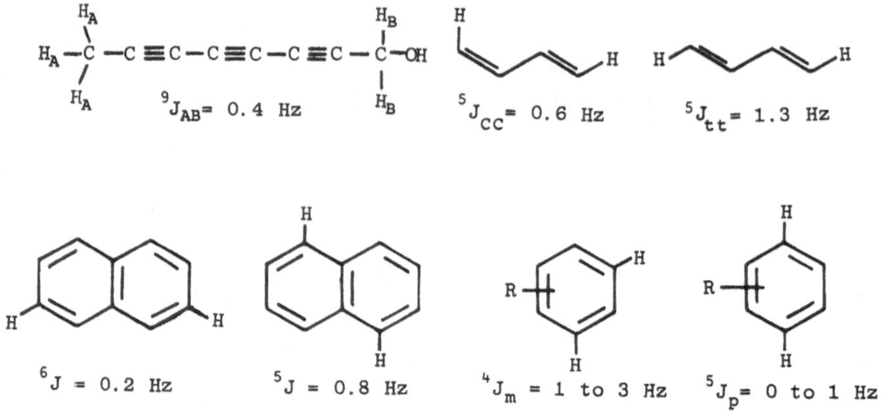

H_A
H_A—C—C≡C—C≡C—C≡C—C—OH
H_A
H_B
H_B
9J_AB = 0.4 Hz
H
H
5J_cc = 0.6 Hz
H
H
5J_tt = 1.3 Hz
H
H
6J = 0.2 Hz
H
H
5J = 0.8 Hz
R
H
H
4J_m = 1 to 3 Hz
R
H
H
5J_p = 0 to 1 Hz

Figure 2.62. Some long-range couplings in polyunsaturated systems.

by the C—H and C—H′ bonds, respectively, with the plane in which the double bond lies, the interaction being greatest when θ and θ' approach 90°.

2.5.1.3 Long-Range Coupling in Other Unsaturated Systems

Long-range couplings have been observed in a number of conjugated unsaturated systems. These appear primarily to be transmitted through the σ framework and the magnitude is largest when there is a zigzag arrangement of atoms.

In polyacetylenes, allenes, and cumulenes, long-range couplings occur primarily through the π framework (Figure 2.62).

2.5.2 Through-Space Coupling

Coupling can sometimes occur directly through space rather than through the intervening σ or π bonds. Thus in the compound shown in Figure 2.63, H_A and H_B are forced close to each other and a coupling constant of 1.1 Hz results. The coupling is probably the result of a direct spin–spin interaction between

Cl
H_a
H_b
Cl
OAc
Cl
Cl
Cl Cl

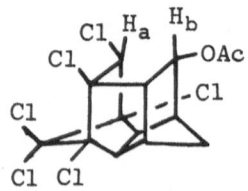

Figure 2.63. H_a and H_b are situated close enough in space so as to afford a coupling constant of 1.1 Hz.

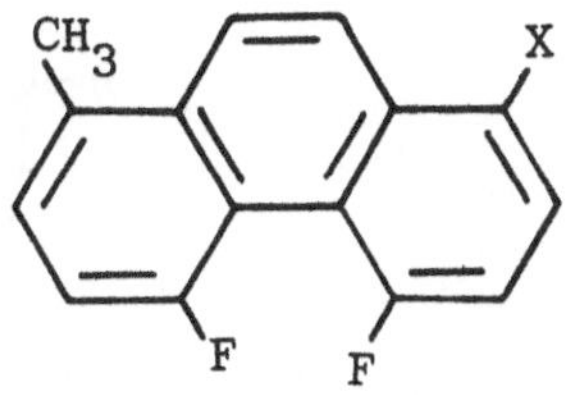

Figure 2.64. Through-space coupling of fluorine atoms.

the two nuclei rather than interaction through the six σ bonds. The "through-space" mechanism receives support from the fact that the F–F coupling in a number of substituted difluoromethylanthracenes (Figure 2.64) is about 167–168 Hz, irrespective of the nature of the group X. If the coupling was occurring via the σ or π bonds, then the electron-withdrawing or -donating nature of X would have been expected to influence its magnitude.

2.5.3 Long-Range Coupling in Saturated Systems

Couplings over 4 or 5 bonds (^4J, ^5J) have sometimes been observed in saturated systems; when the C—H and C—C bonds exist in a zigzag arrangement, such coupling constants may be up to 1.0 Hz. When there are four intervening bonds between the coupled nuclei, then one speaks of an "M" or "W" arrangement of atoms (Figure 2.65). In certain strained bicyclic systems the coupling constants may be fairly large (Figure 2.66). As the orbitals themselves are directed away from one another an extensive overlap of the rear lobes of the bonding orbitals of C—H bonds may occur, which would explain the mechanism of coupling across such long distances.

2.5.4 Dipole–Dipole Interaction

Long-range coupling through space can occur by the direct interaction of nuclear magnetic dipoles of molecules with one another. In liquids this is averaged to zero because of the rapid molecular motions. In solids, however, molecules have a fixed orientation with respect to one another and one can therefore observe dipolar coupling.

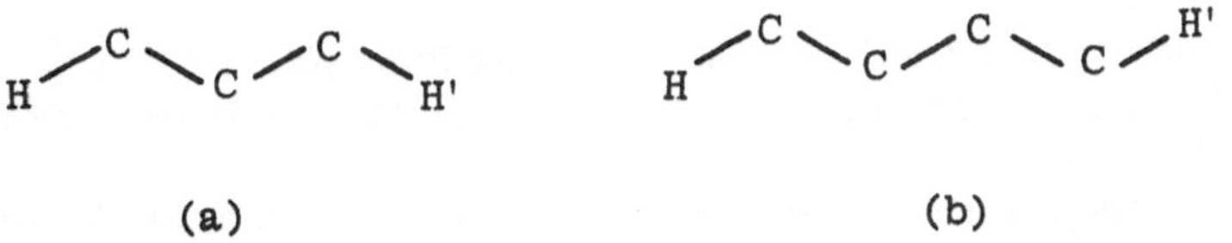

(a)   (b)

Figure 2.65. (a) $^4J_{H,H'}$ with a "W" (or "M") arrangement of atoms. (b) $^5J_{HH'}$.

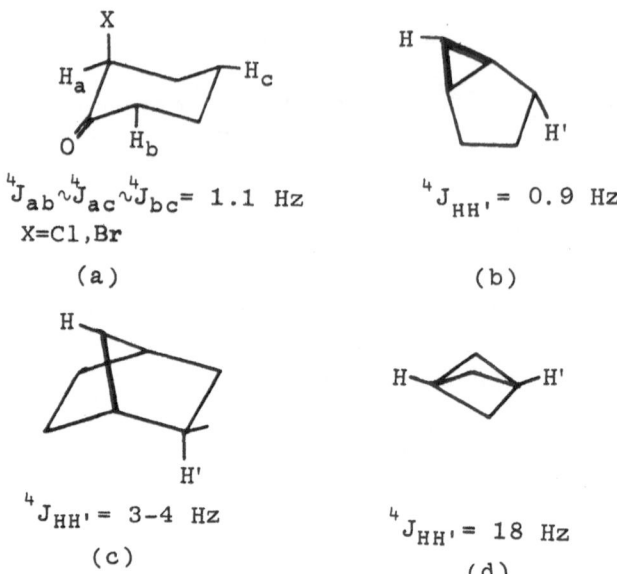

Figure 2.66. Long-range coupling in some saturated systems. Note the large coupling constant in (d).

RECOMMENDED READING

1. F.W. Wehrli and T. Wirthlin, *Interpretation of Carbon-13 NMR Spectra*, Heyden and Son Ltd., London (1978).
2. G.C. Levy, Ed., *Topics in Carbon-13 NMR Spectroscopy*, Vol. 1–3, John Wiley and Sons, New York (1979).
3. E. Breitmaier and W. Voelter, *${}^{13}C$-NMR Spectroscopy*, Verlag Chemie, Weinheim (1978).
4. C. Brevard and P. Granger, *Handbook of High Resolution Multinuclear NMR*, John Wiley and Sons, New York (1981).
5. G.C. Levy, R.L. Lichter, and G.L. Nelson, *Carbon-13 Nuclear Magnetic Resonance Spectroscopy*, John Wiley and Sons, New York (1980).
6. H. Günther, *NMR Spectroscopy*, John Wiley and Sons, New York (1980).
7. L.M. Jackman and S. Sternhell, *Applications of Nuclear Magnetic Resonance Spectroscopy in Organic Chemistry*, International Series in Organic Chemistry, Vol. 10, Pergamon Press, Oxford (1969).
8. F.A. Bovey, *Nuclear Magnetic Resonance Spectroscopy*, Academic Press, New York (1969).
9. E.D. Becker, *High Resolution NMR*, Academic Press, New York (1980).
10. M.L. Martin and G.J. Martin, *Practical NMR Spectroscopy*, Heyden and Son Ltd., London (1980).
11. T. Clerc and E. Pretsch, *Kernresonanz Spektroskopie*, Akademische Verlagsgesellschaft, Frankfurt (1973).
12. A. Carrington and A.D. McLachlan, *Introduction to Magnetic Resonance*, Chapman and Hall, London (1979).

Chapter 3

Experimental Procedures in NMR Spectroscopy

In order to record the proton NMR spectrum of a substance, it must first be dissolved in a suitable solvent in which it is sufficiently soluble (preferably 10–50 mg per ml). The majority of solvents used are deuterated analogues of the common organic solvents (e.g., $CDCl_3$, CD_3COCD_3, C_6D_6, CD_3CN, CD_3OD, C_5D_5N, DMSO-d_6, etc.) to avoid the overlap of large peaks of the protons present in the solvents with the spectrum of the material. Since deuterated solvents are expensive, it is essential that solubility be first checked in the corresponding undeuterated solvents before attempting dissolution in the deuterated materials. Spectroscopic grade CCl_4 is relatively inexpensive and may be used if the solubility is sufficient. Compounds bearing carboxyl groups have a high solubility in pentadeuteropyridine (C_5D_5N) while basic substances can be dissolved in CF_3COOD. A list of some useful NMR solvents is given in Table 3.1. A small amount of a reference compound, usually tetramethylsilane, $(CH_3)_4Si$, is added as an internal standard. For a routine continuous-wave type NMR spectrometer, about 10–20 mg of the compound is required for a good proton NMR spectrum, but in the FT instruments 1–10-mg quantities of sample are usually used. It is possible to record proton NMR spectra using as little as 50 μg of sample but this requires accumulation of several thousand scans, the use of special microcells and high-purity solvents and materials, as even trace impurities in solvents can be significantly magnified and can interfere with the spectrum of the substance being recorded.

For normal proton NMR measurements, 10–15 mg of the compound is dissolved in 0.4 ml of the solvent and the solution filtered into an NMR tube of 5-mm outer diameter through a sintered funnel or a dropping funnel packed with a small amount of cotton wool at the point where its bore narrows. It is important that the solution is completely clear and does not contain any undissolved sample particles such as dust, wool, hair, etc., as these would adversely affect the quality of the NMR spectrum. It is also important that the volume of solution placed in the NMR tube reaches a correct height since it is only a particular volume of the NMR tube which experiences the RF field, and

Table 3.1. Some useful solvents for NMR spectroscopy

Solvent	Liquid temperature range (°C)	Chemical shifts	
		δ_H	δ_C
Acetone-d_6	−95–56	2.17	29.2, 204.1
Acetonitrile-d_3	−44–82	2.00	1.3, 117.7
Benzene-d_6	6–80	7.27	128.4
Methanol-d_4	−98–65	3.4, 4.8	49.3
CCl$_4$	−23–77	—	96.0
CDCl$_2$CDCl$_2$	−44–146	5.94	75.5
CDCl$_3$	−64–61	7.25	76.9
CD$_2$Cl$_2$	−95–40	5.33	53.6
CFCl$_3$	−111–24	—	117.6
CF$_2$BrCl	−140––25	—	109.2
CS$_2$	−112–46	—	192.3
Cyclohexane	6–81	1.43	27.5
Dioxane	12–101	3.7	67.4
DMF	−60–153	2.9, 3.0, 8.0	31.36, 162.4
DMSO-d_6	19–189	2.62	39.6
D$_2$O	0–100	4.70	—
HMPA	7–233	2.60	36.8
Nitrobenzene	6–211	8.2, 7.6, 7.5	149, 134, 129, 124
Nitromethane-d_3	−29–101	4.33	57.3
Pyridine-d_5	−42–115	7.0, 7.6, 8.6	124, 136, 150
Tetrahydrofuran	−108–66	1.9, 3.8	25.8, 67.9
TFA	−15–72	11.3	114.5, 161.5
1,2,4-Trichlorobenzene	17–214	7.1, 7.3, 7.4	133.3, 132.8, 130.7, 130.0, 127.6
Vinyl chloride	−154––13	5.4, 5.5, 6.3	126, 117

if the solution extends above this area, it does not contribute to the NMR signal. Similarly, if the volume of the solution is too small it will result in the production of a vortex when the sample tube is spun on its axis, resulting in the lowering of resolution and the generation of spurious peaks. Special Teflon plugs are often inserted into the NMR tube to a level just touching the upper level of the solution to avoid vortex formation. The spinning of the sample tube on its axis is necessary to reduce field inhomogeneities, and the spinning rate has to be sufficiently high to avoid spinning side bands. These are easily recognized since they are smaller satellite peaks which occur at equal distances on both sides of a large central peak, and their position and intensity are dependent on the spinning rate. Spinning side bands can also arise due to maladjustment of electrical shim coils or the use of poor-quality sample tubes and spinning apparatus.

3.1   PULSED FOURIER TRANSFORM NMR SPECTROSCOPY

The nuclear magnetic resonance spectrometer consists of a permanent magnet or an electromagnet, an RF transmitter, a receiver, and a recording system. An intense magnetic field is applied to a solution of the compound which is placed in a long glass tube in between the poles of the magnet. The application of the magnetic field results in the generation of two energy levels for protons, and transition between these energy levels is stimulated by the RF oscillator present in the transmitter which is tuned to the Larmor frequency (transition frequency) of the nucleus. The resulting signal is received and amplified by the receiver, and recorded on the recorder. A block diagram of an NMR spectrometer and the arrangement of the sample between the poles of the magnet is shown in Figure 3.1.

There are two types of NMR spectrometers: continuous wave (CW) and pulsed Fourier transform (FT). In the continuous-wave instruments, the oscillator frequency is normally kept constant and the magnetic field gradually changed. As the various values of the magnetic field at which transitions of different protons at that particular frequency occur are reached, absorption signals are observed. The disadvantage in this method is that at any one time there is only one field which is being observed. Thus in a proton NMR recorded at 100 MHz covering a range of 1200 Hz (12 ppm), a line of, say, 1-Hz width would be observed for only 1/1200 of the total scan time. For the rest of the time the instrument will be scanning either the baseline or other signals in the spectrum. We could speed up the process by scanning quickly but there is a limitation to the speed at which one can scan. At too high a sweep rate the resolution decreases. One therefore has to sweep at 1 Hz/s or slower. One can enhance the signal-to-noise (S/N) ratio by "signal averaging," i.e., by attaching a computer of average transients (CAT) to the NMR spectrometer which stores the scans in the form of numbers in its memory. The noise, being random, can have a gaussian distribution so that it partly cancels

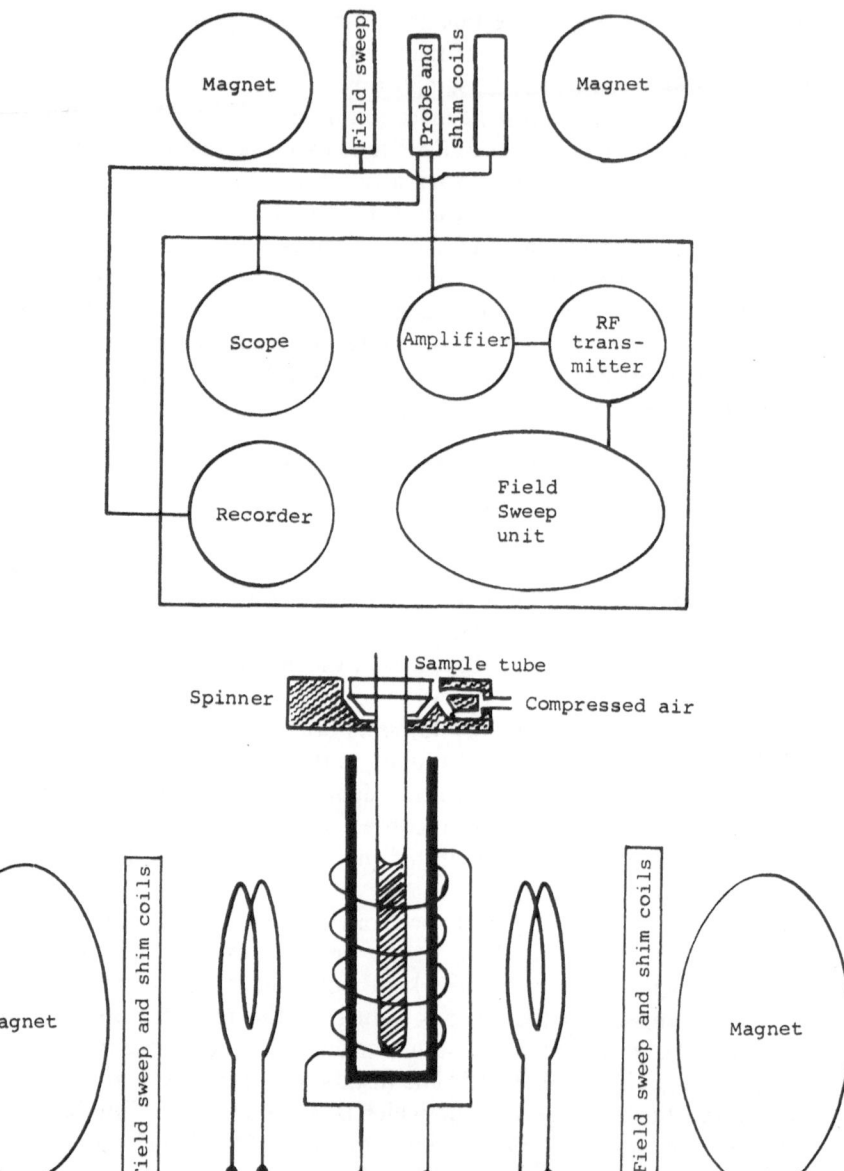

Figure 3.1. Block diagram of NMR spectrometer (cross coil configuration).

out and does not grow at the same rate as the signal. The noise grows at a rate proportional to the *square root* of the scan time (or number of scans) while the signals will grow *linearly* with the scan time. Thus if S represents the signal, B the number of scans, and N the noise, the growth of the signal S with the number of scans B is represented by the equation

$$S = k_1 B$$

where k_1 is a constant. The growth of noise N with the number of scans is represented by the equation

$$N = k_2 B^{1/2}$$

where k_2 is a constant. The signal to-noise ratio, S/N, may therefore be seen to grow with the scan time as follows:

$$\frac{S}{N} = \frac{k_1}{k_2} \cdot B^{1/2} = K \cdot B^{1/2}$$

It can thus be seen that the signal-to-noise ratio, S/N, will grow at a rate proportional to the square root of the number of scans, and after addition of successive scans, the peaks gradually become more intense and emerge out of the noise on accumulation of a sufficiently large number of scans. While this improves the spectrum to some extent, the process is inefficient since it is limited by the long scan time involved in the recording of each spectrum on continuous-wave instruments.

If one could somehow have 1200 oscillators delivering RF signals 1 Hz apart, each connected to its own separate receiver, then one could, in theory, simultaneously excite an entire 1200-Hz wide region of an NMR spectrum, and obtain the spectrum instantaneously. This is clearly impractical. Fortunately, an alternative method of excitation exists which serves the same purpose. This involves applying a short but intense radio frequency pulse which extends over the entire bandwidth of frequencies, and this simultaneously excites all the nuclei falling within that region. The pattern detected by the receiver is not in the form of peaks but is a sum of exponentially decaying sine waves known as a free induction decay (FID). A typical FID is shown in Figure 3.2. Rapid acquisition and accumulation of these FIDs by an analog-to-digital converter in a computer followed by Fourier transformation (a mathematical operation allowing the conversion of the FID from the sinusoidal form to the normal spectral peak shapes) provides the desired NMR spectrum.

3.1.1 Rotating Frame of Reference

As described earlier, if a sample made up of nuclei with nuclear spin $\frac{1}{2}$ is placed in a magnetic field, B_0, the nuclei start to precess around the direction of the applied field with a frequency ω_0, known as the Larmor frequency, and their

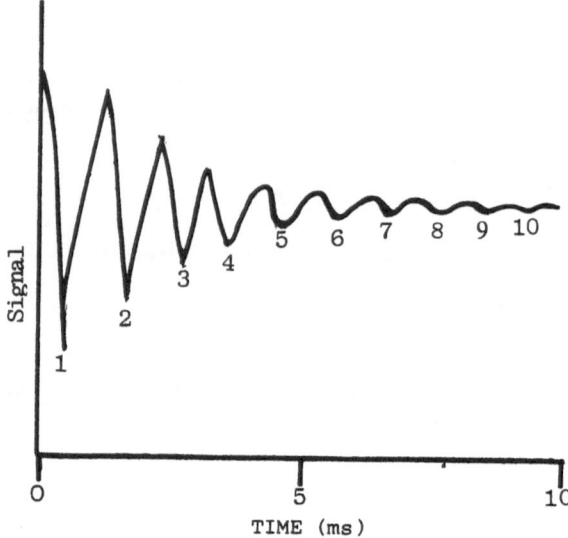

Figure 3.2. Free induction decay (FID)

spins are aligned either with the field or against the field. Since there is a slight excess (Boltzmann excess) of nuclei having spins aligned with the field, they give rise to a resultant magnetization vector M_0 which lies in the same direction as B_0 (see Figure 3.3).

This description of nuclear motions is commonly referred to as the laboratory or "*stationary frame of reference.*" If, however, the observer could himself rotate at the Larmor frequency, then the nuclei would not appear to precess but would appear stationary. The system would then be referred to as a "*rotating frame of reference.*" An analogy to this description is found in the case of communication space satellites which are launched in "geostationary" orbits. Although traveling at high speeds in a circular path around the earth, they "appear" to be stationary to an observer from the earth as their rotation around the earth exactly matches the rotation of the earth on its axis. The description of nuclear motions in a rotating frame of reference greatly simplifies the understanding of the effects produced by radio frequency pulses on the nuclei. Thus, in contrast to the stationary frame of reference in which the x-, y-, and z-axes were stationary, in a rotating frame of reference the x'- and y'-axes rotate around the z-axis at the Larmor frequency, ω_0 (Figure 3.3). If a radio frequency pulse is now applied with a frequency equal to the Larmor frequency, ω_0, along the x-axis (in the stationary frame), in the rotating frame of reference this would be equivalent to applying a *static* field B_1 along the x'-axis (since the x'-axis is *itself* rotating at the Larmor frequency in the rotating frame of reference). This causes the magnetization vector M_0, which was previously stationary and pointing towards B_0 (the z-axis), to tip and start rotating around the x'-axis. The magnitude of the component of M_0 along the y'-axis is given by $M_0 \sin \theta$ (see Figure 3.4). During the subsequent period the

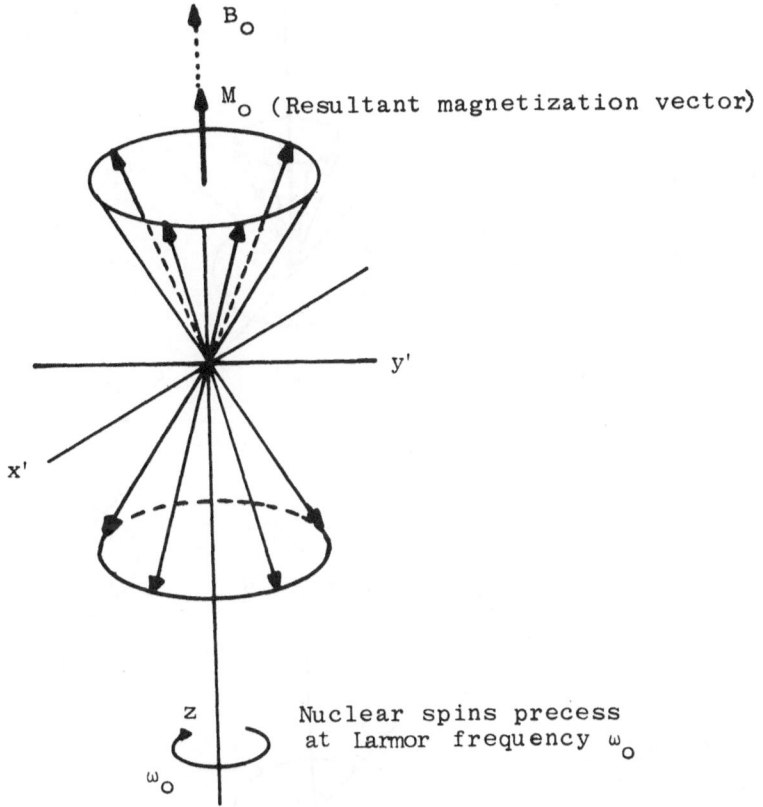

Figure 3.3. Motion of spin $\frac{1}{2}$ nuclei in a magnetic field.

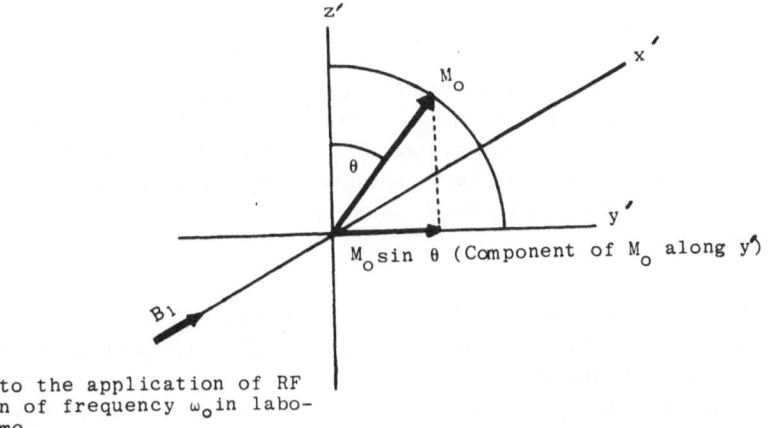

Figure 3.4. Effect of applying RF pulse of frequency ω_0 for time t seconds on bulk magnetization vector M_0.

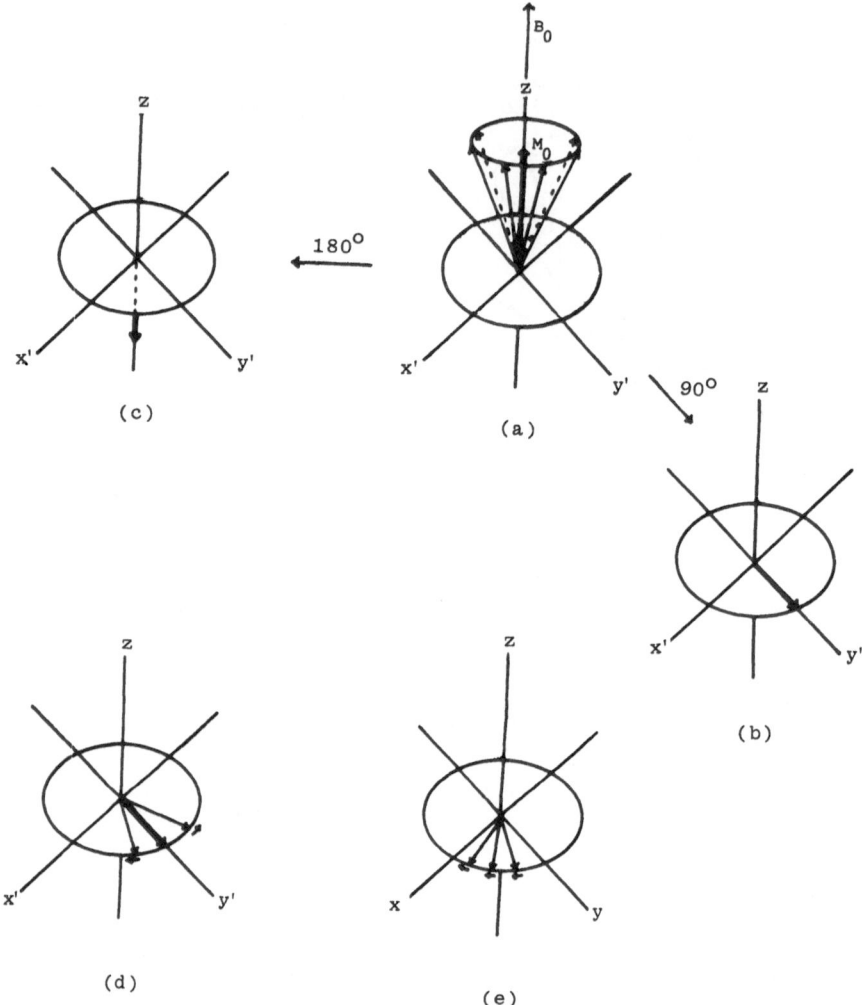

Figure 3.5. (a) Equilibrium position of the nuclei with magnetization M_0 aligned with magnetic field B_0. (b) A 90° pulse tips the magnetization by 90°. (c) A 180° pulse tips the magnetization by 180°. (d) In the rotating frame of reference the slower-moving vector v_2 and the faster-moving vector v_1 move in *opposite* directions with respect to the reference frequency v_0. (e) In the stationary frame of reference the vectors move in the same direction.

vector M_0 splits into a number of component vectors (since different nuclei have different nuclear magnetic moments with corresponding vectors). In the stationary frame of reference these vectors would move in the *same* direction away from the y'-axis [Figure 3.5(e)]. However, in the rotating frame of reference, since the x'- and y'-axes are themselves rotating at the reference frequency v_0, the individual vectors appear to move away from the y'-axis but in *different* directions. The slower-moving vectors move away anti-clockwise while those vectors with precessional frequencies higher than v_0 move away clockwise [Figure 3.5(d)]. Since the spectrometers are normally designed to detect signals along the y'-axis, it is only this component which will be detected and recorded. The angle θ by which M_0 is tipped away from the z-axis is known as the "*pulse angle.*"

It is apparent from the above discussion that when the sample was introduced into the magnetic field, B_0, the population of upper and lower energy levels was equal and there was no polarization of the nuclear spins, i.e., $M = 0$. After placing the sample in the magnetic field, B_0, a slight equilibrium excess (Boltzmann excess) of the nuclei in the lower-energy level is established, which results in a small magnetization, M_0, aligned with the applied field, B_0 [Figure 3.5(a)]. When the sample is then irradiated with a radio frequency field B_1 applied along the x'-axis in the rotating frame of reference (i.e., B_1 and x'-axis are rotating around the z-axis) the net magnetization M is bent away from the z-axis and precesses towards the y'-axis. The longer the time for which the pulse is applied, the more will M be bent towards (or even beyond) the y'-axis. Thus the pulse can be applied for a certain period of time, which can be experimentally determined (usually between 1 and $100\mu s$, to tip M by $90°$ [Figure 3.5(b)]. A *pulse width* twice as long as this time may be employed to tip M by $180°$, i.e., to completely invert M [Figure 3.5(c)]. The latter situation would result in a Boltzmann excess of nuclei in the higher-energy state, and the spin system would then be said to possess a "negative spin temperature." As soon as the excitation pulse is turned off, the relaxation processes begin which allow the nuclei to relax back to their original state shown in Figure 3.5(a). The macroscopic magnetization before the pulse is switched on may be represented as shown in Figure 3.6(a). When the pulse is switched off, the macroscopic magnetization may be represented as in Figure 3.6(b). It is then subjected to two different relaxation processes. The first, termed *spin–spin relaxation*, causes the moments to fan out in the x',y'-plane, as shown in Figure 3.6(c). This results in $M_{y'}$ decreasing to zero [Figure 3.6(d)] with a time constant T_2. The second relaxation process, termed *spin–lattice relaxation*, results from the nuclei losing their energy to the surrounding lattice, and they tip back towards the z-axis forming an increasingly narrowing cone as shown in Figure 3.6(d)–(f), M_z ultimately relaxing to the original equilibrium value of M_0 over a time constant T_1. This may be compared to the closing of an inverted umbrella.

3.1.2   Free Induction Decay

Let us now suppose that one has a substance in which all the protons are identical, e.g., benzene, and which would therefore all come to resonance at the same value of the applied RF field. A $90°$ pulse applied along the x'-axis would cause the magnetization to come to lie along the y'-axis (the transverse magnetization $M_{y'}$ would be along a *fixed* axis in the *rotating* frame). As the signal is detected along the y'-axis, the maximum intensity of the signal is obtained immediately after applying a $90°$ pulse. As soon as the pulse is

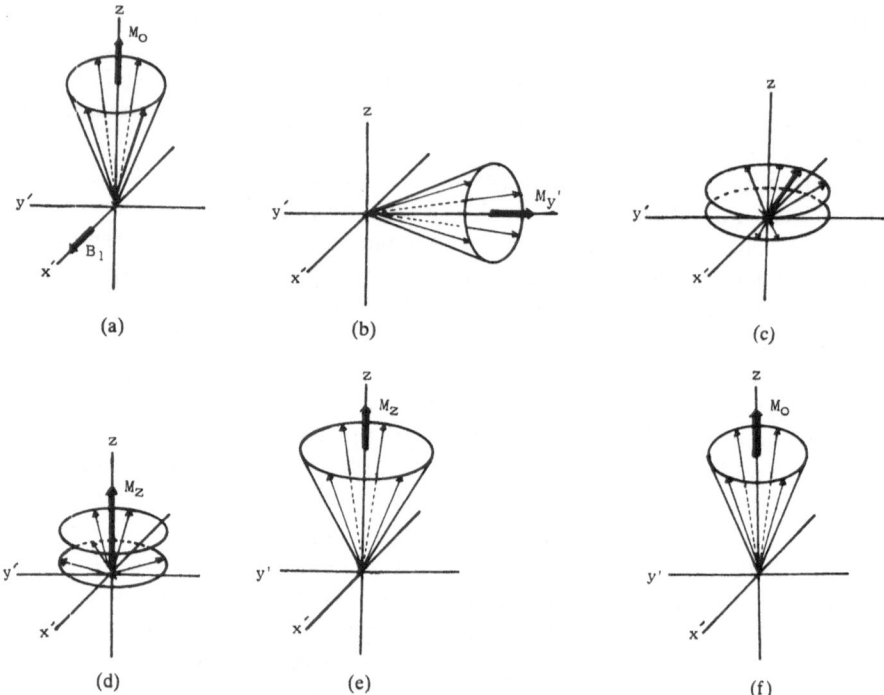

Figure 3.6. (a) Magnetization vector along z-axis. The component vectors precess around the z-axis with a net static magnetic vector pointing in the direction of the z-axis in the rotating frame of reference. (b) Application of 90° pulse tips the magnetization θ towards the y'-axis. (c) Dephasing of nuclear moments by spin–spin relaxation (fanning out of the vectors in the x, y-plane). (d)–(f) Spin–lattice relaxation by the formation of an increasingly narrower cone, resulting ultimately in the establishment of the equilibrium value of M_0 in (f).

switched off, the magnetic moments, M, undergo spin–lattice relaxation (move towards the z-axis) and spin–spin relaxation (fanning out along the x', y'-plane, resulting in the signal gradually becoming weaker and finally becoming zero [Figure 3.6(c)–(f)]. This is the situation when the radio frequency field is exactly equal to the resonance frequency of the protons. If, however, a molecule contains several different types of protons (as is usually the case) which come to resonance at values which are slightly different from that of the applied RF field, the transverse magnetization $M_{y'}$ is no longer fixed but *rotates* relative to the rotating frame, causing $M_{y'}$ and the applied field H_1 to periodically come in phase and then go out of phase with one another. The detector therefore shows not just the exponential decay of M_{xy} caused by relaxation processes, but also the interference effects as the M_{xy} and the applied frequency alternately dephase and rephase. This decaying beat pattern (Figure 3.2) is similar in appearance to the ringing effect (arising from the interference between the nuclear signal and RF) seen in high-resolution NMR spectra recorded on continuous-wave instruments.

3.1.3 Setting Pulse Widths

As mentioned earlier, the maximum signal intensity is obtained with a 90° pulse, and the time for which the pulse must be applied to tip the magnetization vector through 90° onto the y'-axis is known as the 90° pulse time. However, in order for the magnetization vector to return to its equilibrium value along the z-axis, it takes about $5T_1$ seconds (where T_1 is the spin–lattice relaxation time), and ideally one should therefore wait for $5T_1$ seconds between pulses. As T_1 values for ^{13}C nuclei can be fairly long, sometimes exceeding 100 s, it is more practical to apply a smaller pulse angle (determined by a correspondingly smaller pulse duration) of between 30°–50° which would require a much smaller relaxation time and thus allow faster accumulation of scans by reducing the delay times between successive pulses. This usually results in a better signal-to-noise ratio due to the accumulation of a larger number of scans in a given time. A number of modern pulse sequences also employ the use of 90° pulses, at which the signal has maximum amplitude, or of 180° pulses, at which it has minimum amplitude. It is therefore important to be able to accurately determine these pulse widths (in microseconds) for each nucleus or probehead diameter. The 90° pulse is best determined by first finding the 180° pulse by recording a series of computer-controlled spectra of any given sample, at gradually varying pulse widths, and determining the minimum amplitude of signal. Halving this value affords the 90° pulse.

After the application of each pulse, a dedicated computer acquires and stores the free induction decay in a digital form in its memory. Ideally one should continue to acquire ("sample") the FID till its amplitude falls to zero (which would take about $5T_1$ seconds). It may be predicted on theoretical grounds that in order to represent a frequency accurately in a digital form, at least two data points must be acquired in each cycle (Figures 3.7 and 3.8). In a spectrum of, say, 5000 cycles per second width, there must be sampling or a digitization rate of 10,000 words of data storage per second, i.e., one must sample twice as fast as the spectral width. Thus the maximum time for which a signal can be accumulated (called the "acquisition time", AT) is given by the equation

$$AT = \frac{N}{2\Delta}$$

where N is the number of words of data storage memory in the computer and Δ is the spectral width in cycles per second. Thus for a ^{13}C spectrum of 5000-Hz width using a computer with 10,000 words of data memory, the maximum possible acquisition time after each pulse will be given by $10,000/(2 \times 5000) = 1.0$ s. The highest frequency which can be correctly observed for a particular sampling rate is called the Nyquist frequency.

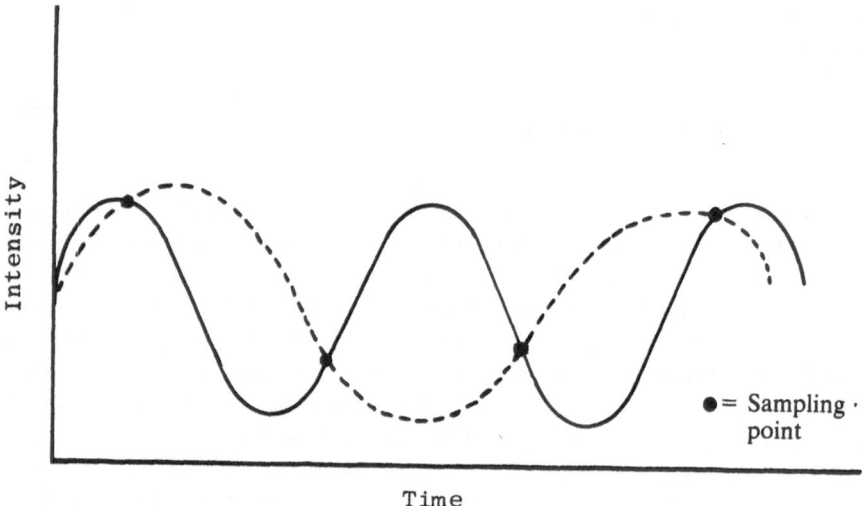

Figure 3.7. Signal sampled less than twice per cycle. More than one curve can be drawn.

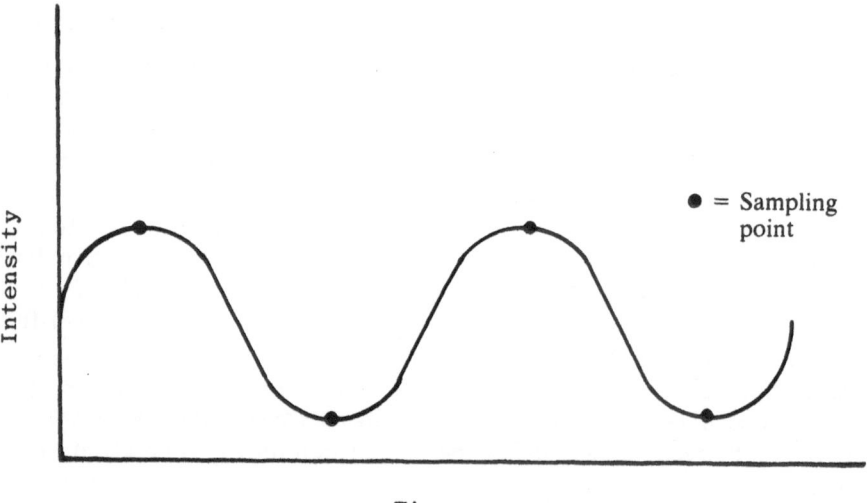

Figure 3.8. Signal sampled twice per cycle. Only one curve can be drawn through the points.

It may be noted that in a continuous-wave (CW) spectrum, the sample is irradiated with a weak field and the energy absorbed by the nuclei is detected. This may therefore be regarded as an absorption spectrum. In the pulsed NMR experiment, the sample is irradiated with a short but powerful burst of a high-energy pulse, and the energy emitted by the system on switching off the pulse is measured. This may therefore be considered as an emission spectrum. Another difference is that in the CW experiment one is measuring signal intensity as a function of frequency, whereas in a pulsed experiment we record the signal intensity as a function of time.

3.1.4 Adjustment of Pulse Frequency

The frequency of the RF pulse can be set outside the Larmor frequency at one end of the spectrum, and the data above this frequency collected. In this "single detector" method, positive and negative frequencies relative to the irradiating frequency cannot be distinguished. If all signals are assumed to be above the irradiating frequency, then only noise lies below the irradiating frequency. The noise "folds over" and comes to lie on the spectrum, thus decreasing the signal-to-noise ratio (Figure 3.9). In the more modern instruments the irradiating frequency is placed in the middle of the spectrum, and the negative frequencies distinguished from the positive frequencies by using two phase-sensitive detectors with their phases 90° apart. This "*quadrature detection*" mode results in improvement of the signal-to-noise ratio by a factor of approximately $2^{1/2}$ (about 40%) since the noise from the "other" side of the irradiating frequency which was being obtained in the single detection mode is eliminated, thus giving twice as much signal per unit time (Figure 3.10).

If too small a spectral width is chosen, a "folding in" of the peaks can occur, in which the peaks lying beyond the highest value of the spectral width chosen can "fold over" and superimpose themselves at a lower frequency value, thus resulting in an unduly complicated spectrum (Figure 3.11).

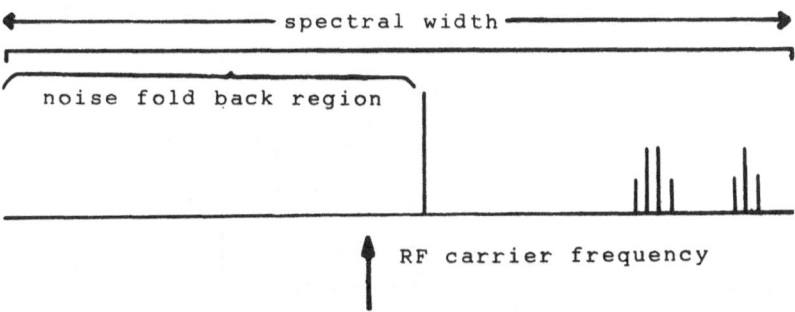

Figure 3.9. In the "single detector" mode, RF carrier is placed outside the spectral region at one end of the spectrum. Setting a large spectral width results in the "folding back" of the noise from the other side of the RF carrier, leading to a reduction in signal-to-noise ratio.

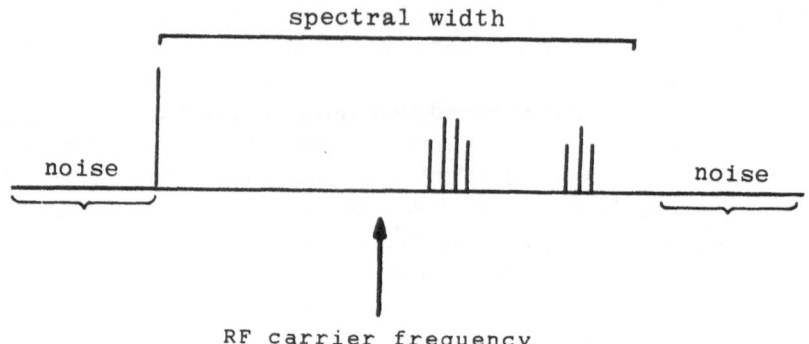

Figure 3.10. In the quadrature detection mode, RF carrier is placed at the center of the spectrum so that the spectral width is reduced. The noise lying beyond the spectrum cannot "fold back" on the spectrum. This results in a greater signal-to-noise ratio.

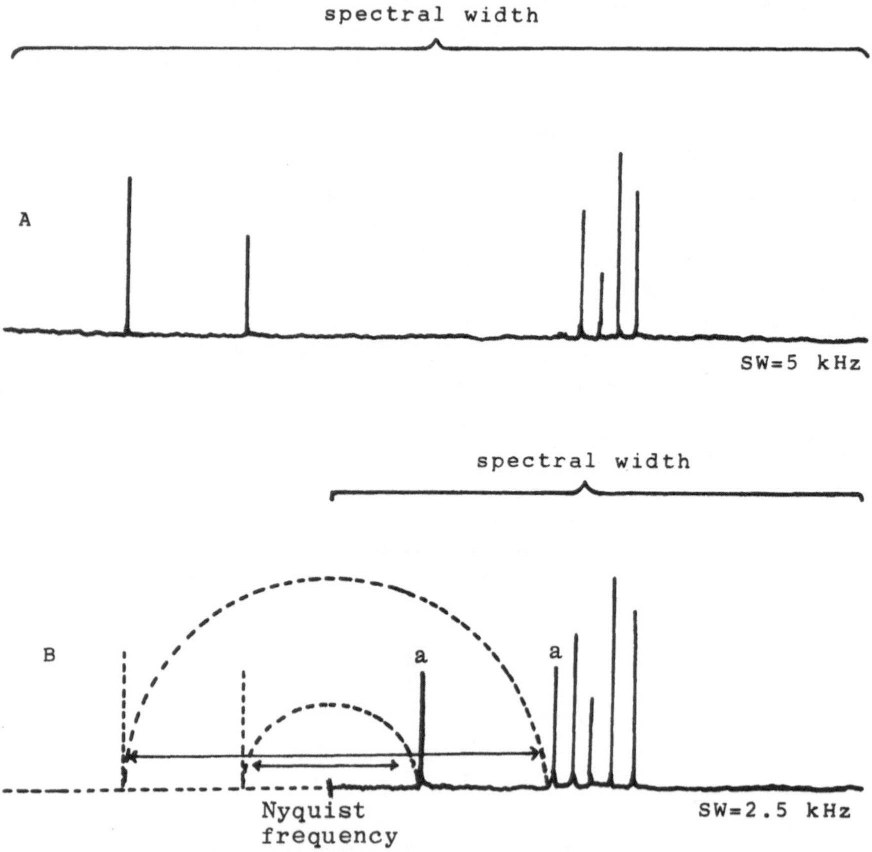

Figure 3.11. (A) Correctly chosen spectral width. (B) Spectral width chosen is too narrow, so that the peaks lying beyond the Nyquist frequency fold over and give rise to erroneous signals marked "a".

3.1.5 Signal Weighting

One way to improve the signal-to-noise ratio or the resolution of a spectrum is by a process of *digital filtering*. This is achieved by multiplying the FID with an exponential factor $e^{\pm at}$. One can thus enhance the signal-to-noise ratio (and somewhat reduce the resolution) by multiplying the FID with a negative exponential factor. The reverse is achieved by multiplication with a positive exponential factor. The reverse is achieved by multiplication with a negative value of the exponential factor. Multiplication of the FID with a slightly affects the initial portion of the FID (which contains most of the desired information) but removes most of the "tail" of the FID by zeroing the last points and reducing the intermediate ones. The effects of multiplying an FID with increasing negative or positive exponential functions are illustrated in Figures 3.12 and 3.13, respectively.

3.1.6 Phase Correction

When a free induction decay is Fourier transformed, some of the peaks may appear distorted or even inverted. They are then said to be out-of-phase with respect to a reference peak. Examples of the quartet of ethylbenzene before and after phase adjustment are given in Figure 3.14. The phase errors may be due to the phase detector setting, delay between pulse and start of data acquisition, or filter settings. The phase is digitally corrected after Fourier transformation.

3.1.7 Double Resonance

The technique of double resonance or spin decoupling involves strongly irradiating one set of coupled nuclei, A, while simultaneously observing the other set of nuclei, B, with which the protons are coupled. The A nuclei are irradiated with the normal weak irradiating frequency while the B nuclei are subjected to a much stronger decoupling frequency–hence the term "double resonance." In the rotating frame of reference the effective fields for the nuclei A and B are quantized along orthogonal axes, and since nuclear spin coupling depends on the product of I_A and I_B the orthogonal quantization of A and B spins causes the coupling J_{AB} to disappear and the A nuclei collapse to a singlet.* This technique thus helps in finding out which nuclei are coupled to which other nuclei in the spectrum. When the nuclei being irradiated and observed are of the same type (e.g., if both are protons), then the technique is called *homonuclear decoupling*, but if the nuclei are different, e.g., if one is irradiating protons while observing ^{13}C nuclei, then the technique is called *heteronuclear decoupling*. Spectra of ethylbenzene with and without

---

* For a more rigorous treatment see R.A. Hoffman and S. Forsen, *Progr. NMR Spectroscopy* **1**, 15 (1966); A.L. Bloom and J.N. Shoolery, *Phys. Rev.* **151**, 102 (1965).

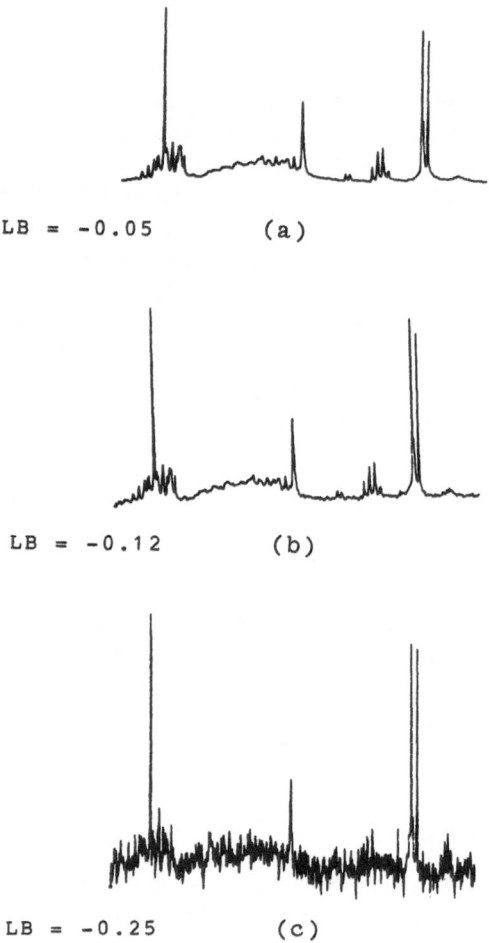

Figure 3.12. Multiplication with increasingly negative exponential factor on going from (a) to (c) causes increased resolution but decreases signal-to-noise ratio.

homonuclear decoupling are shown in Figure 3.15. Irradiation of the methyl protons results in collapse of the methylene protons to a singlet, and vice versa.

In the case of ^{13}C NMR spectra, as the natural abundance of ^{13}C is only 1%, the chances of finding two ^{13}C nuclei bound to one another are $0.01 \times 0.01 = 10^{-4}$, i.e., very low. Therefore $^{13}C-^{13}C$ couplings are not normally observed.* Extensive $^{13}C-^{1}H$ couplings are, however observed. The

---

* See the section on INADEQUATE in a later chapter which describes how $^{13}C-^{13}C$ couplings can be detected and used for structure elucidation.

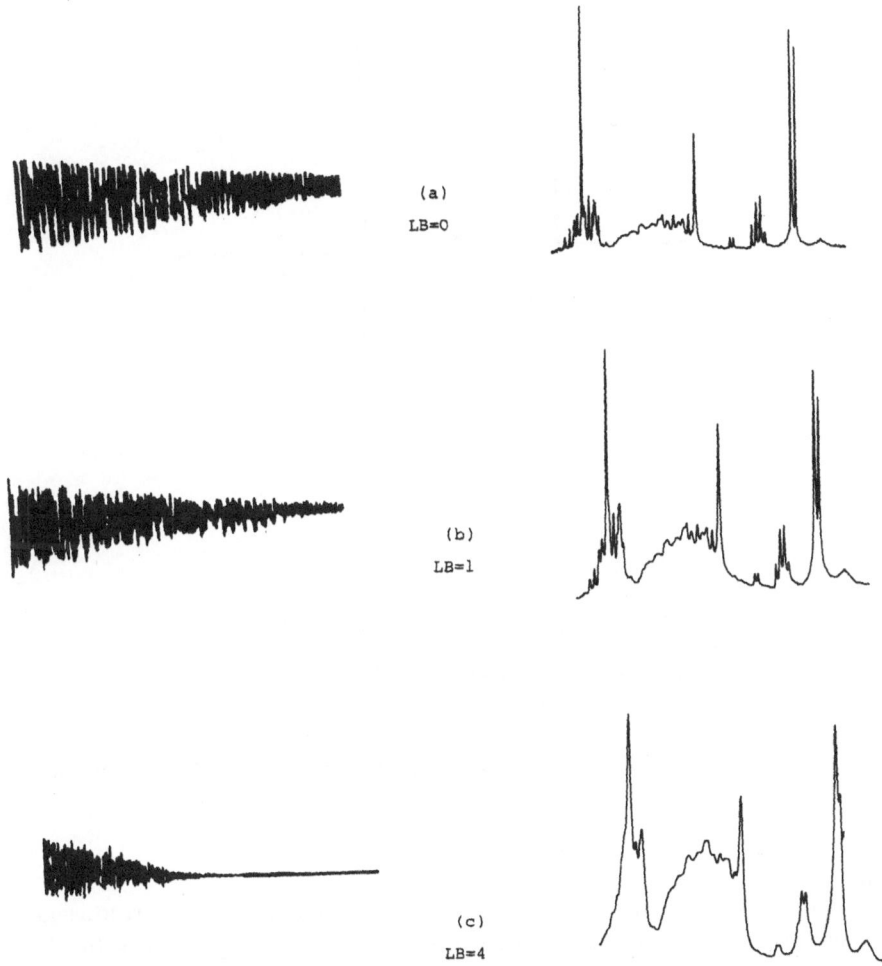

Figure 3.13. Multiplication with increasingly positive exponential factors on going from (a) to (c) improves signal-to-noise ratio but causes peak broadening. LB refers to the value of the line broadening exponential function.

one-bond ($^1J_{CH}$) couplings are large, often in the range of 100–200 Hz (Table 3.2). Additionally, smaller $^{13}C-^1H$ couplings over two or three bonds ($^2J_{CH}$, $^3J_{CH}$) are also observed, with the result that an undecoupled ^{13}C spectrum can be fairly complicated due to extensive overlap of split signals. ^{13}C spectra are therefore usually recorded using proton decoupling techniques. This is usually represented as $^{13}C-\{^1H\}$, the nucleus outside the brackets being the one that is observed while the one inside the brackets is being irradiated.

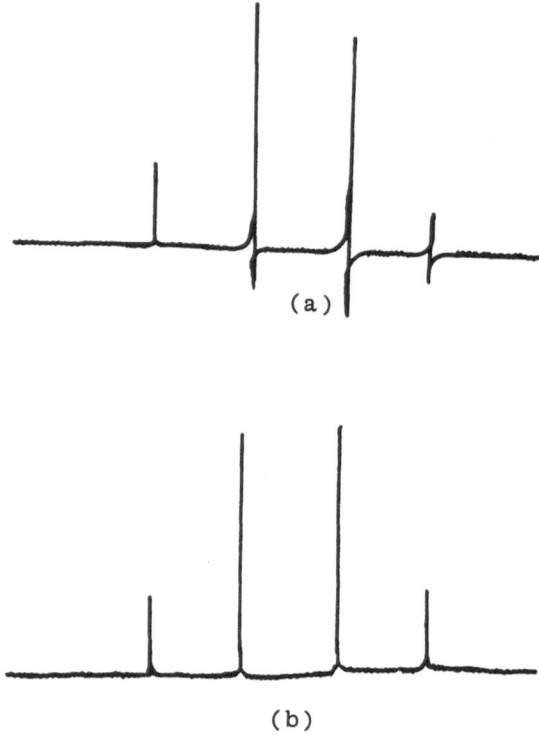

Figure 3.14. (a) Maladjusted phase in a quartet. (b) Correctly adjusted phase.

In the simplest form of heteronuclear decoupling, a particular proton signal is selected and irradiated with a strong RF field. Simultaneously the ^{13}C-NMR spectrum is observed. This results in the collapse of the signal from the carbon atom to which the protons were attached to a singlet, but the remainder of the ^{13}C-NMR spectrum is unaffected. This procedure allows one to correlate the ^{1}H signals with the corresponding ^{13}C signals, and provides valuable structural information. The procedure of sequentially irradiating various ^{1}H signals and observing the ^{13}C-NMR spectra is tedious and time-consuming. A more sophisticated alternative developed in recent years is two-dimensional NMR spectroscopy, which will be discussed in a later section.

The procedure employed to record decoupled ^{13}C-NMR spectra involves placing the decoupler frequency at the center of the proton region and then modulating the entire proton region (1000-Hz area at 24 kG) using a "noise generator." This results in the simultaneous irradiation of every proton and allows the carbon signals to be recorded as singlets. This technique, known as *proton noise decoupling* or *broad-band decoupling*, causes a significant increase

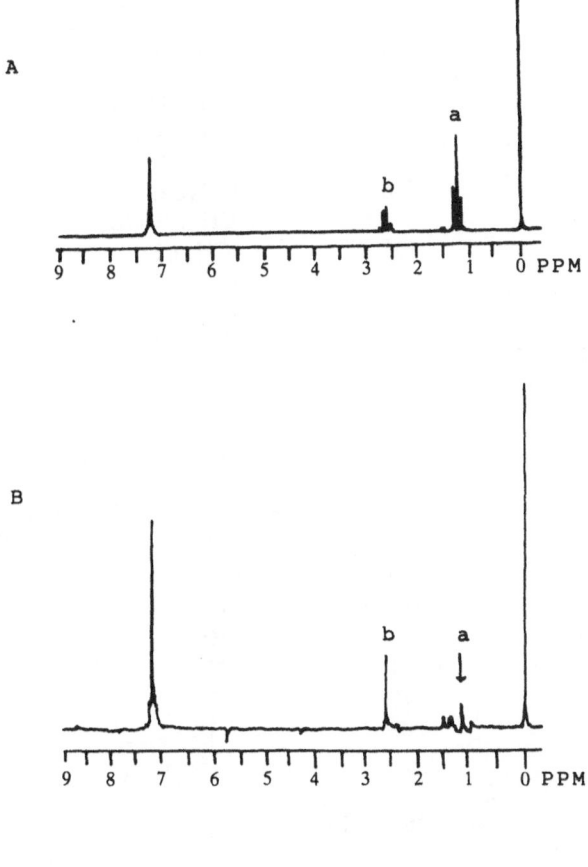

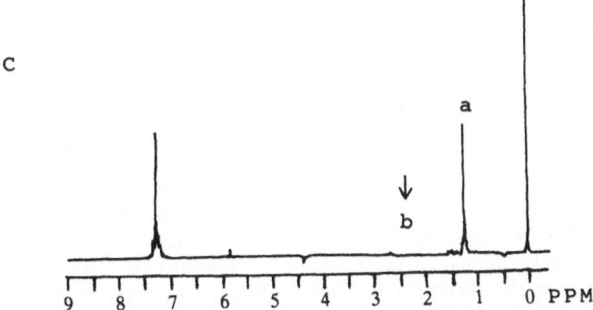

Figure 3.15. (A) ^{1}H-NMR spectrum of ethylbenzene without spin–spin decoupling. (B) Decoupled spectrum of ethylbenzene. Irradiation at the chemical shift of the triplet at "a" results in the quartet at "b" collapsing into a singlet. (C) Decoupled spectrum of ethylbenzene. Irradiation at the chemical shift of the quartet (at "b") results in the triplet at "a" collapsing into a singlet.

Table 3.2. Spin–spin coupling: Characteristic $^1J_{CH}$ couplings[a]

Molecule	Coupling (Hz)	Molecule	Coupling (Hz)
CH_4	125	CH_3NO_2	147
C_2H_6	125	$CH_3{}^+NH_3$	145
C_2H_4	156	CH_3NH_2	133
C_2H_2	248	CH_3OH	141
		$CH_3C{\equiv}CH$	132
cyclopropane (CH_2–CH_2–CH_2)	162	$(CH_3)_4Si$	118
		$\underline{C}H_3CHO$	178
		CCl_3CHO	207
cyclobutane	123	CH_3COCH_3	127
		$\underline{C}H_3CO_2H$	130
		HCO_2H	222
		$Me_2N\underline{C}HO$	191
cyclopentadiene/benzene ring	159	$Me_2{}^+\underline{C}H\bar{S}bF_5Cl$	168
CH_3F	149		
CH_3Cl	150		
CH_3Br	152		
CH_3I	151		
CH_2Cl_2	178		
$CHCl_3$	209		

$CH_2{=}CHX$			
X	α	cis	trans
F	200	159	162
Cl	195	163	161
CHO	162	157	162
CN	177	163	165

pyridine:

C_2	C_3	C_4
170	163	152

fluorobenzene:

C_2	C_3	C_4
155	163	161

X	C_2	C_3
CH_2	170	170
NH	184	170
S	185	167
O	201	175

pyranone:

C_2	C_3
200	169

[a] G.E. Maciel, J.W. McIver, Jr., N.S. Ostlund, and J.A. Pople, *J. Am. Chem. Soc.* **92**, 1,11 (1970); S.R. Johns and R.I. Willing, *Aust. J. Chem.* **29**, 1617 (1976); H. Gunther, H. Seel, and M.-E. Gunther, *Org. Magn. Resonance* **11**, 97 (1978); H. Gunther, H. Seel, and H. Schmickler, *J. Magn. Resonance* **28**, 145 (1977), L. Ernst, V. Wray, V.A. Chertkov, and N.M. Sergeyev, *J. Magn. Resonance* **25**, 123 (1977); V.A. Chertkov and N.M. Sergeyev, *J. Magn. Resonance* **21**, 159 (1976); V. Wray and D.N. Lincoln, *J. Magn. Resonance* **18**, 374 (1975); L. Ernst, D.N. Lincoln, and V. Wray, *J. Magn. Resonance*, **21**, 115 (1976); V. Wray, L. Ernst and E. Lusting, *J. Magn. Resonance* **27**, 1 (1977); L. Ernst and V. Wray, *J. Magn. Resonance* **28**, 373 (1977); W.S. Brey, L.W. Jaques, and H.J. Jakobsen, *Org. Magn. Resonance* **12**, 243 (1979).

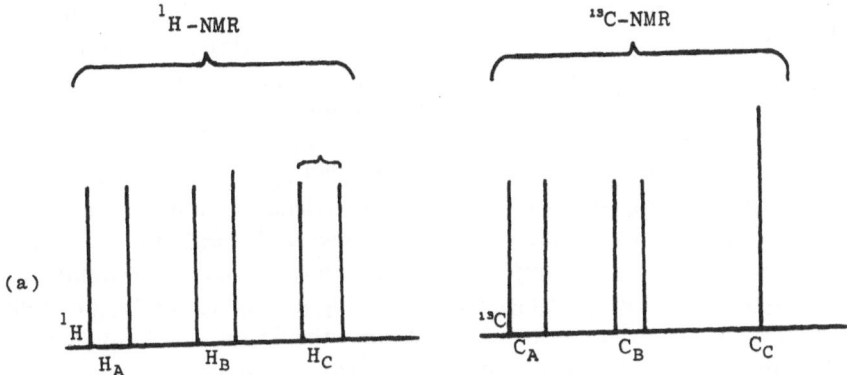

Proton noise decoupling covering

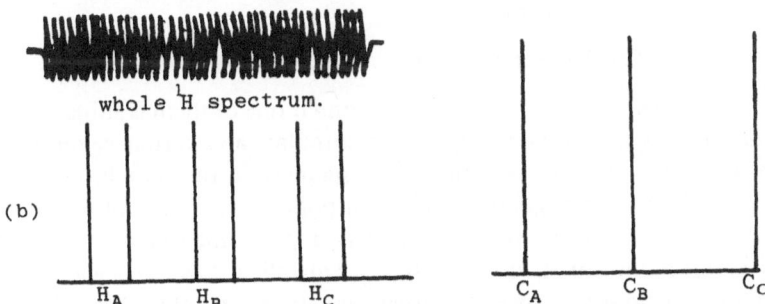

whole ^{1}H spectrum.

Figure 3.16. Proton noise decoupling techniques: (a) single-frequency decoupling; (b) noise decoupling. Irradiation of H_C in (a) results in the collapse of only one carbon, C_C, to a singlet. Proton noise decoupling (b) results in all the carbon atoms appearing as singlets.

in the signal-to-noise ratio due to (a) collapse of the multiplet to a singlet and (b) nuclear Overhauser enhancement (NOE). The NOE effect will be discussed later. Single-frequency decoupling and proton noise decoupling are illustrated in Figure 3.16(a) and 3.16(b), respectively.

3.1.8   Off-Resonance Decoupling

The disadvantage in proton noise decoupling is that it removes all the coupling information and thus one cannot distinguish between methyl, methylene, methine, and quaternary carbon atoms. To overcome this problem, *off-resonance decoupling* is employed. This involves setting the proton decoupling frequency 1000 to 2000 Hz outside the proton region (above TMS) and turning off the noise modulation. This results in the removal of all the two-bond and three-bond $^{13}C-H$ couplings and the larger one-bond ($^1J_{CH}$) couplings are considerably reduced (to 20–25 Hz), causing the primary carbon atoms bearing three hydrogen atoms to appear as quartets, secondary carbon atoms as triplets, tertiary carbon atoms as doublets, and quaternary carbon atoms as singlets. When the off-resonance spectrum is examined in combination with the proton noise-decoupled spectrum, it allows one to assign each carbon atom as a methyl, methylene, methine, or quaternary carbon. It may be noted that the distances between the peaks in, say, a quartet obtained from a methyl group do not represent the coupling constants ($^1J_{CH}$) but are only residual splittings which are proportional to the $^1J_{CH}$ coupling, the power of the decoupler, and the distance the decoupling frequency is set from the protons being decoupled. If there is an extensive overlap of multiplets, this can often be removed by adjusting the position of the decoupling frequency to afford either larger or smaller residual splittings. The nuclear Overhauser enhancement (nOe; see Section 3.1.8.1) is largely retained in the off-resonance spectrum.

3.1.8.1 Nuclear Overhauser Enhancement and Gated Decoupling

When two nuclei are close enough in space, and if one of them is subjected to a strong decoupling field, then this is seen to stimulate an alternative relaxation path for the other nucleus, resulting in an increase in the population of the lower state of that nucleus. This increase in population is accompanied by a corresponding increase in signal intensity, a phenomenon known as the nuclear Overhauser enhancement (NOE). This effect has been widely used in structure elucidation since it allows one to probe the stereochemical disposition of close-lying protons, and also results in an enhanced signal intensity of carbon nuclei in broad-band decoupled spectra. Thus when a proton noise-decoupled spectrum is recorded and the intensity compared with that of the corresponding coupled spectrum, it is seen that the increase in intensity of the singlets in the noise-decoupled spectrum is significantly greater than would be expected simply from the collapse of the multiplet structure into a single line. This nuclear Overhauser enhancement is the result of the decoupling of the hydrogen spins from the carbon spins, which results in a transfer of spin polarization to the carbon atoms and hence results in an increase of the carbon intensities.

When a ^{13}C nucleus is directly bonded to a hydrogen atom, it experiences a small field due to the dipolar interaction with the protons. In a rapidly tumbling molecule in solution, this field generated from the proton will fluctuate and have a large number of components of different frequencies; some of these components may be close to the Larmor frequency of the carbon atom and thus afford it an opportunity to give up some of its energy to the proton and relax to the lower-energy state. This dipolar relaxation mechanism is dependent on the rapidity with which the molecules tumble in solution and is proportional to r^{-6}, where r is the distance between the interacting nuclei. Since this interaction decreases rapidly with distance, it is usually observed only for protons spatially close to other protons, or for carbon atoms bonded directly to hydrogen atoms. This explains why in substituted benzenes, the resonances of the carbon atoms carrying substituents are usually much less intense than those of carbons carrying protons, since the former do not have access to the dipolar relaxation pathway available to their protonated partners.

The dipole–dipole relaxation mechanism is responsible for the nuclear Overhauser effect. When a proton bonded to carbon is irradiated, the homonuclear relaxation processes are not sufficient to restore the equilibrium population of the protons, and they therefore give up some of their energy via the dipole–dipole relaxation mechanism to the ^{13}C atoms to which they are bound. The additional amount of energy thus received by the carbon atoms results in their behaving as if they had been irradiated and causes them to relax, thereby increasing the population of carbon atoms in the ground state and producing an enhanced signal.

Let us consider two nuclei A and X which are sufficiently close in space that they can influence the relaxation processes of one another. They do not need to be coupled but only to be sufficiently close in space to show NOE effects. Each of these nuclei can exist in two different spin states, α or β. By convention, the α state is regarded as the lower-energy one. The two nuclei can thus exist in four different spin combinations—$\alpha\alpha$, $\alpha\beta$, $\beta\alpha$, and $\beta\beta$. These are shown in the energy level diagram in Figure 3.17. Only transitions between adjacent levels are allowed, i.e., from $\alpha\alpha$ to $\alpha\beta$ or from $\alpha\beta$ to $\beta\beta$ but not from $\alpha\alpha$ to $\beta\beta$. The energy level diagram shows that there are four transitions which can take place, corresponding to two different energy level differences, E^H and E^C. Transition between levels 1 and 4, corresponding to a double energy jump, E^2, is not allowed except as a relaxation process. At equilibrium (i.e., before application of the saturation pulse) the population of levels 2 and 3 will be equal and may be represented by the symbol N, the population of level 1 will

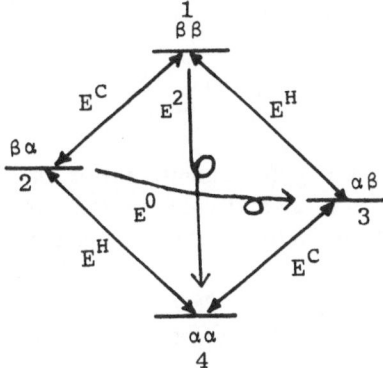

Figure 3.17. ^1H and ^{13}C spin transitions. Relaxation can occur through the forbidden pathway, E^2, involving a "double jump" resulting in a positive NOE. Large molecules can relax through pathway E^0, giving rise to a negative NOE.

be slightly lower $(N - d)$ while that of level 4 will be slightly higher $(N + d)$. The population difference between levels 1 and 4 will thus be $2d$.

When a high-power decoupling field is applied to one of the two nuclei, it results in an equalization of the populations of levels 1 and 3, as well as of levels 2 and 4. What this means, in effect, is that half the excess (d) of nuclei in levels 3 and 4 are promoted to levels 1 and 2. The populations of the levels at equilibrium and immediately after application of the decoupling field are shown in Table 3.3.

The difference in population between levels 1 and 4 was $2d$ at equilibrium but after application of the saturation pulse it is only d. This difference is restored to some extent by the relaxation pathway E^2 involving direct relaxation from level 1 to level 4, primarily a dipole–dipole interaction between the coupled nuclei. If the number of nuclei relaxed by this path from level 1 to level 4 is x, then the intensity of the lines (which is dependent on the difference in populations between levels 1 and 2, or between levels 3 and 4) will be increased by an amount x due to this additional relaxation, and a corresponding effect will be observed.

This situation, giving rise to a positive NOE, prevails in small rapidly tumbling molecules with a molecular weight of say 100–300. In the case of larger molecules with a molecular weight of say 6000, the tumbling motions may be a thousandfold slower ($\sim 10^8$ s^{-1} as compared to $\sim 10^{11}$ s^{-1} in small molecules). An alternative nonradiative transition, E^0, associated with a smaller energy difference is then promoted, resulting in an increasing number of transitions between levels 2 and 3. This results in a decrease in the difference in population between levels 2 and 1 (or 4 and 3), giving rise to a corresponding decrease in the intensity of the lines, or a *negative* NOE effect.

In an A–{X} experiment (i.e., irradiate X, observe A), the NOE is dependent on the gyromagnetic ratio of A and X nuclei, provided that the intramolecular dipolar relaxation mechanism is the predominant pathway through which nuclei A are relaxing. Thus the maximum NOE factor $f_A(X)$ for the A signal in an A–{X} experiment is given by

$$f_A(X) = \frac{\gamma_X}{2\gamma_A}$$

where γ_X and γ_A are the gyromagnetic ratios of the X and A nuclei, respectively. Since $\gamma_{^{13}C} = 6.726$ and $\gamma_{^1H} = 26.752$ rad s^{-1} gauss^{-1}, in a ^{13}C–{^{1}H} experiment the NOE factor for the ^{13}C nuclei would be given by

$$f_{^{13}C}(^1H) = \frac{26.752}{2(6.726)} = 1.988$$

Since the increase in signal-to-noise ratio is 1 plus the NOE factor, the ^{13}C signals will be increased by $1 + 1.988 = 2.988$ (i.e., up to a threefold increase will be observed at maximum NOE).

Table 3.3. Populations of two close nuclei at equilibrium and after saturation of one

Level	Population at equilibrium	Difference in population at equilibrium	Population after saturation	Difference in population after saturation	Population after saturation and E^2 relaxation	Difference in population after E^2 relaxation
1	$N-d$		$N-\dfrac{d}{2}$		$N-\dfrac{d}{2}-x$	
		$\longleftrightarrow d \longleftrightarrow$ $\quad$ $\longleftrightarrow 2d \longleftrightarrow$		$\longleftrightarrow$ Nil $\longleftrightarrow$ $\quad$ $\longleftrightarrow d \longleftrightarrow$		$\longleftrightarrow x \longleftrightarrow$ $\quad$ $\longleftrightarrow d+2x \longleftrightarrow$
2	N		$N+\dfrac{d}{2}$		$N+\dfrac{d}{2}$	
3	N		$N-\dfrac{d}{2}$		$N-\dfrac{d}{2}$	
		$\longleftrightarrow d \longleftrightarrow$		$\longleftrightarrow$ Nil $\longleftrightarrow$		$\longleftrightarrow x \longleftrightarrow$
4	$N+d$		$N+\dfrac{d}{2}$		$N+\dfrac{d}{2}+x$	

Table 3.4. Maximum values of the nuclear Overhauser enhancement for various nuclei

Irradiated nucleus	^{1}H	^{1}H	^{1}H	^{1}H	^{1}H	^{1}H
Observed nucleus	^{1}H	^{13}C	^{15}N	^{19}F	^{29}Si	^{31}P
Maximum NOE (η_{max})^a	0.5	1.99	−4.93	0.53	−2.52	1.24

^{a} η has a negative value when a nucleus with a negative gyromagnetic ratio is observed while ^{1}H is decoupled.

In the corresponding experiment in which ^{13}C are being irradiated and protons are being observed, ^{1}H–{^{13}C}, the NOE for ^{1}H would be

$$f_{^1H}(^{13}C) = \frac{6.726}{2(26.752)} = 0.126$$

It is thus clear that maximum sensitivity enhancements are obtained when the nucleus with a high gyromagnetic ratio (e.g., proton) is irradiated and the nucleus with a low gyromagnetic ratio (e.g., ^{13}C) is observed. The maximum NOE effects for different nuclei observed when ^{1}H is irradiated are shown in Table 3.4.

NOE difference measurements can be used to estimate distances between protons. This can be done by following the buildup of NOE till it reaches a steady state (usually over 5 s). The rate at which the NOE grows is proportional to r^{-6}, where r is the distance between the two nuclei. If the constant of proportionality can be determined, then the rate of buildup can be used to calculate the internuclear distance. This method works particularly well in rigid systems and where distances to be estimated are below 3 Å. A plot of the size of NOE against time affords an exponential curve (Figure 3.18) (provided only direct NOE effects are involved, i.e., there is no third proton intervening between the irradiated and observed protons, and a simple two-spin approach is applicable), whereas a plot of $\ln(h_\infty - h_t)$ against t (where h_t is the size of NOE at time t and h_∞ is the final steady-state value of NOE after it has fully built up) gives a straight line of slope $-k$ (where $k \propto r^{-6}$) (Figure 3.19). The relationship between k and internuclear distance has been determined by measuring NOE buildup in protons with known internuclear distances (*ortho* aromatic protons and CH_3—CH groups) which showed that for $k = 4.5$, $r = 2.15$ Å. Thus, by determining the value of k from the slope of the $\ln(h_\infty - h_t)$ versus t plot, it is possible to calculate the internuclear distance. This method has been used for the measurement of intermolecular as well as intramolecular distances in investigations of the binding of the antibiotics vancomycin and ristocetin A to peptide cell wall analogues.*

---

* G. Wagner and K. Wutrich, *J. Magn. Resonance* **33**, 675 (1981).

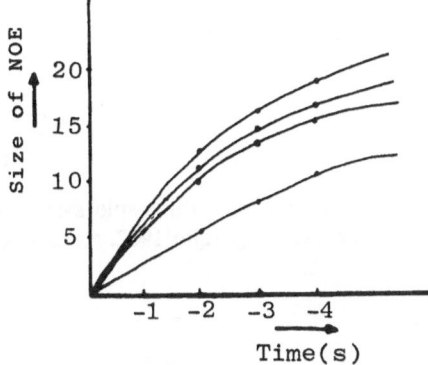

Figure 3.18. Buildup of NOE after irradiation of bound alanine methyl protons in ristocetin A–tripeptide complex. The four curves represent four different protons. (Reprinted with permission from *J. Am. Chem. Soc.*, Copyright 1983, American Chemical Society.)

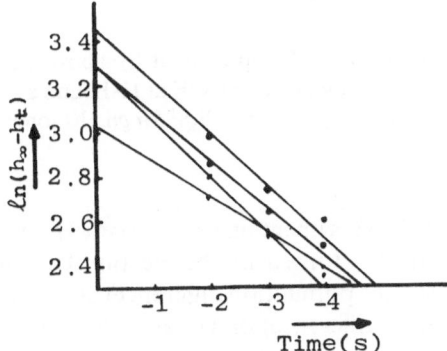

Figure 3.19. The data in Figure 3.18, plotted as $\ln(h_\infty - h_t)$ vs. time. The slopes of the straight lines give the rate at which the NOE is building up, which is proportional to r^{-6}. (Reprinted with permission from *J. Am. Chem. Soc.*, Copyright 1983, American Chemical Society.)

The above discussion has been concerned only with direct NOEs, which are positive except when the tumbling is slowed down, as in macromolecules or in small molecules dissolved in viscous solvents.* In certain geometrical arrangements, protons can also give rise to *negative* NOEs by an indirect mechanism.[†] Let us first consider the direct effects in a three-spin system, shown in

---

* M.P. Williamson and D.H. Williams, *J. Chem. Soc., Chem. Commun.*, 165 (1981); D. Neuhaus, H.S. Rzepa, R.N. Sheppard, and I.R.C. Bick, *Tetrahedron Lett.* 2933 (1981).
[†] J.H. Noggle and R.E. Schirmer, *The Nuclear Overhauser Effect*, Academic Press, New York (1971).

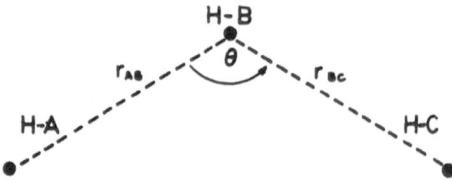

Figure 3.20. Schematic representation of a three-spin system. (Reprinted with permission from *Org. Magn. Resonance*, Copyright 1982, Heyden & Sons, Ltd.

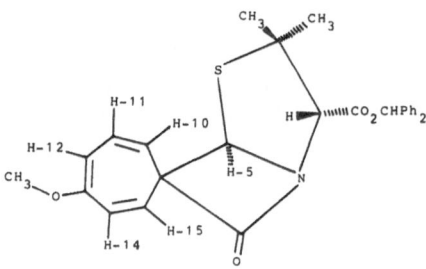

Figure 3.21. Indirect negative NOE represented by above example. Irradiation of OCH_3 protons results in a strong positive NOE at H-12 and a small *negative* NOE at H-11. (Reprinted with permission from *Org. Magn.* Resonance, Copyright 1982, Heyden & Sons, Ltd.

Figure 3.20. In the NOE experiment one proton (say H_A) is irradiated for some time and the irradiation is turned off before the RF pulse. This results in decrease of the intensity (or population difference) of H_A and an increase in the intensity (or population difference) of H_B as well as a small increase in the intensity of H_C. Since a *decrease* in the population difference of H_A is causing an *increase* in the intensity of H_B, the reverse should be equally possible: the increase in the population difference of H_B should cause a decrease in the intensity of H_C (i.e., give rise to a negative NOE). Thus H_C would experience two opposing effects, a *small* increase (due to the larger internuclear distance) caused by direct NOE interaction between H_A and H_C, and a larger negative contribution caused by the indirect propagated effect through H_B, resulting in an overall negative NOE at H_C. This negative NOE will depend on the size of NOE at H_B caused by the irradiation of H_A (since it is this effect which is being transmitted) as well as on the extent to which this NOE effect is transmitted (in a negative mode) to H_C. The larger the value of the individual NOE effects, the more discernible will be the negative NOE effect at H_C.

   An example of the indirect negative NOE effect is provided by the molecule shown in Figure 3.21.* Irradiation of the methoxyl group gives rise to a

---

* J.D. Mersh and J.K.M. Sanders, *Org. Magn. Resonance* **18**(2), 122 (1982).

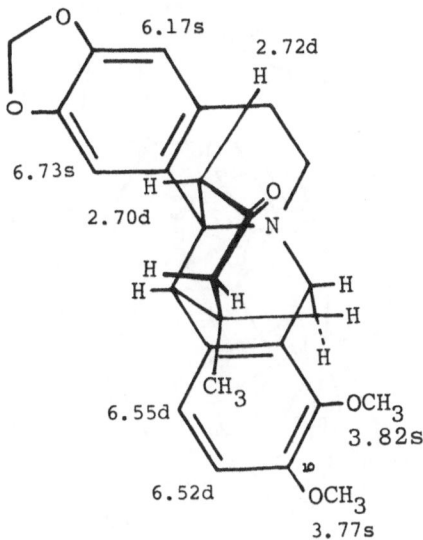

Figure 3.22. Karachine.

positive 19% NOE at H-12, which in turn leads to a 3% negative effect at H-11 and a further propagated effect of 0.5% at H-10. This method is useful in establishing otherwise inaccessible connectivities in complex spectra and provides an effective probe for establishing the optimum conformations of molecules in solution.

The use of NOE in structure elucidation may be illustrated by the example of karachine (Figure 3.22), a bridged protoberberine alkaloid isolated by Atta-ur-Rahman and coworkers.* The substitution pattern in the two aromatic rings was established by NOE difference measurements. These involved successively recording a normal (unenhanced) spectrum, followed by irradiation of a particular proton or group of protons and recording the spectra again while irradiating. Subtraction of the former spectrum from the latter yields *NOE difference* spectra, which are stored in the computer memory. After a few hundred accumulations, the sum of the NOE difference spectra is recorded. This results in the elimination of all peaks which do not show any enhancement, and only those peaks which have undergone some increase (or decrease, as in negative NOE effects) are recorded. Even small enhancements at different positions in the spectrum, which may otherwise be lost under other peaks, can thus be recorded.

Karachine (Figure 3.22) contains two methoxyl groups which appear as singlets at δ 3.82 and δ 3.77. Irradiation of the C-10 methoxyl singlet at δ 3.77 resulted in an overall 11.6% increase in the area of the double doublets for the

---

* For details see G. Blasko, N. Murugesan, A.J. Freyer, M. Shamma, A.A. Ansari, and Atta-ur-Rahman, *J. Am. Chem. Soc.* **104**, 2039 (1982).

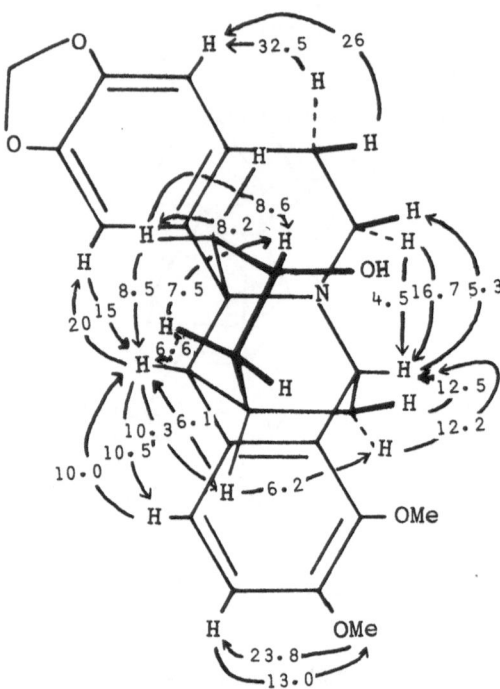

Figure 3.23. Nuclear Overhauser enhancements between protons in valachine. (Reprinted with permission from *J. Chem. Soc. Commun.*, Copyright 1984, Royal Society of Chemistry.)

two aromatic protons which resonate at δ 6.52 and δ 6.55. This showed that the OCH$_3$ group was located in the vicinity of these two aromatic protons. Irradiation of the H-1 singlet at δ 6.17 resulted in a 2.8% increase in the area of the δ 2.70 and δ 2.72 double doublets. Examination of the structure of karachine shows why this should be so. The two protons α to the C=O group, which give rise to the two doublets at δ 2.70 and δ 2.72, lie fairly close above the H-1 proton and therefore an NOE enhancement is to be expected on irradiation of H-1. Irradiation at δ 6.17 and δ 6.73 (H-1 and H-4 singlets) did not result in enhancement of the OCH$_3$ signals, suggesting that no —OCH$_3$ groups are located near the protons giving rise to these resonances. These arguments substantially contributed to the elucidation of the structure of karachine.

The structure of another closely related alkaloid, valachine, has also been established using NOE difference measurements. The NOE effects (% increases) observed between protons lying in close proximity to one another are shown in Figure 3.23.

3.1.8.1.1 Gated Decoupling. As described above, there can be an almost threefold increase in ^{13}C signal intensity under conditions of proton decoupling. However, this effect is no longer visible when the proton decoupling is not being carried out and the peaks become significantly smaller because of (a) loss of NOE and (b) splitting of the singlets into multiplets, with a corresponding division of peak area and reduction in intensity of the signals, when the decoupler is turned off. As the coupling information is very useful in distinguishing methyl, methylene, methine, and quaternary carbons from one another, it is highly desirable to find some way of recording coupled spectra without losing the signal-to-noise enhancement produced by the nuclear Overhauser effect. This is achieved by a technique known as *gated decoupling*. This utilizes the fact that while the decoupling effect appears or vanishes instantaneously on switching the decoupler on and off, the nuclear Overhauser effect depends on spin–lattice relaxation time and takes many seconds to evolve and decay. Thus if the decoupling power is applied prior to the application of the pulse, the populations of the spin states will be disturbed and they will take some time to return to their equilibrium values. The decoupler is then switched off and the pulse is applied. Since the decoupler is off during the pulse, coupling information is retained; moreover, since the spin populations are approximately the same as under conditions of proton decoupling (this is so because the pulse is applied for a very short period and the disturbed spin populations take much longer to return to their equilibrium values) the nuclear Overhauser enhancement is largely retained. This sequence of events is shown in Figure 3.24. A pulse delay of τ seconds is inserted with the decoupler switched on between successive pulses in order to allow the decoupler sufficient time to build up NOE spin populations.

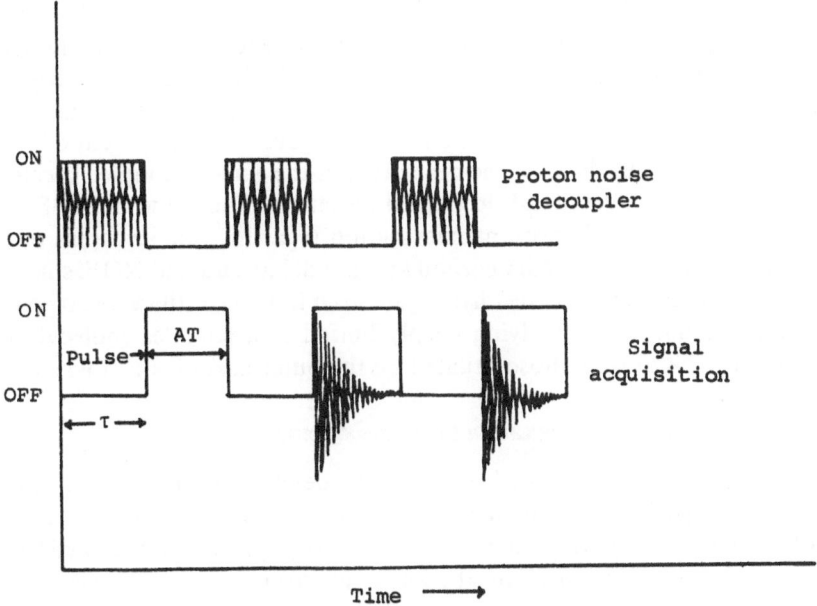

Figure 3.24. Sequence for gated decoupling, producing a coupled spectrum with NOE.

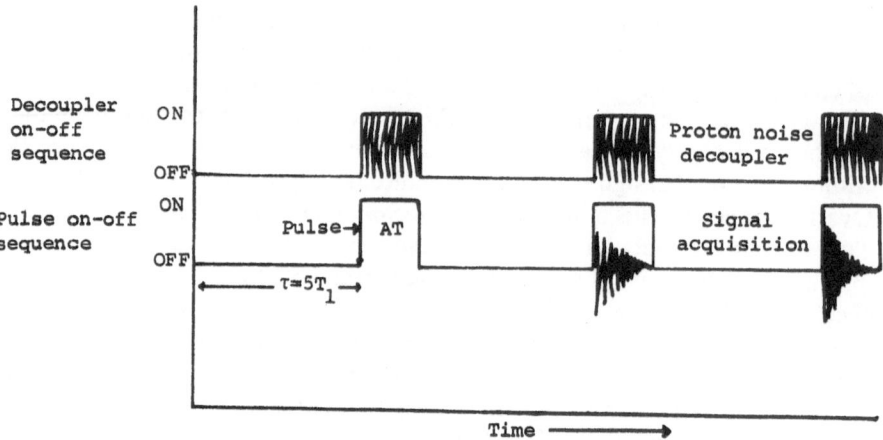

Figure 3.25. Sequence for gated decoupling to afford decoupled spectrum without NOE.

If one wishes to record a coupled spectrum without NOE, then the decoupler is not switched on at any stage. To record a decoupled spectrum but without NOE, the decoupler is switched on as soon as the pulse is applied and it remains on during the signal acquisition so that a decoupled spectrum is obtained. However, since the spin states are perturbed and NOE builds up during the time that the decoupler is on, a sufficient delay has to be introduced before the next pulse is applied in order to allow the equilibrium populations to be reestablished and the NOE to disappear. This delay is usually $5T_1$, where T_1 is the relaxation time of the nucleus. This sequence of events is shown in Figure 3.25. The gated and noise-decoupled ^{13}C-NMR spectra of isopropanol are shown in Figure 3.26. Similarly, the coupled ^{13}C spectra of dioxane, with and without gated decoupling, are shown in Figure 3.27. The extent of NOE can be determined by dividing the peak areas of the gated decoupled spectrum by those of the conventional decoupled spectrum. The value of NOE thus obtained can provide important information regarding the structure of the substance. Thus all quaternary carbon atoms will have a small NOE, since the distance in space to the nearest hydrogen atom influences the value of NOE. Those quaternary carbons lying deeply buried in a complex molecule will have smaller NOE than those situated on the outer part of the molecule.

3.1.8.2   Spin–Lattice Relaxation (T_1) Measurements

There are a number of methods for the determination of spin–lattice relaxation time, T_1. Only the *inversion recovery technique*, which is the one most commonly used, will be described here. A 180° pulse is first applied which causes the magnetization along the z-axis to invert to the $-z$-axis (Figure 3.28). As soon as the pulse is switched off, the magnetization vector begins to relax back along the z-axis by the spin–lattice relaxation processes (Figure

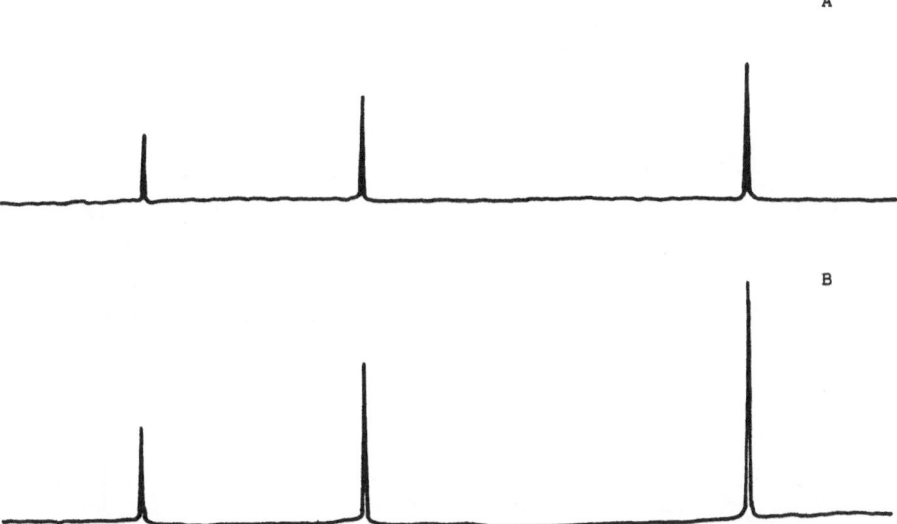

Figure 3.26. (A) Gated decoupled spectrum of isopropanol. (B) Noise-decoupled spectrum of isopropanol. The NOE enhancement in (B) can be determined by dividing the peak area in the lower peak by the peak area in the corresponding upper peak.

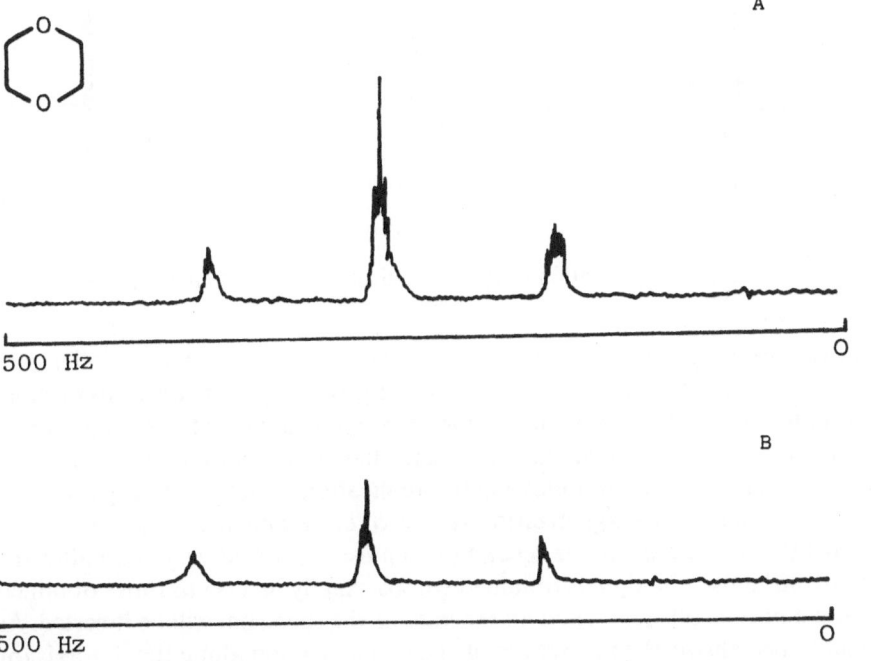

500 Hz

500 Hz

Figure 3.27. Undecoupled ^{13}C spectrum of dioxane: (A) with gated decoupling to retain NOE; (B) without gated decoupling.

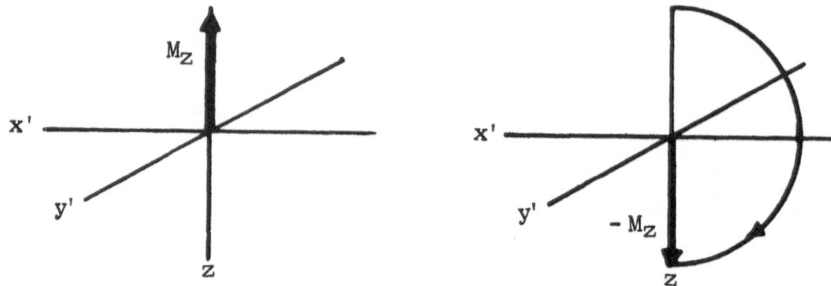

Figure 3.28. Inversion of magnetization vector by the application of a 180° pulse.

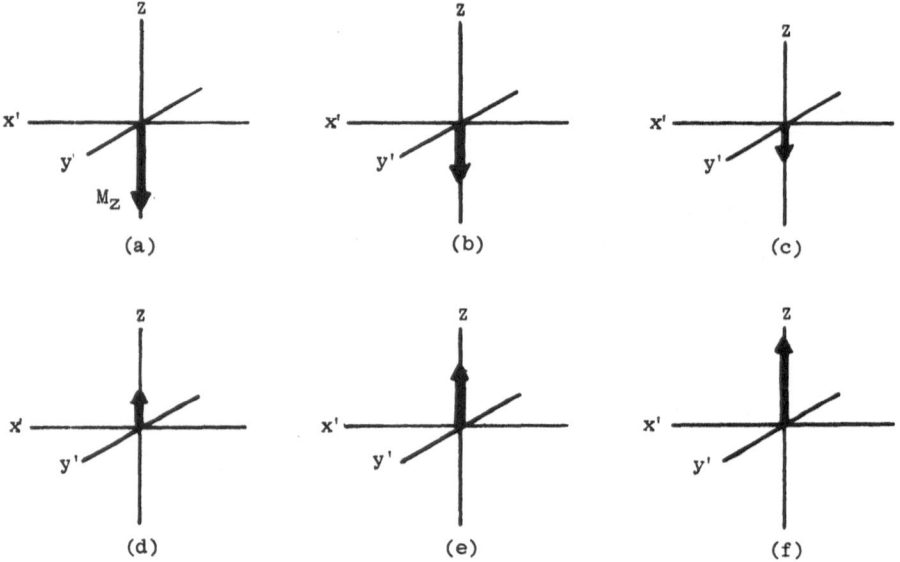

Figure 3.29. Relaxation of the magnetization vector along the z-axis.

3.29). The magnetization vector along the $-z$-axis becomes smaller and smaller, passes through zero, and then starts growing back towards its original equilibrium value. As most spectrometers are not capable of detecting signals along the z-axis, the magnetization vector has to be subjected to a 90° pulse after a delay of τ seconds following the application of the first 180° pulse. This 90° pulse causes the magnetization vector to rotate a further 90° (i.e., a total of 270°) to the $-y$-axis. If the second (90°) pulse is applied very soon after the first 180° pulse, the signals obtained will be exactly inverse to those obtained from a normal 90° pulse, and all the lines of the spectrum will be inverted. As mentioned above, the magnitude of the magnetization along the $-z$-axis (or $+z$-axis) will depend on the delay interval τ between the pulses, the signals being strongly negative (due to the large negative value of the magnetization along the $-z$-axis) if τ is very small and the magnetization vector has had insufficient time to relax; the negative values of the signals decrease with longer and longer delay values, τ, and when the delay is greater than $T_1 \ln 2$, the signals become positive.

Thus if we consider a second case in which the second 90° pulse is applied after the nuclei have had some time to relax (i.e., the magnitude of the magnetization along the $-z$-axis has somewhat decreased), it will rotate the vector to the $-y$-axis and the signals recorded will still be negative, but less intense than when the delay τ was very small.

In a third situation, the delay τ is kept fairly long so that the magnetization has relaxed back up the $-z$-axis through zero and partly up the $+z$-axis (but has not attained the original equilibrium value of the magnetization vector). A 90° pulse would then rotate the vector along the y-axis and positive signals (though somewhat weaker in intensity than the signals from a normal 90° pulse) would be recorded.

Thus the full pulse sequence in the inversion recovery method is given by

$$(180° \text{ pulse–delay } \tau\text{–}90° \text{ pulse–}T)_n$$

where τ is a variable delay time and T is a long pulse delay inserted between successive pulses to allow the magnetization vector to fully relax back to its equilibrium value before a new pulse is applied. This delay T is usually kept to a value of $5T_1$ seconds (after $5T_1$ seconds, magnetization vector $M_z = 0.993 M_z^0$ where M_z^0 is its equilibrium value). The whole sequence is repeated n times, the computer executing the pulses and adjusting the values of the variable delays between the 90 and 180° pulses and the fixed delays T between successive pulse sequences automatically. Figure 3.28 shows the inversion of the magnetization M_z^0 after the application of the first 180° pulse. Figure 3.29 shows how this magnetization vector relaxes back along the $-z$-axis and then up the $+z$-axis. Figure 3.30 illustrates the effects of recording spectra with different τ values. The interrelationship between the spin–lattice relaxation time T_1 and the change in intensity of the magnetization vector is given by the equation

$$\ln(M_z - M_z^0) = -\ln(2M_z^0) - \frac{\tau}{T_1}$$

where M_z is the magnetization vector along the z-axis τ seconds after the application of the 180° pulse, M_z^0 is the equilibrium value of the vector, τ is the variable delay between the 180° and 90° pulses, and T_1 is the spin–lattice relaxation time. Hence a plot of $\ln(M_z - M_z^0)$ against τ will afford a straight line, with a slope of $-1/T_1$ from which T_1 can be determined. The value of M_z^0 can be determined by keeping τ very long ($>5T_1$). Alternatively, one can record a series of "stacked plots" and the point at which the signals go from

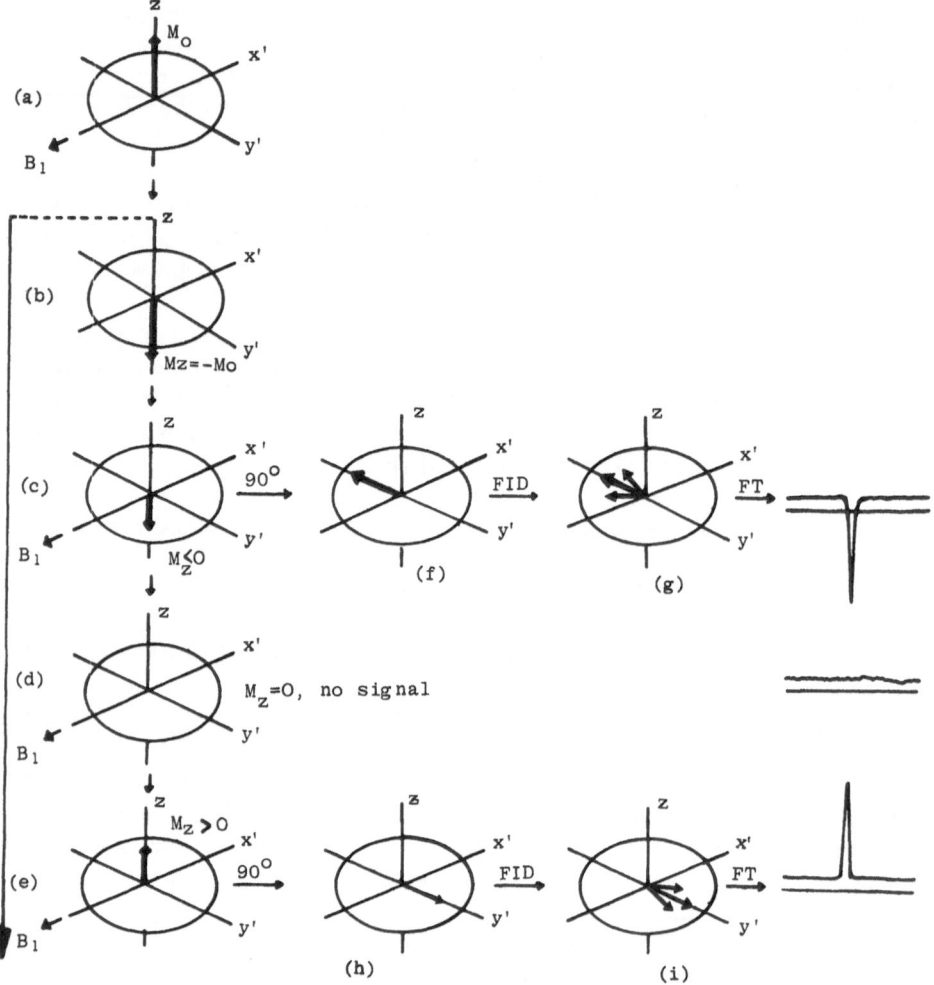

Figure 3.30. Movement of magnetization vector M_0 during an inversion recovery experiment.

the negative side through the zero point can be determined. This zero transition time, τ_0, is related to T_1 by the equation: $T_1 = \tau_0/\ln 2 = \tau_0/0.69$. Hence determining the value of τ (in seconds) at which the signal intensity for a particular carbon is zero and dividing by 0.69 affords its spin–lattice relaxation time. The stacked plots obtained for ethylbenzene are shown in Figure 3.31.

Other methods used for determining T_1 include the *saturation recovery method* and the *progressive saturation method*. Their discussion is beyond the scope of this text.

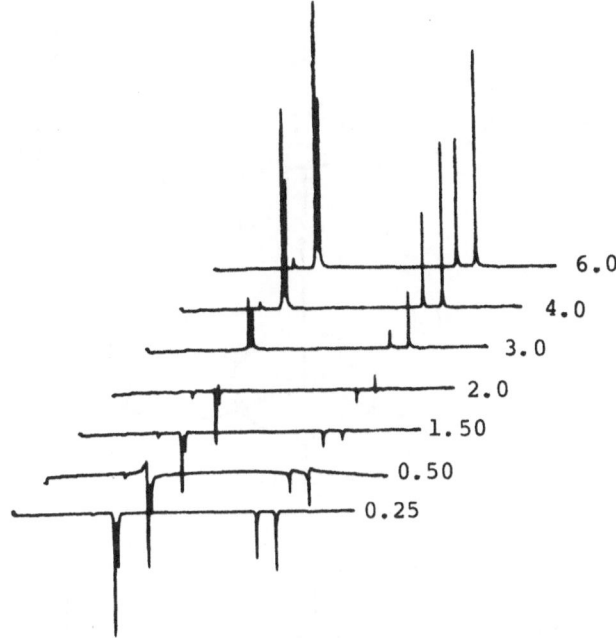

6.0

4.0

3.0

2.0

1.50

0.50

0.25

Figure 3.31. Stacked plots for ethylbenzene; T_1 measurements are made by this procedure.

3.1.8.3   Spin–Spin Relaxation (T_2) Measurements

The Carr, Purcell, Meiboom, and Gill spin-echo (or CPMGSE) method is employed to determine spin–spin relaxation times. The basic method involves the application of a 90° pulse which causes the magnetic vector to bend away from the z-axis, and come to lie on the y-axis. The spins then begin to lose phase coherence and start "fanning out" in the x, y-plane because of field inhomogeneities and spin–spin relaxation (Figure 3.32). The magnetic vectors formed by the fanning-out process travel outwards at different speeds. Thus nucleus "a" in Figure 3.32A, (ii) precesses slightly faster and nucleus "b" precesses slightly slower than the average value. If after a time τ a 180° pulse is applied along the y-axis, all the vectors are rotated about the y-axis by 180°. This causes the faster-moving vector of the "a" nucleus to be positioned *behind* the slower-moving vector of the "b" nucleus [Figure 3.32A, (iv)]. Since the direction of dephasing after the 180° pulse remains the same, the faster-moving nuclei will "catch up" with the slower-moving nuclei after a time 2τ seconds. This "rephasing" results in the production of a *spin-echo* after 2τ seconds [Figure 3.32A, (v)]. The FID of the echo is stored in the computer memory, and the focused vector is again allowed to dephase for a further τ seconds before the application of another 180° pulse which results in

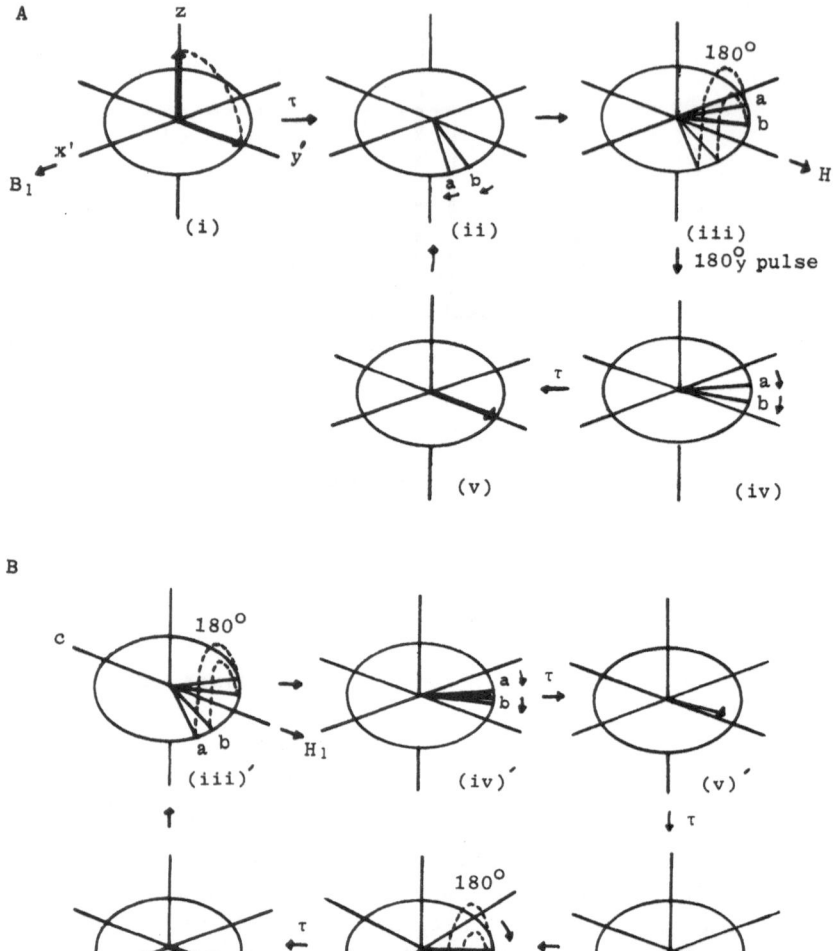

Figure 3.32. (A) (i) 90° pulse tips magnetization by 90°; (ii) nucleus "a" precesses away faster than nucleus "b"; (iii) 180° y pulse causes magnetization vectors "a" and "b" to adopt mirror image positions across the y-axis; (v) nucleus "a" catches up with "b" producing a spin-echo at time 2τ after the original 90° pulse. (B) If pulse angle is not adjusted correctly, the error in setting the 180° pulse causes the two new nuclei to refocus slightly above the y-axis [see (v)']. After subsequent time period τ a second 180° pulse (which is similarly maladjusted) is applied. This causes an "equal and opposite error" [see (vii)'] which results in production of a correctly focused spin-echo in (viii)'.

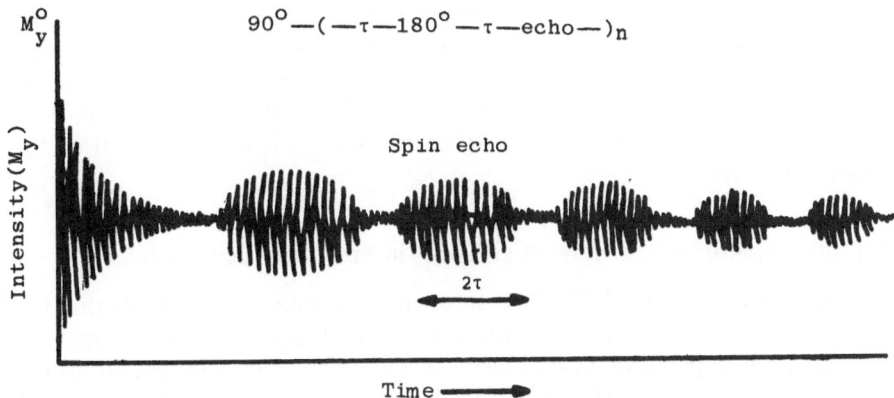

Figure 3.33. Generation of an echo after every 2τ seconds.

rephasing of the vectors and the production of another spin-echo. Thus the pulse sequence $90°-(-\tau-180°-\tau-\text{echo}-)_n$ results in the generation of an echo every 2τ seconds (Figure 3.33). Measurements of the signal intensity, M_y, are made after each echo and a plot of $\ln(M_y - M_y^0)$ against t gives a straight line with a slope of $1/T_2$ from which the spin–spin relaxation time can be obtained. M_y^0 is the intensity of the signal measured immediately after applying the $90°$ pulse.

A variation in the procedure usually employed is to apply the $180°$ pulses in *opposite* directions. This is to neutralize any errors in the timings of the pulses. Thus if the first pulse deviates from the desired $180°$ by a certain error, the application of a second pulse in the opposite direction, having the same fixed error, results in the correct focusing of phases and production of a true echo. By this procedure, signals which contain the error and signals which are error-free are produced alternatively [Figure 3.32, (iii)′–(viii)′]. By using only the even (error-free echoes) and ignoring the odd (error-containing) echoes, the correct value of T_2 can be determined. The sequence used is

$$90°-(-\tau-180°-\tau-\text{echo}-\tau-(-180°)-\tau-\text{echo}-)_n$$

$$\qquad\qquad\;\uparrow\qquad\qquad\qquad\qquad\;\;\uparrow$$

$$\text{With error}\qquad\qquad\text{Without error}$$

These spin-echo Fourier transform experiments are known as the *SEFT* methods.

The method described is only suitable for determining T_2 values of single lines. For more complex spectra, approximations of T_2 values can be made by determining the spin–lattice relaxation times in the rotating frame, $T_{1(\text{rot})}$. In liquids, $T_{1(\text{rot})}$ is approximately equal to the spin–spin relaxation time T_2. $T_{1(\text{rot})}$ times are determined by the *spin-locking* Fourier transformation experiment. This involves the application of a strong field B_1 along the y-axis immediately after the $90°$ pulse continuously for t seconds. During this time, t, the spins are forced to precess in phase with the y'-axis of the rotating frame (hence the name, "spin locking") and dephasing due to magnetic field inhomogeneities is avoided. When the field B_1 is turned off after t seconds, the developing transverse magnetization produces an FID signal. A plot of $\ln(M_y - M_y^0)$ against t gives a straight line, and T_2 is the reciprocal of the slope of this line.

3.1.8.4   Significance of Relaxation Times in Structure Elucidation

When one examines the ^{13}C-NMR spectrum of a molecule, one finds that the signals of the various carbon atoms are not of equal size but may vary considerably, depending on the efficiency with which they can relax from the higher-energy state to the lower-energy state. If one records the NMR spectrum before the nuclei have had a chance to relax, some signals may appear inverted while others may have passed the baseline and start to appear positive. Generally, quaternary carbon atoms, particularly those which are not near any protons, relax very slowly and thus afford only small signals as compared to the proton-bearing carbon atoms. This provides a very useful additional tool in distinguishing the quaternary carbon atoms from other carbon atoms in a complex molecule.

In solution, molecules are tumbling rapidly and these random tumblings cannot be readily considered in terms of their rotational, vibrational, and translational components. The molecular motions are therefore measured in terms of the average time taken by a molecule for one vibration or rotation, and this is known as the *molecular correlation time* τ_c. T_1 measurements provide information on relatively fast molecular reorientations ($\sim 10^{-7}$ s) while T_2 measurements shed light on the slower reorientations ($\sim 10^{-4}$ s). Little work has been done on T_2 measurements, so at the present time T_1 values are more useful in structural assignments.

The effect of the molecular correlation time τ_c on T_1 and T_2 is shown in Figure 3.34. The top two curves show the variation of T_1 values with the correlation time. Most molecules lie either at the minima of the curves or to the left of the minima. Moreover, as the molecules become bigger they can tumble less freely and the molecular correlation times increase while the spin–lattice relaxation times, T_1, decrease. Thus in general it may be said that T_1 *values decrease with increasing molecular weight.* This generalization is not applicable, however, to very large molecules with long molecular correlation times which can lie to the right of the minima, where the T_1 values start increasing again. At higher magnetic fields, the minimum lies toward shorter correlation times, corresponding to higher "frequencies" of molecular motion. Also evident from Figure 3.34 is the fact that T_2 values lie close to T_1 values for molecules with small correlation times. An exception to this are carbon atoms bound to quadrupolar nuclei, which affect T_1 values much more than T_2 values. o-Dichlorobenzene serves to illustrate this dramatically, the carbon atoms bonded to chlorine having much longer T_1 values, while the T_1 and T_2 values of the other carbons are similar (Table 3.5).

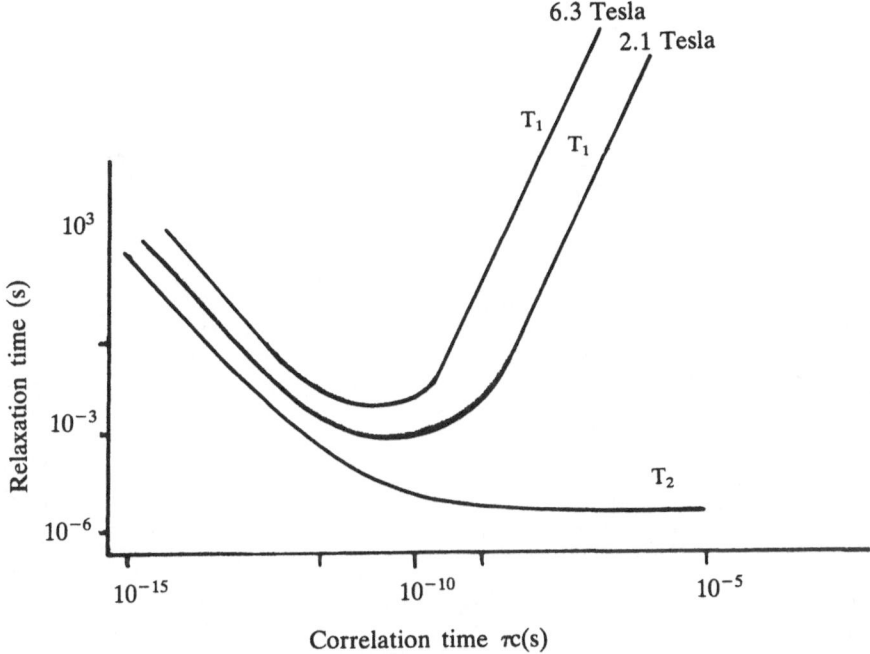

Figure 3.34. Variation of T_1 and T_2 with the molecular correlation time (τ_c). In stronger magnetic fields, the T_1 minimum shifts toward smaller correlation times, corresponding to higher frequencies of molecular motion. (Reprinted with permission from *J. Am. Chem. Soc.*, Copyright 1979, The American Chemical Society.)

Molecular correlation times are calculated on the assumption that the molecule is tumbling or rotating in all directions equally (isotropically). In practice, molecules may rotate in one direction more frequently than others and this will influence the T_1 values of the carbon atoms. Thus in nitrobenzene the preferred axis of rotation lies along the length of the molecule. A comparison of the T_1 values of the proton-bearing carbons shows that the *para* carbon has a shorter T_1 than the *ortho* or *meta* carbon atoms. This is due to the fact that the *para* carbon lies along the axis of the molecule and its correlation time is considerably larger (and T_1 therefore smaller) than that of the *ortho* or *meta* protons. A much larger T_1 value is obtained for C-1 because of its quaternary nature.

Table 3.5. T_1 and T_2 values in *o*-dichloro-benzene

	$T_1(s)$	$T_2(s)$
C_1	66	4.2
C_2	7.8	7.7
C_3	6.3	6.4

Table 3.6. Characteristic ^{13}C T_1 values (s) of some compounds[a]

Compound	T_1
$\underline{C}H_3OH$	13
$\underline{C}H_3I$	13
$\underline{C}H_3Br$	8.8
$\underline{C}HBr_3$	1.6
$\underline{C}HCl_3$	32.4
$H\underline{C}N$	18
$\underline{C}H_3COOH$	10.5
$CH_3\underline{C}OOH$	35
$(CH_3)_2\underline{C}O$	36
$Ph.\underline{C}O.CH_3$	34
Benzene	28
Cyclohexane	20(18)
Cycloalkanes, $(CH_2)_n$	
$n = 3$	37
$n = 4$	36
$n = 5$	29
$n = 6$	20
$n = 7$	16
$n = 8$	10
$n = 10$	5
$\underline{C}S_2$	19
$\underline{C}Cl_4$	6
$\underline{C}_6D_6$	22
$\underline{C}H_3CN$	13
$CH_3\underline{C}N$	5
$H\underline{C}OOH$	10

[a] R.J. Abraham and P. Loftus, *Proton and Carbon-13 NMR Spectroscopy*, Heyden and Son, London (1979); G.C. Levy, *Acc. Chem. Res.* 6, 161 (1973); S. Berger, F.R. Kreissl, and J.D. Roberts, *J. Am. Chem. Soc.* 96, 4348 (1974); G.C. Levy, *J. Chem. Soc., Chem. Commun,* 47 (1972).

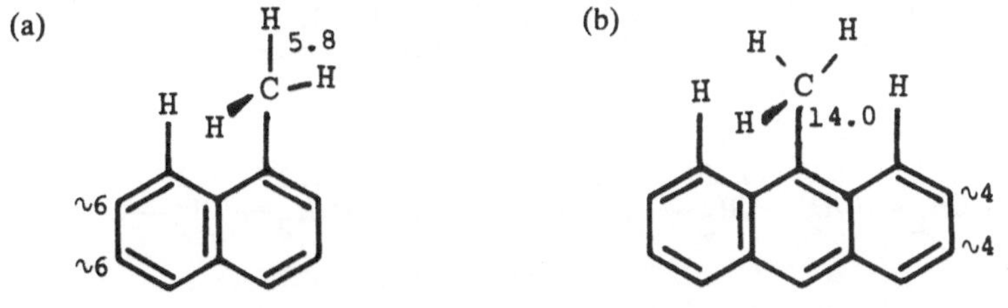

Figure 3.35. (A) 1-Methylnaphthalene. (B) 9-Methylanthracene. (Reprinted with permission from *Accounts of Chemical Research*, Copyright 1973, American Chemical Society.)

The characteristic T_1 value of carbons of many different classes of compounds are given in Table 3.6. T_1 values can vary considerably; generally for molecules with a molecular weight below 1000 they lie between 0.1–300 s (for protonated carbons the values are between 0.1–10 s while for quaternary carbons or for carbons in small symmetrical molecules they lie between 10–300 s). As a broad generalization it may be stated that *the intensity of ^{13}C signals is inversely proportional to the relaxation time*, though this is not always true since certain other factors may also effect signal intensities. In carbons which are part of a rigid skeleton and which are relaxed predominantly by the dipolar mechanism, the T_1 value is inversely proportional to the number of protons directly bound to those carbon atoms. Quaternary carbons which have no protons in their immediate neighborhood exhibit very long relaxation times, and it is usually possible to distinguish various quaternary carbons by correlating their T_1 values with the number of protons which lie in their neighborhood.

Any steric hindrance of rotations of methyl protons also results in decrease of T_1 values. This effect is illustrated by comparison of the T_1 value of the methyl groups in 1-methylnaphthalene (Figure 3.35A) with the corresponding value for 9-methylanthracene (Figure 3.35B). In the former case the *peri* proton forces the methyl group to adopt a preferred conformation, thus restricting its movement, while in the latter case two equivalent *peri* interactions occur so that the methyl does not adopt any preferred conformation.

In molecules which have a rigid skeleton bearing a flexible side chain, the movement in the flexible part will be greater than in the rigid portion. The correlation times for the flexible portions will therefore be shorter and the T_1 values greater, leading to reduction in the intensity of corresponding signals. In the case of a long chain, the ends of the chain will move more freely and the T_1 values will therefore be greater for the end carbons; as one moves away from the ends the T_1 values become progressively smaller because of the greater restrictions of rotational freedom. Attachment of bulky groups to the end of the chain reduces its freedom of movement and thus decreases the T_1 values of the carbon atoms located at the end of the chain. These effects are illustrated by the T_1 values of decane and 1-bromodecane (Figure 3.36).

$$
\begin{array}{ccccccccc}
8.74 & 6.64 & 5.71 & 4.95 & 4.78 & 4.36 & 4.95 & 5.71 & 6.64 & 8.79
\end{array}
$$

$$CH_3-CH_2-CH_2-CH_2-CH_2-CH_2-CH_2-CH_2-CH_2-CH_3$$

(a)

$$
\begin{array}{ccccccccc}
2.5 & 2.7 & 1.9 & 2.0 & 2.0 & 2.1 & 2.2 & 3.1 & 3.9 & 5.3
\end{array}
$$

$$Br-CH_2-CH_2-CH_2-CH_2-CH_2-CH_2-CH_2-CH_2-CH_2-CH_3$$

(b)

Figure 3.36. T_1 values of decane and 1-bromodecane. (Reprinted with permission from *J. Chem. Soc. Chem. Commun.*, Copyright 1973, Royal Society of Chemistry.)

T_1 values are of considerable significance in probing the mechanisms of a number of important biological processes. Thus conformational changes occurring in proteins (helix–random coil transitions), segmental mobility in phospholipids which determines molecular transport, and permeability through membranes, etc., can all be studied by determining T_1 values of suitable carbon atoms.

3.1.8.5   Solvent Elimination

A problem often encountered in recording NMR spectra is the presence of large solvent peaks which may overlap with those of the sample. Since most solvents have much lower molecular weights than those of the samples, their molecular correlation times are usually much shorter and T_1 values correspondingly longer. This fact can be utilized by using the inversion recovery method to eliminate the solvent peaks selectively. Application of a 180° pulse along the z-axis inverts the magnetization along this axis. A delay τ is then introduced which allows the magnetization to recover. The delay τ is selected so that it is equal to $0.693\,T_{1\,\text{solvent}}$, since this is the time at which the solvent peak will be passing from its inverted state upwards through zero. Since the sample will have shorter T_1 values, the sample signal will have relaxed back to its original value by this time. At this point a second 90° pulse is employed to bend the magnetization to the y-axis where it can be detected; since the solvent peak will have zero intensity, only the sample peaks will be recorded. The procedure may be difficult to use if the T_1 value of the solvent is not significantly different from the T_1 values of the protons to be measured. However, it may in many cases be better than the alternative procedure of applying a decoupling field at the solvent chemical shift, as that tends to remove the sample signals which may be lying under the solvent signals. The pulse sequence employed is

$$(-180°-\tau-90°-PD-)_n$$

where the pulse delay (PD) of magnitude $5\,T_{1\,\text{solvent}}$ is introduced in order to allow the solvent peak to return to its equilibrium value before a second pulse is applied. This procedure can also be used to selectively eliminate signals of one of two different protons which differ significantly in their T_1 values.

3.2 STUDY OF DYNAMIC EFFECTS BY NMR SPECTROSCOPY

NMR spectroscopy offers a method for the study of various dynamic processes encountered in organic chemistry. A classical example of an equilibrium process between two different structures of the same molecule is provided by N,N-dimethylformamide. The bond between the carbonyl carbon and nitrogen in this molecule has significant double bond character [Figure 3.37(a)], which results in a differentiation between the two methyl groups. This effect is visible if the rate of rotation of the $N(CH_3)_2$ group is sufficiently slow, since at any given time one of the methyl groups would be *cis* to the carbonyl group while the other would be *trans* to it [Figure 3.37(b)]. It is, therefore, not surprising that the two methyl groups of dimethylformamide give two separate signals. On heating, these signals tend to broaden as the rate of rotation of the $N(CH_3)_2$ group increases, and above 120°C they coalesce to a single peak. This is shown in Figure 3.38.

The *position* of an equilibrium, $A \rightleftharpoons B$, between two states A and B in equilibrium is determined by the free energy, ΔG, of the process, and it may be described by the equation

$$n_B/n_A = \exp(-\Delta G/RT)$$

provided $n_A + n_B = 1$, where n_A and n_B are the mole fractions of the two states A and B, R is the universal gas constant (8.31 J K^{-1}), and T is the absolute temperature, K.

The *rate*, k, of the interconversion between the states A and B is determined by the free energy of activation, $\Delta G^{\ddagger}$, as given by the equation

$$k = \frac{RT}{Nh}\exp(-\Delta G^{\ddagger}/RT)$$

The position at which the peaks of the methyl groups A and B of dimethylformamide just merge into each other is called the "coalescence

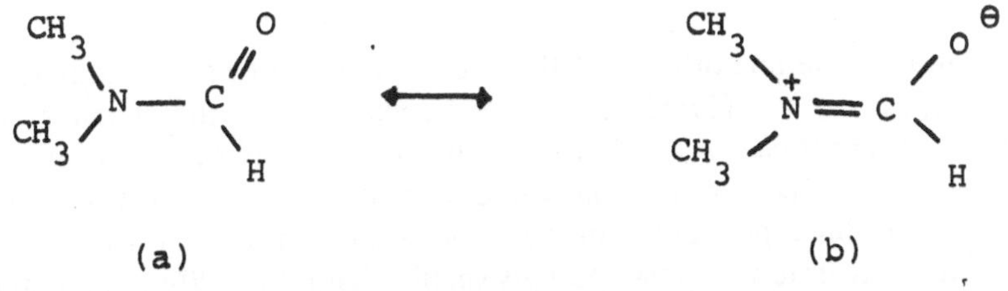

(a)   (b)

Figure 3.37. *N,N*-dimethylformamide. The double bond character in the C—N bond results in the two *N*-methyl groups appearing as separate signals. On heating, the rotation increases, resulting in the coalescence of the *N*-methyl signals to a single peak.

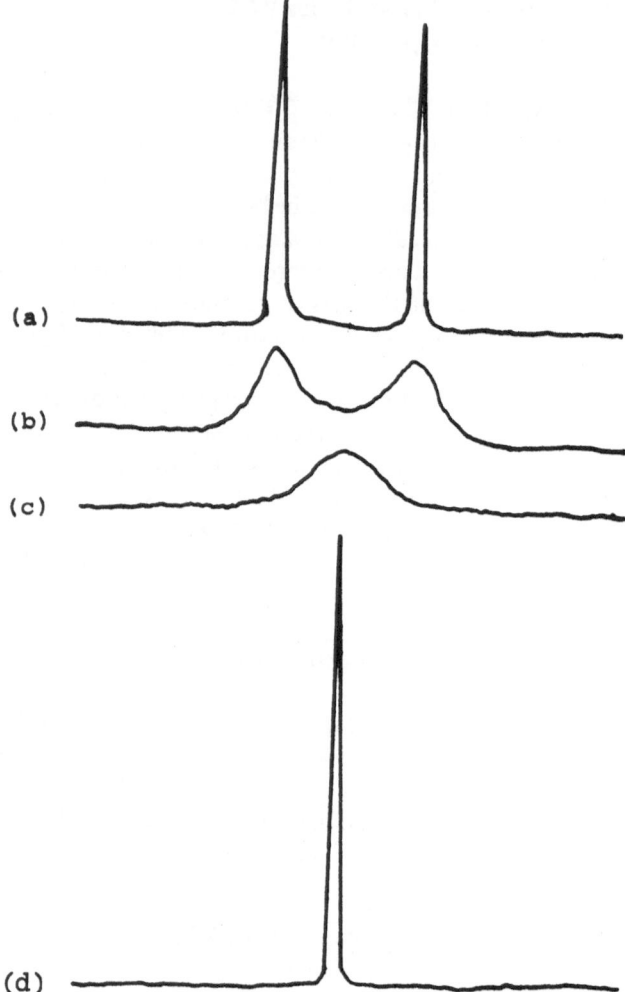

Figure 3.38. (a) Methyl groups of dimethylformamide appear as two singlets but with increasing temperature they broaden (b), coalesce (c), and appear as *one* sharp singlet (d).

point", and the lifetime of each of the two methyl groups at the coalescence point is given by $\tau = [2/(\pi\delta v)]^{1/2}$ seconds, where δv is the chemical shift difference in Hz between A and B. In equilibrating systems, δv for protons may be between 0 and 100 s so that τ may be about 10^{-1} s while for ^{13}C nuclei the chemical shift differences between the nuclei are larger and the value of τ at coalescence is therefore correspondingly smaller. Thus for a 20-ppm difference in chemical shifts between nuclei A and B in equilibrium the lifetime of the nuclei at the coalescence point would be approximately 1×10^{-3} s. The rate constant at coalescence ($k_{\text{coal.}}$) is given by the relation: $k_{\text{coal.}} = \pi \delta v/(2)^{1/2} = 2.22\delta v$. Separate signals for the two nuclei in equilibrium will be observed if $k/\delta v > 2.22$ but an averaged signal at $(v_A + v_B)/2$ will be recorded if $k/\delta v < 2.22$. The free energy of activation $\Delta G^{\ddagger}$ can be obtained from the equation $\Delta G^{\ddagger} = 19.14T_c(9.97 + \log\ T_c/\delta v)$ where T_c is the temperature of coalescence. The *energy barrier* that is obtained by this equation is temperature dependent because $\Delta G^{\ddagger}$ depends on the entropy of the system.

One can broadly distinguish three rate types in exchange processes:

a. *Slow rate of exchange.* In this process the species A and B are inter-converting slowly and two separate spectra for the two molecular species will be observed. The relative intensities of the signals would give the mole fractions of A and B (i.e., n_A and n_B) from which the free energy ΔG can be obtained.

b. *Fast rate of exchange.* If the interconversion of A and B is fast, then only one time-averaged spectrum is observed. The frequency v_{AV} at which this averaged signal appears is given by $v_{AV} = n_A v_A + n_B v_B$, while the coupling is given by $J_{AV} = n_A J_A + n_B J_B$.

c. *Medium rate of exchange.* In this situation broad lines are observed. Peak broadening due to medium rate of exchange should be distinguished from other factors causing broadening of peaks such as presence of quadrupole nuclei (e.g., ^{14}N), presence of paramagnetic species or solid impurities, etc.

In the following sections the equilibria which can be studied by NMR spectroscopy are categorized.

3.2.1 Hindered Internal Rotation

The example of partial double bond character hindering internal rotation in amides has already been discussed above for dimethylformamide. Other examples of hindered rotation are presented in Table 3.7.

The *tert*-butyl group offers another case of hindered rotation. When attached to cycloalkanes, the *tert*-butyl group experiences significant hindrance to rotation, resulting in one of its methyl groups appearing at a different chemical shift than the other two at low temperatures (Figure 3.39).

3.2.2 Keto–Enol Tautomerism

Keto–enol tautomerism encountered in ethyl acetoacetate (Figure 3.40) is another slow exchange process. It results in a small contribution of the enol tautomer of ethyl acetoacetate to the spectrum, the enol form being partly stabilized due to conjugation of the double bond of the enol with the α-carbonyl group of the ester moiety.

It may be noted that one can sometimes "lose" signals for ^{13}C nuclei due to equilibration between tautomeric forms. This is so because the signal-to-noise ratio in ^{13}C-NMR spectra is sometimes poor, resulting in weak signals. If an equilibrium process is simultaneously occurring and the temperature at which

Table 3.7. Examples of hindered rotation with activation energies in kcal mol^{-1}

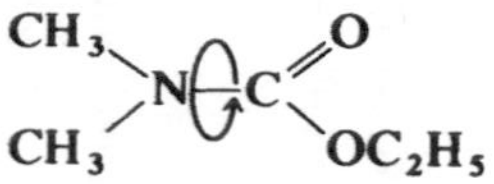

$\Delta G^{\ddagger}_{270} = 16.0$

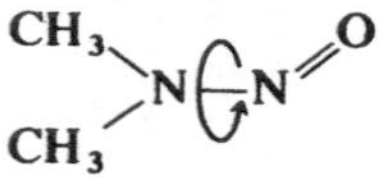

$E_a = 23$

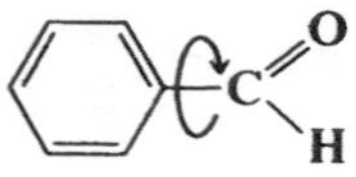

$\Delta G^{\ddagger}_{146} = 7.9$

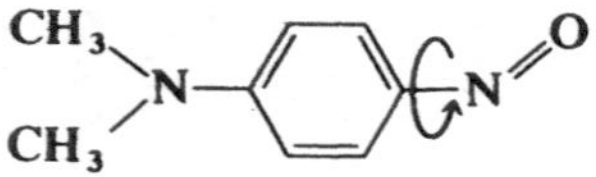

$\Delta H^{\ddagger} = 14.9$

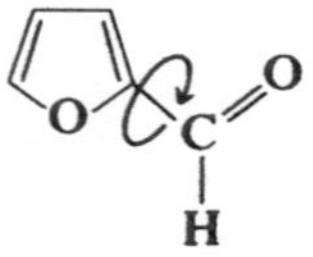

$\vec{E}_a = 12.0$

$\overleftarrow{E}_a = 11.0$

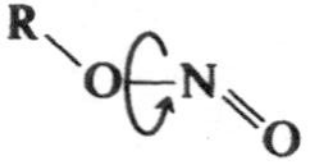

$E_a = 10$

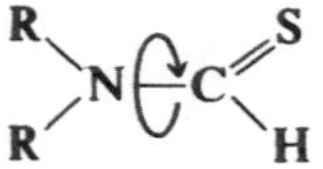

$E_a = 25$

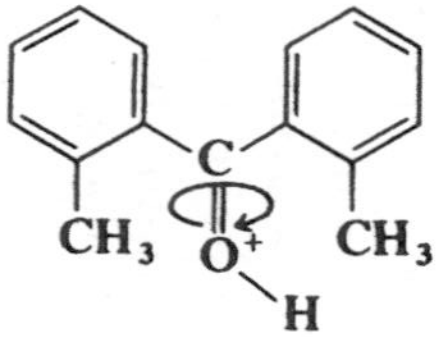

$\Delta G^{\ddagger}_{298} = 13.6$

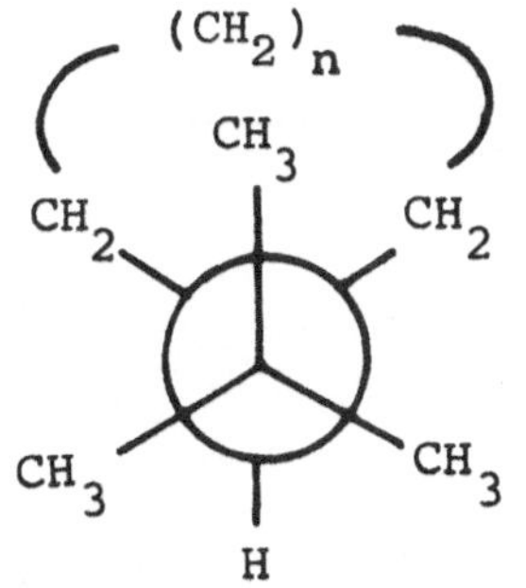

Figure 3.39. A Newman projection of the *tert*-butyl group in a substituted cycloalkane. One of the methyl groups lies in a different steric environment from the other two, and the restricted rotation caused by steric hindrance therefore results in two separate signals being observed for the three methyl groups.

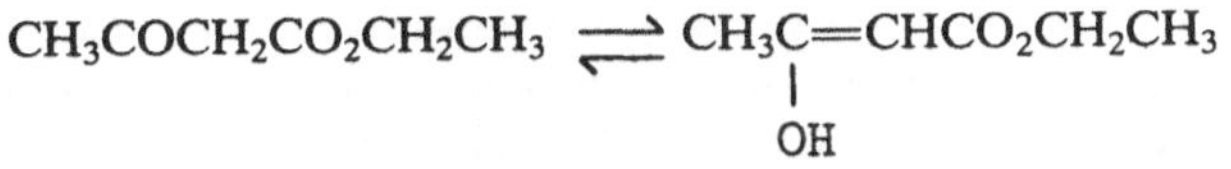

$$CH_3COCH_2CO_2CH_2CH_3 \rightleftharpoons CH_3C{=}CHCO_2CH_2CH_3$$
$$\underset{OH}{|}$$

Figure 3.40. Keto–enol tautomerism in ethyl acetoacetate.

the NMR spectrum is being recorded is near the coalescence point, then these weak signals may broaden out so as to disappear in the noise. An example of this is provided by the ^{13}C-NMR spectrum of nigellicine, a new alkaloid isolated by Atta-ur-Rahman and coworkers from seeds of *Nigella sativa*. It can be seen from Figure 3.41 that some of the signals for the ^{13}C atoms are absent. Since the structure was confirmed by X-ray crystallography, the missing signals may be attributed to exchange occurring between tautomeric forms as well as to the low intensity of NMR signals of nonprotonated carbon atoms.

3.2.3   Inversion of Configuration

The inversion of configuration in nitrogen compounds has been studied by NMR spectroscopy in the case of many amines, such as dibenzylmethylamine, aziridines, and a number of other cyclic amines. Similarly, the conformational mobility of ring systems, such as cyclohexane, has been extensively studied. The inversion of one chair form of cyclohexane to another would bring axial protons to equatorial positions and vice versa. Since different chemical shifts would be expected for axial and equatorial protons, it should be possible to study this equilibrium. This has been done using cyclohexane-d_{11} and decoupling the H–D coupling effects so that only one proton in axial and equatorial positions could be observed. At room temperature cyclohexane-d_{11} gives a sharp single line but on cooling to $-60°$, the line broadens due to the slowing of the equilibrium. At $-68°$ it becomes split into two broad lines for the axial and equatorial protons, which sharpen at $-89°$.

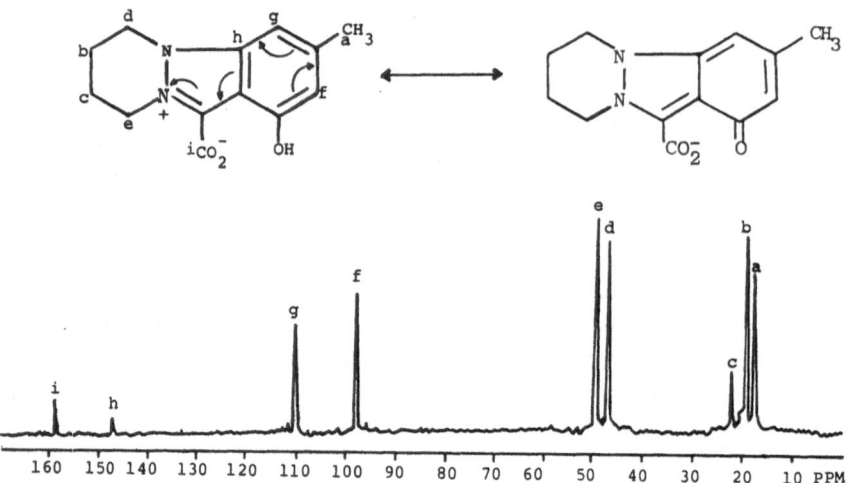

Figure 3.41. ^{13}C-NMR spectrum of nigellicine; note that certain carbon atoms are not visible. This is accounted for by the low intensity of quaternary carbons, and the contribution of tautomeric structures, which results in the lowering of the amount of any one particular contributing species; hence the apparent "loss" of certain carbon atoms. (Reprinted with permission from *Tetrahedron Lett.*, Copyright 1985, Pergamon Press.)

3.2.4 Proton Exchange Equilibria

There are a number of proton exchange equilibria which can be studied by NMR spectroscopy. A simple example is the NMR spectrum of ethanol which, in the absence of moisture or traces of acid, shows coupling of the hydroxyl proton with the adjacent methylene protons (Figure 3.42A). With increasing amounts of acid, the rate of exchange of the hydroxylic proton increases so that ultimately the methylene protons "see" the hydroxylic proton in a single time-averaged environment because of the very fast exchange (Figure 3.42B).

The interconversion of the N—H protons in the pyrrole rings of the porphyrin molecule provides another interesting example of proton exchange equilibrium. The equilibrium structures shown in Figure 3.43 illustrate that in *meso*-tetraphenylporphyrin (TPP)* there are two types of β-hydrogens attached to the two different types of five-membered heterocyclic rings. However, the NMR spectrum of TPP shows only one signal for the β-hydrogens at room temperature but on cooling to −60°, the signal is split into a doublet ($J = 17$ Hz). At room temperature one is therefore looking at a fast rate of exchange which is slowed down by cooling to afford a doublet for the β-hydrogens.

---

* In porphyrin systems there are three types of carbons, conventionally named α-carbon, β-carbon, and *meso*-carbon.

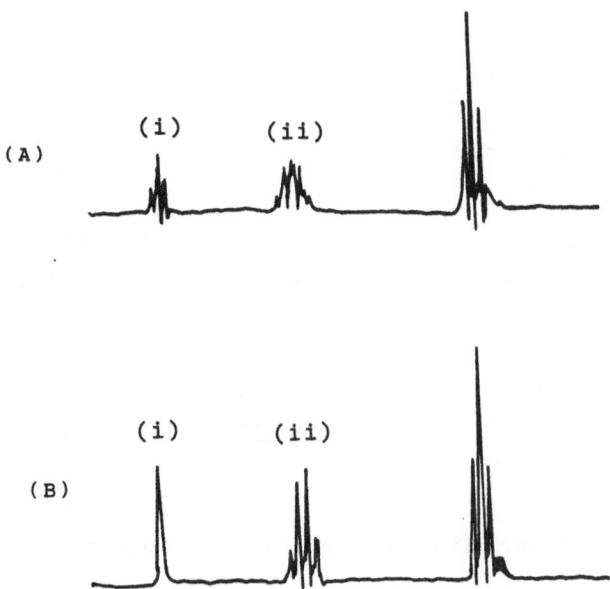

Figure 3.42. (A) Slow rate of exchange of —OH proton results in coupling between hydroxyl (i) and methylene (ii) protons in ethanol. (B) Fast rate of exchange of —OH proton.

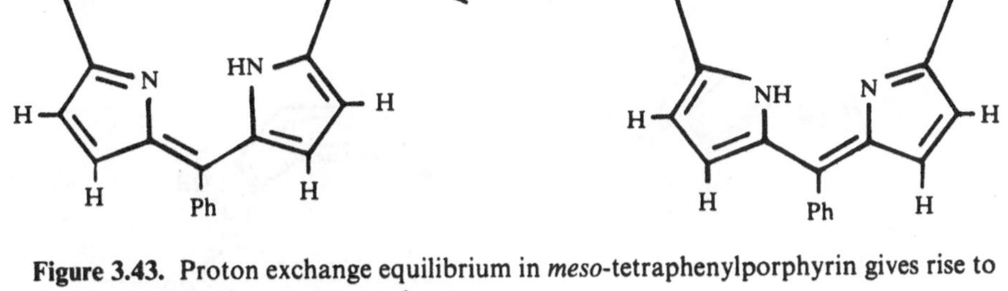

Figure 3.43. Proton exchange equilibrium in *meso*-tetraphenylporphyrin gives rise to the two contributing structures shown.

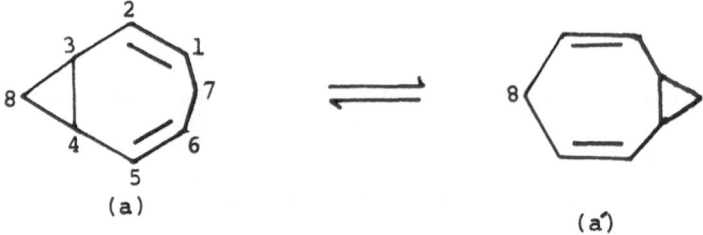

Figure 3.44. Valence tautomerism in 3,4-homotropilidene.

Figure 3.45. Benzene oxide–oxepine isomerization.

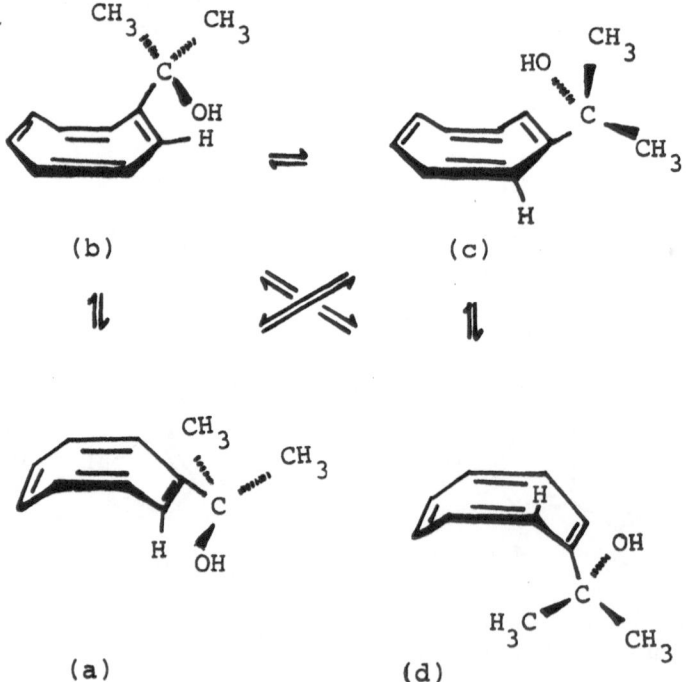

Figure 3.46. Structural and conformational isomers of cyclooctatetraene derivatives.

3.2.5 Valence Tautomerism

Another type of equilibrium which has been studied by NMR spectroscopy is valence tautomerism. This is illustrated by the example of 3,4-homo-tropilidene (Figure 3.44) in which bond migration leads to the cyclopropane ring switching between two alternative positions. Thus C-8 changes from α cyclopropyl to an allylic carbon. By lowering the temperature to below −40°C, it is possible to freeze the equilibrium sufficiently to record the spectrum of a single structure rather than that of the equilibrium mixture.

Other equilibria which have been studied by NMR spectroscopy include those of oxepine–benzene oxide isomerization (Figure 3.45), structural and conformational isomers of cyclooctatetraene derivatives (Figure 3.46), carbocations, and organometallic compounds.

RECOMMENDED READING

1. J.W. Cooper, *Spectroscopic Techniques for Organic Chemists*, John Wiley and Sons, New York (1980).
2. D. Shaw, *Pulsed Fourier Transform NMR Spectroscopy*, Elsevier Scientific Publishing Co., Amsterdam (1979).
3. F.W. Wehrli and T. Wirthlin, *Interpretation of Carbon-13 NMR Spectra*, Heyden and Son Ltd., London (1978).
4. G.C. Levy, Ed., *Topics in Carbon-13 NMR Spectroscopy*, Vol. 1-3, John Wiley and Sons, New York (1979).
5. E. Breitmaier and W. Voelter, *^{13}C-NMR Spectroscopy*, Verlag Chemie, Weinheim (1978).
6. C. Brevard and P. Granger, *Handbook of High Resolution Multinuclear NMR*, John Wiley and Sons, New York (1981).
7. G.C. Levy, R.L. Lichter, and G.L. Nelson, *Carbon-13 Nuclear Magnetic Resonance Spectroscopy*, John Wiley and Sons, New York (1980).
8. H. Günther, *NMR Spectroscopy*, John Wiley and Sons, New York (1980).
9. L.M. Jackman and S. Sternhell, *Applications of Nuclear Magnetic Resonance Spectroscopy in Organic Chemistry*, International Series in Organic Chemistry, Vol. 10, Pergamon Press, Oxford (1969).
10. F.A. Bovey, *Nuclear Magnetic Resonance Spectroscopy*, Academic Press, New York (1969).
11. E.D. Becker, *High Resolution NMR*, Academic Press, New York (1980).
12. M.L. Martin and G.J. Martin, *Practical NMR Spectroscopy*, Heyden and Son Ltd., London (1980).
13. T. Clerc and E. Pretsch, *Kernresonanz Spektroskopie*, Akademische Verlagsgesellschaft, Frankfurt (1973).
14. A. Carrington and A.D. McLachlan, *Introduction to Magnetic Resonance*, Chapman and Hall, London (1979).
15. J.H. Noggle and R.E. Shirmer, *The Nuclear Overhauser Effect; Chemical Applications*, Academic Press, New York (1971).

Chapter 4

Chemical Shifts and Spin–Spin Couplings in ^{13}C-NMR Spectroscopy

4.1 CHEMICAL SHIFTS IN ^{13}C-NMR SPECTROSCOPY

The range of ^{13}C chemical shifts is normally 0–230 ppm. In a proton-decoupled spectrum they are found as sharply resolved single peaks, and they are measured with reference to TMS taken as zero, although addition of TMS is often not carried out since the position of the solvent signal with reference to TMS is known, and the signals may be calibrated with reference to the solvent signal. The chemical shifts of common solvents are presented in Table 4.1.

4.1.1 Factors Affecting Chemical Shifts

As discussed earlier for protons, there are a number of factors which affect chemical shifts of ^{13}C atoms. The extent of shielding which a ^{13}C atom experiences can be considered in terms of the shielding constant σ_i, which is made up of a number of contributing components:

$$\sigma_i = \sigma^{\text{dia}} + \sigma^{\text{para}} + \sigma^{\text{N}}$$

The *local diamagnetic shielding term*, σ^{dia}, is indicative of the shielding contribution due to the isotropic circulation of electrons around the nucleus. The greater the electronic density around the nucleus, the more shielded it will be and this will cause corresponding upfield shifts of the resonance lines. The magnitude of the diamagnetic term is inversely proportional to the distance between the nucleus and the circulating electrons. Thus s electrons, which lie closer to the nucleus, will cause greater shielding than p electrons which are farther away. The *local paramagnetic term*, σ^{para}, represents the contribution of the anisotropic (nonspherical) electron circulation effects. It is a deshielding effect and thus acts in opposition to σ^{dia}. For nuclei other than ^{1}H, such as ^{13}C, σ^{para} is the dominant effect. The *neighbor anisotropy term*, σN represents the

Table 4.1. ^{13}C data for common deuterated solvents and their ^{1}H isotopes

Compound	C–D signal multiplicity	^{13}C shifts (ppm)	Isotopic ^{1}H compound ^{13}C shifts (ppm)
CD_2Cl_2	quintet	53.1	53.8
$CDCl_3$	triplet	77.0 ± 0.5	$78.0 + 0.5$
CD_3CN	septet	1.3	1.7
	*	118.2	118.2
CD_3OD	septet	49.0	49.9
C_2D_5OD	septet	15.8	16.9
	quintet	55.4	56.3
$C_2D_5OC_2D_5$	septet	13.4	14.5
	quintet	64.3	65.3
1,4-Dioxane-d_8	quintet	66.5	67.6
	quintet	25.2	26.2
Tetrahydrofuran-d_8	quintet	67.4	68.2
CD_3COCD_3	septet	29.8	30.7
	*	206.5	206.7
DMSO-d_6	septet	39.7	40.9
CD_3COOD	septet	20.0	20.9
	*	178.4	178.8
DMF-d_7	septet	30.1	30.9
	septet	35.2	36.0
	triplet	167.7	167.9
HMPA-d_{18}	septet	35.8	36.9
cyclohexane-d_{12}	quintet	26.4	27.8
C_6D_6	triplet	128.0 ± 0.5	128.5 ± 0.5
Pyridine-d_5	triplet (3, 5)	123.5	124.2
	triplet (4)	135.5	136.2
	triplet (2, 6)	149.2	149.7

* Long-range multiplet, not resolved.

fields caused by electron circulation around neighboring atoms. It therefore depends on molecular geometry and the nature of the neighboring atoms.

Chemical shifts in ^{13}C spectroscopy are dependent on the factors described in the following sections.

4.1.1.1 Hybridization of the Observed Nucleus

The range within which ^{13}C signals occur is greatly influenced by the state of hybridization of the observed nucleus. Thus sp^3-hybridized carbons resonate between -20 and 100 ppm and sp^2 carbons resonate between 120 and 240 ppm while sp hybridized carbons resonate between 70 and 110 ppm.

Table 4.2. ^{13}C chemical shifts induced by replacement of a terminal hydrogen atom by an electronegative substituent in n-alkanes

$$X-\overset{\alpha}{C}H_2-\overset{\beta}{C}H_2-\overset{\gamma}{C}H_2-CH_2\cdots$$

		Shift in ^{13}C resonance		
Substituent X	Electronegativity	α	β	γ
H	2.1	0	0	0
CH$_3$	2.5	+9	+10	−2
NH$_2$	3.0	+29	+11	−5
Cl	3.0	+31	+11	−4
F	4.0	+68	+9	−4
SH	2.5	+11	+12	−6

4.1.1.2 Electronegativity of Substituents

The electronegativities of substituents have a pronounced effect on ^{13}C chemical shifts. This is best seen by comparing the chemical shifts of α-, β-, and γ-carbon atoms in halogenated n-alkanes to those of the corresponding unsubstituted compounds (Table 4.2). On theoretical grounds it can be predicted that the inductive effects will be propagated along the chain in such a manner that alternating $\delta+$ (deshielding) and $\delta-$ (shielding) effects will be produced on alternate carbon atoms, and the effect should decrease with the inverse cube of the distance (r^{-3}) from the substituent:

$$\underset{X}{\overset{\delta-}{}}\!\!-\!\!\underset{C}{\overset{\delta+}{}}\!\!-\!\!\underset{C}{\overset{\delta\delta-}{}}\!\!-\!\!\underset{C}{\overset{\delta\delta\delta+}{}}\!\!-$$

The effects at the β- and γ-carbons should thus be much smaller than are actually observed, suggesting that some other factors are operative simultaneously.

When certain second-row heteroatoms such as N, O, or F are present in a conformational array such that the carbon is anti-periplanar to the hetero-atom (i.e., X=N, O, or F in Figure 1(c) or (d)), then the γ carbon is seen to undergo a significant upfield shift (in comparison to compounds which have a CH$_3$, CH$_2$, or a third-row heteroatom present in an anti-periplanar arrangement to the γ carbon atom). This γ effect can be several ppm in magnitude and it may be important in stereochemical assignments. It has been attributed to the hyperconjugative transfer of charge from the electron pairs on the heteroatom through the $C_\alpha C_\beta$ bond to the γ carbon atom (Figure 4.1(d)).* It is

---

* E.L. Eliel, W.F. Bailey, L.D. Kopp, R.L. Willer, D.M. Grant, R. Bertrand, K.A. Christensen, D.K. Dalling, M.W. Duch, E. Wenkert, F.M. Schell and D.W. Cochran, *J. Am. Chem. Soc.*, **97**, 322 (1975).

not observed when the heteroatom is located at the bridgehead of a bicyclic compound.

4.1.1.3 Steric and van der Waals Effects

Steric crowding of proton-bearing carbon atoms can result in their being shielded. Steric perturbation of C—H bonds results in the electrons moving away from the hydrogen atoms and towards the carbon atoms, thus resulting in the carbon atoms being shielded (and the protons deshielded). This effect is most pronounced when the proton-bearing carbon atoms and the substituents are in a γ-gauche disposition with respect to one another, resulting in steric crowding of the hydrogen atoms (Figure 4.1(b); Table 4.3). Gauche hetero-atoms are generally more effective in producing upfield shifts than methyl or methylene groups.

4.1.1.4 Electron Deficiency

Electron deficiency at a carbon atom (e.g., by carbonium ion formation) results in marked downfield shifts. The chemical shifts of various carbonium ions shown in Table 4.4 illustrate the shifts that can occur.

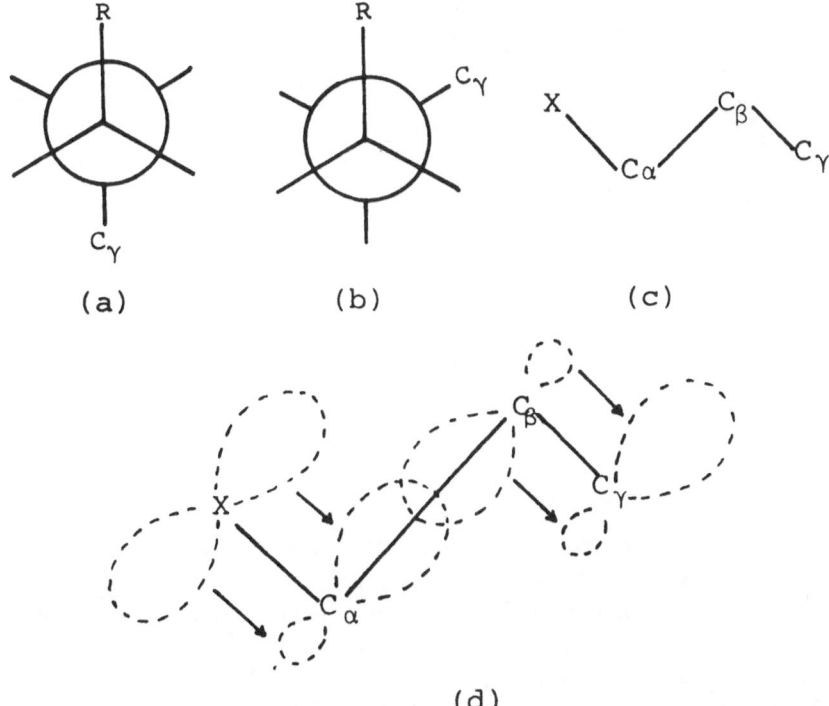

Figure 4.1. (a), (b): C_γ shown in *trans* and *gauche* positions; (c) (d): anti-periplanar arrangement of C_γ with respect to heteroatom.

Table 4.3. Summary of observed ^{13}C chemical shift increments resulting from *gauche* interactions

Interaction	Shift induced at indicated carbon (in ppm)	Interaction	Shift induced at indicated carbon (in ppm)
	-3.97		-8.51
	-2.05		-17.79
	-2.58		-5.10
	-6.20		-10.75
	-11.44		-5.48
	-8.37		-6.32
	-4.22		

4.1.1.5 Mesomeric Effects

Electron-donating groups attached to benzene rings shield the *ortho* and *para* carbon atoms while electron-withdrawing groups cause deshielding effects. This effect is manifested in the chemical shift values of aniline and benzonitrile (Figure 4.2). In α,β-unsaturated carbonyl compounds, the olefinic carbon atom β to the carbonyl group experiences a marked deshielding effect.

Another example is provided by the upfield shifts experienced by carbonyl groups when a double bond is introduced in conjugation. Thus the carbonyl carbon in methyl ethyl ketone resonates at δ 207 but in methyl vinyl ketone the

Table 4.4. Chemical shifts for carbocation centers[a]

Carbon	Chemical shift (δ)	Temperature (°C)	Carbon	Chemical shift (δ)	Temperature (°C)
$(CH_3)_2\overset{+}{C}{-}H$	320.6	-80^b	$(Ph)_2\overset{+}{C}{-}H$	200.2	-60^c
$(CH_3)_2\overset{+}{C}{-}CH_3$	335.2	-80^b	$(Ph)_2\overset{+}{C}{-}Ph$	212.7	-60^c
$(CH_3)_2\overset{+}{C}{-}Et$	335.4	-80^b	$H{-}\overset{+}{C}(OH)_2$	177.0	-30^d
cyclopropyl$\overset{+}{C}(CH_3)_2$	281.5	-60^c	$CH_3{-}\overset{+}{C}(OH)_2$	196.2	-30^d
$Ph{-}\overset{+}{C}(CH_3)_2$	255.7	-60^c	$HO{-}\overset{+}{C}(OH)_2$	166.6	-50^d
$H_2\overset{+}{C}{-}OH$	223.8	-50^d	$Ph{-}\overset{+}{C}{=}O$	154.6	
			$CH_3{-}\overset{+}{C}{=}O$	152.0	
$(CH_3)_2\overset{+}{C}{-}OH$	250.3	-50^d	$FCH_2\overset{+}{C}{=}O$	145.3	
$CH_3(H)\overset{+}{C}{-}OH$	237.2	-50^d	$ClCH_2\overset{+}{C}{=}O$	146.4	
cyclohexyl$\overset{+}{C}{-}OH$	251.1	-115	$BrCH_2\overset{+}{C}{=}O$	147.1	
			$ICH_2\overset{+}{C}{=}O$	150.6	

[a] G.A. Olah and D.J. Donovan, *J. Am. Chem. Soc.* **99**, 5026 (1977); G.A. Olah, A. Germain, and H.C. Lin, *J. Am. Chem. Soc.* **97**, 5481 (1975).
[b] Solvent: SO_2ClF-SbF_5.
[c] Solvent: SO_2-SbF_5.
[d] Solvent: SO_2-FSO_3H-SbF_5.

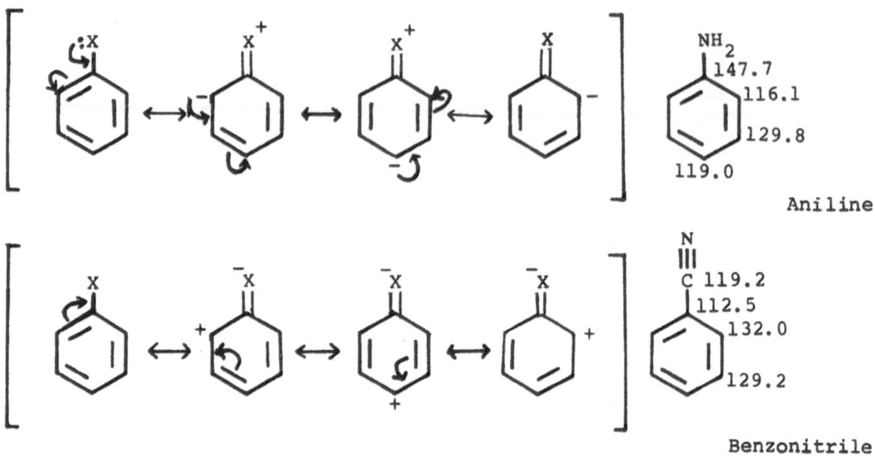

Aniline

Benzonitrile

Figure 4.2. Electron-releasing substituents, e.g., NH$_2$, shield the *ortho* and *para* carbon atoms of monosubstituted benzenes; the opposite effect is seen with electron-withdrawing substituents, e.g., CN.

corresponding resonance is shifted upfield to δ 194. This is because the partial positive charge on the carbonyl carbon atom can be delocalized on to the β-carbon atom when a conjugating double bond is present, thus causing an upfield shift of the carbonyl carbon (see Figure 4.3).

4.1.1.6   Intramolecular Anisotropic Effects

The neighbor anisotropic contribution, σ^N, to the shielding effect is exemplified by the ring current effects seen in aromatic compounds. The effect is more marked in ^{1}H-NMR than in ^{13}C-NMR, where the shielding effect is <2 ppm. A comparison of the chemical shifts of methyl groups in methyl cyclohexene and toluene serves to illustrate this (Figure 4.4).

4.1.1.7   Heavy Atom Effect

Attachment of heavy atoms results in a significant upfield shift of the substituted carbon atoms. Thus when halogens are attached to a carbon atom,

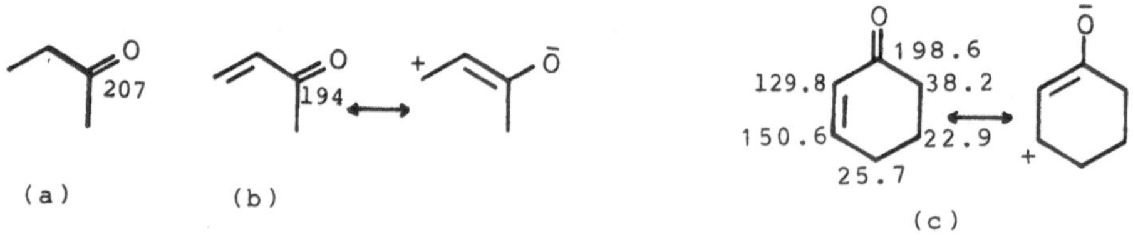

Figure 4.3. (a) Methyl ethyl ketone. (b) Methyl vinyl ketone; introduction of the double bond in (b) causes an upfield shift of the resonance of the carbonyl carbon. (c) Cyclohexenone; the carbon β to the carbonyl group resonates at a lower field because of the electron-withdrawing influence of the carbonyl group.

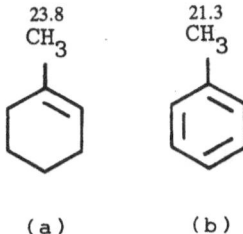

Figure 4.4. (a) Methyl cyclohexene. (b) Toluene. The chemical shift of the methyl group in toluene is 2.5 ppm upfield from that of the methyl group in methyl cyclohexene due to the ring current effect in toluene.

the diamagnetic shielding effect caused by the introduction of the heavy atoms is more dominant than the inductive effects.

4.1.1.8 Conjugation

Introduction of another double bond in conjugation with an existing double bond results in a shielding effect. This is shown in Figure 4.5.

4.1.1.9 Isotope Effect

Upfield shifts of up to 1.5 ppm may be experienced by carbon atoms when hydrogen atoms attached to them are replaced by deuterium atoms. Deuterium isotope effects are dependent on the degree of deuteration.

4.1.1.10 Hydrogen Bonding

Carbonyl carbon atoms experience deshielding when hydrogen bonds are formed with the carbonyl oxygens. This effect is shown in Figure 4.6.

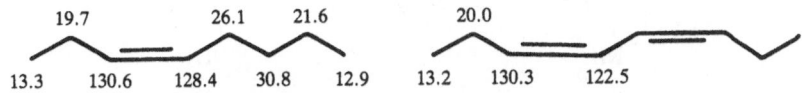

Figure 4.5. (a) Insertion of additional double bond in conjugation with existing double bond causes a shielding effect at the nearest olefinic carbon (140.2 ppm → 137.2 ppm). (b) Similar effect in another compound (128.4 ppm → 122.5 ppm). (Reprinted with permission from Pretsch et al., *Structureaufklarung Organischen Verbindungen.* Copyright 1981, Springer-Verlag, Heidelberg.)

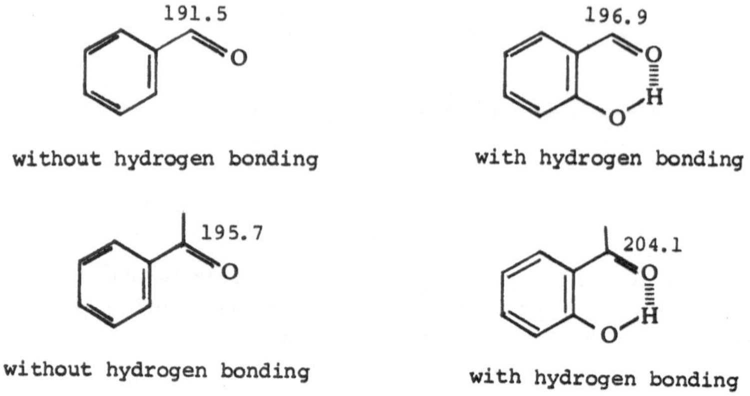

191.5

without hydrogen bonding

196.9

with hydrogen bonding

195.7

without hydrogen bonding

204.1

with hydrogen bonding

Figure 4.6. Hydrogen bonding causes deshielding of the carbonyl carbon atoms.

4.1.1.11   Effect of pH

Protonation of carboxylate anions, amines, and α-aminocarboxylic acids results in upfield shifts at the α, β, and γ carbon atoms; the effect is understandably most marked at the carbon atoms α and β to the carbonyl group and decreases down the chain (Figure 4.7).

4.1.1.12   Solvent Shifts

Downfield shifts are observed when going from nonpolar solvents such as carbon tetrachloride to polar solvents, particularly those in which hydrogen

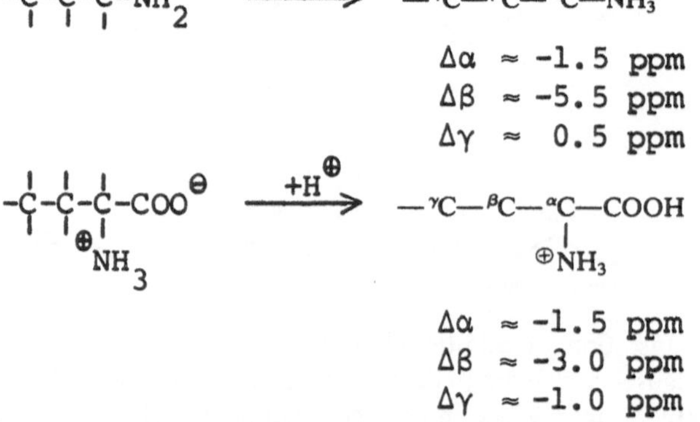

$$-\overset{|}{\underset{|}{C}}-\overset{|}{\underset{|}{C}}-\overset{|}{\underset{|}{C}}-COO^{\ominus} \quad \xrightarrow{+H^{\oplus}} \quad -^{\gamma}C-^{\beta}C-^{\alpha}C-COOH$$

$\Delta\alpha \approx -4.5$ ppm
$\Delta\beta \approx -1.5$ ppm
$\Delta\gamma \approx -0.3$ ppm

$$-\overset{|}{\underset{|}{C}}-\overset{|}{\underset{|}{C}}-\overset{|}{\underset{|}{C}}-NH_2 \quad \xrightarrow{+H^{\oplus}} \quad -^{\gamma}C-^{\beta}C-^{\alpha}C-\overset{\oplus}{N}H_3$$

$\Delta\alpha \approx -1.5$ ppm
$\Delta\beta \approx -5.5$ ppm
$\Delta\gamma \approx 0.5$ ppm

$$-\overset{|}{\underset{|}{C}}-\overset{|}{\underset{\underset{\oplus}{N}H_3}{C}}-\overset{|}{\underset{|}{C}}-COO^{\ominus} \quad \xrightarrow{+H^{\oplus}} \quad -^{\gamma}C-^{\beta}C-^{\alpha}C-COOH$$
$$\underset{\oplus NH_3}{}$$

$\Delta\alpha \approx -1.5$ ppm
$\Delta\beta \approx -3.0$ ppm
$\Delta\gamma \approx -1.0$ ppm

Figure 4.7. Protonation of amines, carboxylate anions, and α-aminocarboxylic acids results in shielding of adjacent carbon atoms.

bonding can occur. Upfield shifts ranging from 0.5 to 7.0 ppm can occur on dilution of a solution but after a certain dilution is reached, no further shifts take place.

4.1.2 Additivity of Substituent Effects

An important fact about ^{13}C chemical shifts, particularly in aliphatic compounds, is that substituent effects are additive, and the chemical shifts can be empirically predicted by adding the contributions of α, β, γ, and δ substituents. The downfield shift at the α-carbon is dependent mainly on the electronegativity of substituents. The substituent effects at the β-carbon atom are also downfield but less prominent than those at the α-carbon atom. Sterically induced polarizations of C—H bonds can cause *upfield* shifts of carbons atoms separated from a substituent by two other carbon atoms. The effect is maximum in rigid systems when the γ-carbon atom is *gauche* to the substituent (γ-gauche effect, discussed earlier) and minimum when the two groups are *trans* to each other (except when the substituent is N, O, or F; the upfield shift at the γ-carbon is then maximum in a *trans* disposition, as explained earlier in Figure 4.1). The γ-effect due to diaxial interaction can be of diagnostic value. Thus in a steroid or triterpenoid molecule the number of Me:H interactions when they are in a 1,3-diaxial relationship often govern their chemical shifts; the larger the number of such interactions with a given methyl group, the farther upfield it will usually appear. The effect when the ^{13}C being observed and the substituent are separated by three other carbon atoms is negligible unless the substituent and the δ-carbon are in a synaxial conformation (Figure 4.8) when the spatial proximity results in a *downfield* shift of the δ-carbon. It is curious that this effect is in the opposite direction to that found in γ-gauche interactions and indicates that when a substituent is brought very close to a carbon atom, some other mechanism must come into play. Second-order field effects induced by fluctuating local fields may be responsible for it.

A useful equation which allows one to calculate the chemical shifts of aliphatic carbon atoms is

$$\delta = -2.3 + \sum_i Z_i + S + \sum_j K_j \tag{4.1}$$

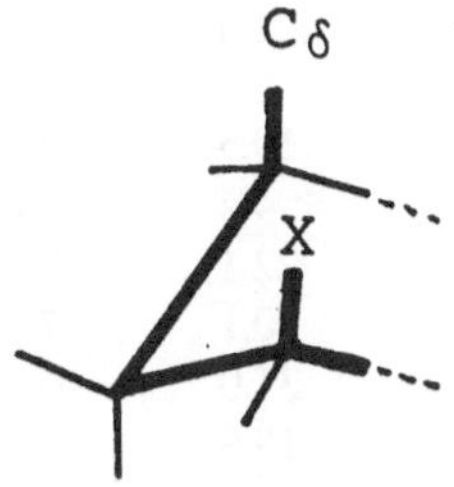

Figure 4.8. Synaxial conformation of C_δ with respect to the substituent X.

where Z_i represents the increments to be added for each α, β, and γ carbon atom, S is the steric correction factor, and K_j is the conformational correction factor for γ-substituted carbon atoms in preferred steric environments. Table 4.5 presents the values of Z_i, S, and K which should be taken into consideration when calculating chemical shifts.

As an illustration let us consider how to calculate the chemical shifts of N-(tert-butoxycarbonyl)alanine:

$$
\begin{array}{c}
\text{(iii)} \\
\overset{\text{O}}{\underset{\|}{}}\quad \overset{\text{CH}_3}{\underset{|}{}} \\
(\text{CH}_3)_3\text{C}-\text{O}-\text{C}-\text{NH}-\text{CH}-\text{CO}_2\text{H} \\
\text{(iv)}\quad\text{(ii)}\qquad\qquad\text{(i)}
\end{array}
$$

To calculate the chemical shift of carbon (i), we proceed as follows:

− 2.3	common factor to be added, from Equation (4.1)
9.1	increment for one α-carbon from Table 4.5
20.1	increment for one α-CO$_2$H from Table 4.5
28.3	increment for one α-NH group from Table 4.5
2.0	increment for one β-COO group from Table 4.5
0.3	increment for one δ-carbon from Table 4.5
− 3.7	increment for steric correction factor S for tertiary carbon 2 from Table 4.5
53.8	= calculated value for carbon (i)

Table 4.5. Parameters for calculation of ^{13}C chemical shifts of aliphatic carbon atoms

$$\delta = -2.3 + \sum_i Z_i + S + \sum_j K_j$$

Substituent		Z_i for substituent at position			
		α	β	γ	δ
—H		0.0	0.0	0.0	0.0
—C—	(*)	9.1	9.4	− 2.5	0.3
▽ (O)	(*)	21.4	2.8	− 2.5	0.3
—C=C—	(*)	19.5	6.9	− 2.1	0.4
—C≡C—		4.4	5.6	− 3.4	− 0.6
—Phenyl		22.1	9.3	− 2.6	0.3
—F		70.1	7.8	− 6.8	0.0
—Cl		31.0	10.0	− 5.1	− 0.5
—Br		18.9	11.0	− 3.8	− 0.7
—I		− 7.2	10.9	− 1.5	− 0.9
—O—	(*)	49.0	10.1	− 6.2	0.0
—O—CO—		56.5	6.5	− 6.0	0.0

Table 4.5. (*Continued*)

Substituent		Z_i for substituent at position			
		α	β	γ	δ
—O—NO		54.3	6.1	−6.5	−0.5
—N<	(*)	28.3	11.3	−5.1	0.0
—$^+$N≤	(*)	30.7	5.4	−7.2	−1.4
—NH$_3^+$		26.0	7.5	−4.6	0.0
—NO$_2$		61.6	3.1	−4.6	−1.0
—NC		31.5	7.6	−3.0	0.0
—S—	(*)	10.6	11.4	−3.6	−0.4
—S—CO—		17.0	6.5	−3.1	0.0
—SO—	(*)	31.1	9.0	−3.5	0.0
—SO$_2$Cl		54.5	3.4	−3.0	0.0
—SCN		23.0	9.7	−3.0	0.0
—CHO		29.9	−0.6	−2.7	0.0
—CO—		22.5	3.0	−3.0	0.0
—COOH		20.1	2.0	−2.8	0.0
—COO$^-$		24.5	3.5	−2.5	0.0
—COO—		22.6	2.0	−2.8	0.0
—CON<		22.0	2.6	−3.2	−0.4
—COCl		33.1	2.3	−3.6	0.0
—CSN		33.1	7.7	−2.5	0.6
—C=NOH *syn*		11.7	0.6	−1.8	0.0
—C=NOH *anti*		16.1	4.3	−1.5	0.0
—CN		3.1	2.4	−3.3	−0.5
—SN⟨		−5.2	4.0	−0.3	0.0

	Steric correction factor S			
Type of carbon being observed	Number of non-hydrogen substituents on the most branched α-carbon (valid only for those substituents marked with an asterisk)			
	1	2	3	4
Primary	0.0	0.0	−1.1	−3.4
Secondary	0.0	0.0	−2.5	−7.5
Tertiary	0.0	−3.7	−9.5	−15.0
Quaternary	−1.5	−8.4	−15.0	−25.0

Conformation of γ-substituent	Conformational correction K
Synperiplanar	−4.0
Synclinal	−1.0
Anticlinal	0.0
Antiperiplanar	2.0
Not fixed	0.0

* Asterisks are meant to identify those substituents for which the steric correction factor 's' has to be added.

This is fairly close to the experimentally determined value of δ 49.0 for this carbon.

Similarly, calculations for carbons (ii), (iii), and (iv) can be made. The calculation for carbon (ii) is presented below as a further illustration.

Common increment factor	-2.3 (see Table 4.5)
3 α-carbons (3 × 9.1)	27.3
1 α-OCO group	56.5
1 γ-NH group	-5.1
1 δ-carbon	0.3
S (quaternary, 1)	-1.5
Calculated value	75.2
Experimentally determined value	78.1

The reader should attempt to calculate the chemical shifts for carbons (iii) and (iv) [calculated values for carbons (iii) and (iv) should come to 16.2 and 28.7, which are close to the experimentally observed values of 17.3 and 28.1, respectively].

4.1.3 Chemical Shifts of Organic Molecules

4.1.3.1 Alkanes

The chemical shifts of aliphatic carbon atoms lie in the range of -2 to $+50$ ppm. Increasing substitution of a given carbon atom causes downfield shifts of about $+9$ ppm for each substitution. This is known as the α-effect, and is illustrated by the chemical shifts of the carbon atoms marked with an asterisk in the following series:

$$\overset{*}{C}H_4 \qquad \overset{*}{C}H_3{-}CH_3 \qquad CH_3{-}\overset{*}{C}H_2{-}CH_3$$
$$-2.1 \qquad\quad 5.9 \qquad\qquad\quad 16.1$$

$$CH_3{-}\overset{*}{C}H(CH_3)_2$$
$$25.2$$

Increasing the number of substituents at the position β to a given carbon also causes downfield shifts of about 9 ppm. This β-effect is shown in the chemical shifts for asterisked carbon atoms:

$$\overset{*}{C}H_3{-}CH_3 \qquad \overset{*}{C}H_3{-}CH_2{-}\overset{\beta}{C}H_3$$
$$5.9 \text{ (no } \beta\text{-C)} \qquad\quad 15.6 \text{ (1 } \beta\text{-C)}$$

$$\overset{*}{C}H_3{-}CH(\overset{\beta}{C}H_3)_2$$
$$24.3 \text{ (2 } \beta\text{-C)}$$

Increasing the number of substituents at the γ-position also causes changes in the chemical shift of about 2.5 ppm but these are in the opposite direction, i.e., upfield instead of downfield. The γ-effect is illustrated below for asterisked carbon atoms:

$$\overset{*}{C}H_3-\overset{\alpha}{C}H_2-\overset{\beta}{C}H_3 \qquad \overset{*}{C}H_3-\overset{\alpha}{C}H_2-\overset{\beta}{C}H_2-\overset{\gamma}{C}H_3$$

$$\text{15.6 (no } \gamma\text{-C)} \qquad\qquad\qquad \text{13.2 (1 } \gamma\text{-C)}$$

$$\overset{*}{C}H_3-\overset{\alpha}{C}H_2-\overset{\beta}{C}H(\overset{\gamma}{C}H_3)_2 \qquad \overset{*}{C}H_3-\overset{\alpha}{C}H_2-\overset{\beta}{C}(\overset{\gamma}{C}H_3)_3$$

$$\text{11.5 (2 } \gamma\text{-C)} \qquad\qquad\qquad \text{8.7 (3 } \gamma\text{-C)}$$

These observations have been amalgamated in the equation

$$\delta(k) = -2.1 + \sum n_{ik}A_i$$

where $\delta(k)$ is the chemical shift of a paraffin carbon k, n_{ik} is the number of carbon atoms at the ith position, and A_i is the term to be added for each substituent at the ith position. A_1 (for an α-carbon) is $+9.1$ ppm, A_2 (for a β-carbon) is $+9.4$ ppm, and A_3 (for a γ-carbon) is -2.5 ppm. Thus for n-decane the chemical shifts for the carbon atoms can be calculated as follows (experimental values are given in parentheses):

$$(\overset{1}{C}H_3-\overset{2}{C}H_2-\overset{3}{C}H_2-\overset{4}{C}H_2-\overset{5}{C}H_2)_2$$

$$\delta_1 = -2.1 + 9.1 + 9.4 + (-2.5) = 13.9 \ (13.9)$$
$$\delta_2 = -2.1 + 2(9.1) + 9.4 + (-2.5) = 23.0 \ (22.8)$$
$$\delta_3 = -2.1 + 2(9.1) + 2(9.4) + (-2.5) = 32.4 \ (32.2)$$
$$\delta_4 = -2.1 + 2(9.1) + 2(9.4) + 2(-2.5) = 29.9 \ (29.7)$$
$$\delta_5 = -2.1 + 2(9.1) + 2(9.4) + 2(-2.5) = 29.9 \ (30.1)$$

For longer or branched alkanes, a more complex equation is employed, which takes into consideration the nature of the α-carbon atom (i.e., whether it is CH_3, CH_2, or CH) and includes the effect of a carbon atom at the δ-position. The chemical shift of a carbon k is then given by

$$\delta(k) = A_n + \sum_{m=0}^{2} N_m^\alpha \alpha_{nm} + N^\gamma \gamma_n + N^\delta \delta_N$$

where $n =$ the number of hydrogens on carbon k, $m =$ the number of hydrogens on the α-carbon, $N_m^\alpha =$ the number of CH_m groups at the α-position (0–2; α-methyl groups are not considered), $N^\gamma =$ the number of γ-carbons, and $N^\delta =$ the number of δ-carbons. The constants are given in Table 4.6. As an example, we can apply this equation in the case of 2,2,3-trimethylbutane to

Table 4.6. Parameters for estimating chemical shifts

n	A_n	m	α_{nm}	γ_n	δ_n
0	27.77	2	2.26	0.86	0
		1	3.96		
		0	7.35		
1	23.46	2	6.60	−2.07	0
		1	11.14		
		0	14.70		
2	15.34	2	9.75	−2.69	0.25
		1	16.70		
		0	21.43		
3	6.80	2	9.56	−2.99	0.49
		1	17.83		
		0	25.48		

calculate the chemical shift of, for instance, C-3:

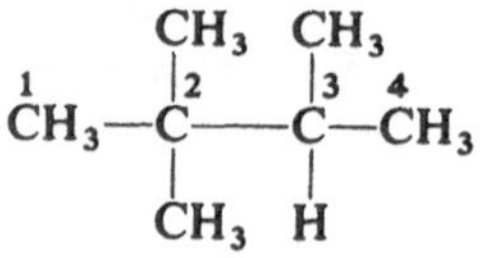

$$
\underset{}{CH_3}-\overset{CH_3}{\underset{CH_3}{\overset{|}{\underset{|}{C}}}}-\overset{CH_3}{\underset{H}{\overset{|}{\underset{|}{C}}}}-CH_3
$$

Since C-3 has one hydrogen, the value of A_n from Table 4.6 is 23.46. There is one quaternary α-carbon (C-2), giving a value of 1(14.70), and two α-methyl groups which are ignored. There are no γ or δ carbon atoms. Hence the calculated value of C-3 would be

$$\delta = 23.46 + 1(14.70) + 0(-2.07) + 0(0) = 38.16$$

This is very close to the experimentally determined value of 38.1.

4.1.3.2 Alkenes

Olefinic carbon atoms normally resonate between 100–165 ppm. Terminal methylene groups resonate about 24 ppm upfield from the other olefinic carbon. In long-chain alkenes, the olefinic carbon atom bearing the longer side chain appears at lower field.

A series of additive parameters have been developed, as in alkanes, which allow calculation of chemical shifts of the olefinic carbon atoms. The parameters are shown below for the olefinic carbon marked with an asterisk:

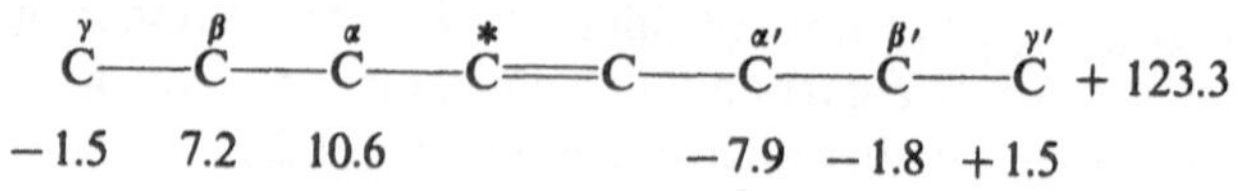

$$
\overset{\gamma}{C}-\overset{\beta}{C}-\overset{\alpha}{C}-\overset{*}{C}=C-\overset{\alpha'}{C}-\overset{\beta'}{C}-\overset{\gamma'}{C} \quad +123.3
$$

$-1.5 \quad 7.2 \quad 10.6 \qquad\qquad -7.9 \quad -1.8 \quad +1.5$

Thus to calculate the chemical shift of C-2 in 2-hexene,

$$\overset{1\alpha}{C}H_3-\overset{*}{C}H=CH-\overset{1\alpha'}{C}H_2-\overset{1\beta'}{C}H_2-\overset{1\gamma'}{C}H_3$$

we proceed as follows:

$$\delta C_2 = 123.3 + 1\alpha + 1\alpha' + 1\beta' + 1\gamma'$$
$$= 123.3 + 10.6 - 7.9 - 1.8 + 1.5 = 125.7$$

If the olefin is *trans*, then no further correction term is needed but if it is *cis* oriented then a factor of -1.1 is added. The effect of introducing various substituents at α, β, and γ or α', β', and γ' carbon atoms is given by the following equation:

$$\delta_c(k) = 123.3 + \sum_i A_{ki}(R_i) + \sum_i A_{ki'}(R_{i'}) + \text{correction factors}$$

$$\overset{\gamma}{C}-\overset{\beta}{C}-\overset{\alpha}{C}-\overset{*}{C}=C-\overset{\alpha'}{C}-\overset{\beta'}{C}-\overset{\gamma'}{C}$$

$$A_{ki}(R_i) \qquad\qquad A_{ki'}(R_{i'})$$

The empirical parameters for calculation of the alkene chemical shifts are given in Table 4.7. Chemical shifts of olefinic carbon atoms when they bear

Table 4.7. Empirical parameters for the calculation of alkene chemical shifts[a]

	Parameters $A_{ki}(R_i)$					
	—C—	—C—	—C=$\overset{*}{C}$—C—		—C—	—C—
R_i	γ'	β'	α'	α	β	γ
C	+1.5	−1.8	−7.9	+10.6	+7.2	−1.5
OH		−1	—	—	+6	
OR		−1	−39	+29	+2	
OAc			−27	+18		
COCH$_3$			+6	+15		
CHO			+13	+13		
COOH			+9	+4		
COOR			+7	+6		
CN			+15	−16		
Cl		+2	−6	+3	−1	
Br		+2	−1	−8	~0	
I			+7	−38		
C$_6$H$_5$			−11	+12		

[a] Correction terms: α, α' (*trans*), 0; α, α' (*cis*), -1.1; α, α, -4.8; α', α', $+2.5$; β, β, $+2.3$; all other interactions, ~ 0.

one or two substituents are given in Tables 4.8 and 4.9. Chemical shifts of olefinic carbon atoms in some α, β-unsaturated ketones, substituted styrenes, and allenes are presented in Tables 4.10, 4.11, and 4.12, respectively.

4.1.3.3   Alkynes

Alkynes have chemical shifts between 63 and 95 ppm. The ^{13}C chemical shifts for some alkynes are presented in Table 4.13. The effect of carbon substituents on alkyne chemical shifts are presented in Table 4.14. It may be noted that the resonance of the non-hydrogen-bearing carbon in alkynes is less intense due to lower NOE and is therefore usually readily recognized.

4.1.3.4   Aromatic Carbons

Since aromatic carbon atoms are not affected by the paramagnetic effects (deshielding effects) caused by ring currents which deshield protons, aromatic

Table 4.8. ^{13}C shifts of monosubstituted alkenes[a]

$$\begin{array}{c} C-2\,C-1 \\ CH_2=CH-X \end{array}$$

Substituent X	C—1	C—2
Cl	126.0	117.3
Br	115.5	122.0
I	85.3	130.4
CO$_2$Et	129.7	130.4
CH$_2$Br	133.1	117.6
CH$_2$OEt	135.7	114.6
OCOCH$_3$	141.6	96.3
OCH$_3$	153.2	84.1
O—CH$_2$—CH(CH$_3$)CH$_3$	152.9	85.0
O—CH(CH$_3$)CH$_3$	151.4	87.5
O—C(CH$_3$)(CH$_3$)CH$_3$	146.8	90.2
SO$_2$CH=CH$_2$	137.7	131.3

[a] G.E. Maciel, *J. Phys, Chem.* **69**, 1947 (1965); K. Hatada, K. Nagata, and H. Yuki, *Bull Chem. Soc. Jap.* **43**, 3195, 3267 (1970); G. Miyajima, K. Takahashi, and K. Nishimoto, *Org. Magn. Resonance* **6**, 413 (1974).

Table 4.9. ^{13}C chemical shifts of cis- and trans-disubstituted ethylenes, XHC=CHX[a]

Substituent X	δ_{cis}	δ_{trans}	Δ
Cl	121.3	119.4	−1.9
Br	116.4	109.4	−7.0
I	96.5	79.4	−17.1
CH$_3$	123.3	124.5	1.2
CH$_2$CH$_3$	131.2	131.3	0.1
CN	120.8	120.2	−0.6
CO$_2$CH$_3$	128.7	132.4	3.7
CO$_2$C$_2$H$_5$	130.5	134.1	3.8
CO$_2$C$_4$H$_9$	129.0	132.8	3.8

[a] G.E. Maciel, P.D. Ellis, J.J. Natterstad, and G.B. Savitsky, *J. Magn. Resonance* **1**, 589 (1969).

Table 4.10. Alkene chemical shifts in α, β-unsaturated ketones[a]

[a] D.H. Marr and J.B. Stothers, *Can. J. Chem.* **43**, 596 (1976); M. H. Loots, L. R. Weingarten, and R.H. Levin, *J. Am. Chem. Soc.* **98**, 4571 (1976).

Table 4.11. ^{13}C chemical shifts of substituted styrenes[a]

X	Y	C-1	C-2
H	H	136.7	113.2
o-Me	H	136.4	114.6
p-Me	H	137.5	114.0
p-OMe	H	136.5	111.5
o-Br	H	137.5	117.2
p-Br	H	136.2	115.1
m-NO$_2$	H	134.9	116.6
p-NO$_2$	H	135.9	119.0
p-NMe$_2$	H	137.3	108.7
H	Me	143.1	112.4
H	Et	150.0	110.9

[a] D.S. Khami and J.B. Stothers, *Can. J. Chem.* **43**, 510 (1965); G.K. Hamer, I.R. Peat, and W.F. Reynolds, *Can. J. Chem.* **51**, 915, 2596 (1973); W.F. Reynolds, I.R. Peat, M.H. Freedman, and J.R. Lyerla, *Can. J. Chem.* **51**, 1857 (1973).

Table 4.12. ^{13}C chemical shifts of allenes[a]

R_1	R_2	R_3	R_4	C_1	C_2	C_3
H	H	H	H	74.8	213.5	74.8
Me	H	H	H	84.4	210.4	74.1
Me	Me	H	H	93.4	207.3	72.1
Me	H	Me	H	85.4	207.1	85.4
Me	Me	Me	Me	92.6	200.2	92.6
Me	SMe	H	H	99.9	203.6	80.1
Ph	Ph	Ph	Ph	113.6	209.5	113.6
OMe	H	H	H	123.1	202.0	90.3
Br	H	H	H	72.7	207.6	83.8
CN	H	H	H	80.5	218.7	67.2

[a] J.K. Crandall and S.A. Sojka, *J. Am. Chem. Soc.* **94**, 5084 (1972); R. Steur, J. van Dongen, M. DeBie, and W. Drenth, *Tetrahedron Lett.*, 3307 (1971); W. Runge, W. Kosbahn, and J.Winkler, *Ber. Bunsenges. Phys. Chem.* **79**, 381 (1975); W. Runge and J. Firl, *Ber. Bunsenges. Phys. Chem.* **79**, 913 (1975); C. Charrier, D.E. Dorman, and J.D. Roberts, *J. Org. Chem.* **38**, 2644 (1973).

Table 4.13. ^{13}C chemical shifts of alkynes[a]

Compound	C-1	C-2	C-3	C-4	C-5	C-6	C-7	C-8
1-Butyne	67.0	84.7						
2-Butyne		73.6						
1-Hexyne	67.4	82.8	17.4	29.9	21.2	12.9		
2-Hexyne	1.7	73.7	76.9	19.6	21.6	12.1		
3-Hexyne	14.4	12.0	79.9					
1-Heptyne	67.4	82.9	17.7	28.1	30.7	22.4	14.0	
2-Heptyne	2.3	74.2	77.6	17.7	31.2	21.8	13.5	
3-Heptyne	13.7	12.0	78.3	80.2	20.2	22.5	13.1	
1-Octyne	69.0	84.0	18.4	29.2	29.2	32.0	23.4	14.9
2-Octyne	2.9	74.8	78.5	18.4	29.5	31.4	22.7	14.5
3-Octyne	15.1	13.0	79.7	81.0	18.7	31.9	22.8	14.6
4-Octyne	12.7	21.4	19.2	79.0				
Dodeca-5,7-diyne					66.7	77.4		

[a] J.K. Becconsall and P. Hampson, *J. Mol. Phys.* **10**, 21 (1965); D.D. Traficante and G.E. Maciel, *J. Phys. Chem.* **69**, 1348 (1965); S. Rang, T. Pehk, E. Lippma, and O. Eisen, *Eesti NSV Tead. Akad. Toim. Keem. Geol.* **16**, 346 (1967) and **17**, 294 (1968); N. Muller and D.E. Pritchard, *J. Chem. Phys.* **31**, 768 and 1471 (1959).

Table 4.14. Alkyne substituent parameters[a,b]

$$\overset{\alpha'}{C}-\overset{\beta'}{C}-\overset{\gamma'}{C}-\overset{\delta'}{C}-C\equiv C-\overset{\delta}{C}-\overset{\gamma}{C}-\overset{\beta}{C}-\overset{\alpha}{C}$$

$\alpha' = -5.69$	$\alpha = 6.93$
$\beta' = 2.32$	$\beta = 4.75$
$\gamma' = -1.31$	$\gamma = -0.13$
$\delta' = 0.56$	$\delta = 0.51$

[a] W. Hobold, R. Radeglia, and D. Klose, *J. Prakt. Chem.* **318**, 519 (1976).
[b] Base = 72.8 ppm.

carbons fall in the same range as alkenes, i.e., 110–135 ppm. In substituted benzenes, the substituent-bearing carbon atoms are often readily recognized by their lower-intensity signals. The effect of substituents on the chemical shifts of aromatic carbon atoms are shown in Table 4.15 (values to be added to δ 128.5, the chemical shift of benzene carbons). The *ortho* and *para* carbon atoms are shielded while the *meta* carbon atoms are not affected when the substituent is an electron-donating group such as methoxy. The carbon to which the group is attached (the *ipso* carbon) is deshielded. Tables 4.16–4.19 present some typical values for the chemical shifts of polynuclear aromatics

Table 4.15. Substituent effects on ^{13}C resonances of monosubstituted benzenes[a]

Substituent	C-1	ortho	meta	para	Remarks
H	0.0	—	—	—	10% in CCl$_4$
F	+34.8	−12.9	+1.4	−4.5	10% in CCl$_4$
Cl	+6.2	+0.4	+1.3	−1.9	10% in CCl$_4$
COPh	+9.4	+1.7	−0.2	+3.6	10% in CCl$_4$
COOH	+2.1	+1.5	0.0	+5.1	10% in CCl$_4$
CHO	+8.6	+1.3	+0.6	+5.5	10% in CCl$_4$
COCl	+4.6	+2.4	0.0	+6.2	10% in CCl$_4$
COCH$_3$	+9.1	+0.1	0.0	+4.2	10% in CCl$_4$
COCF$_3$	−5.6	+1.8	+0.7	+6.7	10% in CCl$_4$
C≡C—H	−6.1	+3.8	+0.4	−0.2	10% in CCl$_4$
CN	−15.4	+3.6	+0.6	+3.9	10% in CCl$_4$
CH$_3$	+8.9	+0.7	−0.1	−2.9	10% in CCl$_4$
CF$_3$	+2.6	−3.3	−0.3	+3.2	10% in CCl$_4$
Br	−5.5	+3.4	+1.7	−1.6	10% in CCl$_4$
NCO	+5.7	−3.6	+1.2	−2.8	10% in CCl$_4$
NH$_2$	+18.0	−13.3	+0.9	−9.8	10% in CCl$_4$
NH$_3^+$	+0.1	−5.8	+2.2	+2.2	10% in CCl$_4$
NO$_2$	+20.0	−4.8	+0.9	+5.8	10% in CCl$_4$
OCH$_3$	+31.4	−14.4	+1.0	−7.7	10% in CCl$_4$
OH	+26.9	−12.7	+1.4	−7.3	10% in CCl$_4$
Ph	+13.1	−1.1	+0.4	−1.2	10% in CCl$_4$
SH	+2.3	+1.1	+1.1	−3.1	10% in CDCl$_3$
SCH$_3$	+10.2	−1.8	+0.4	−3.6	10% in CDCl$_3$
SO$_2$NH$_2$	+15.3	−2.9	+0.4	+3.3	10% in CDCl$_3$
O$^-$	+39.6	−8.2	+1.9	−13.6	Neat
OPh	+29.2	−9.4	+1.6	−5.1	Neat
OAc	+23.0	−6.4	+1.3	−2.3	Neat
NMe$_2$	+22.6	−15.6	+1.0	−11.5	Neat
NEt$_2$	+19.9	−15.3	+1.4	−12.2	Neat
NHAc	+11.1	−9.9	+0.2	−5.6	Neat
CH$_2$OH	+12.3	−1.4	−1.4	−1.4	Neat
I	−32.0	+10.2	+2.9	+1.0	Neat
SiMe$_3$	+13.4	+4.4	−1.1	−1.1	Neat
SnMe$_3$	+13.1	+7.2	−0.4	−0.4	Neat
CH=CH$_2$	+9.5	−2.0	+0.2	−0.5	Neat
CO$_2$Me	+1.3	−0.5	−0.5	+3.5	Neat
COCl	+5.8	+2.6	+1.2	+7.4	Neat
CHO	+9.0	+1.2	+1.2	+6.0	Neat
COEt	+7.6	−1.5	−1.5	+2.4	Neat
CO-iPr	+7.4	−0.5	−0.5	+4.0	Neat
CO—tBu	+9.4	−1.1	−1.1	+1.7	Neat

Table 4.15. (*Continued*)

Substituent	C-1	*ortho*	*meta*	*para*	Remarks
Li	−43.2	−12.7	+2.4	+3.1	†
MgBr	−35.8	−11.4	+2.7	+4.0	†
▽	+15.1	−3.3	−0.6	−3.6	10% in CDCl$_3$
HN▽	+12.2	−2.9	−0.3	−1.8	10% in CDCl$_3$
O▽	+9.1	−0.1	−3.0	−0.5	10% in CDCl$_3$
PMe$_2$	+13.6	+1.6	−0.6	−1.0	10% in CDCl$_3$

a D.F. Ewing, *Org. Magn. Resonance* **12**, 499 (1979); G.L. Nelson and E.A. Williams, *Progr. Phys. Org. Chem.* **12**, 229 (1976); G. Barbarella, P. Dembech, A. Garbesi, and A. Fava, *Org. Magn. Resonance* **8**, 108 (1976); V. Formacek, L. Desnoyer, H.P. Kellerhals, and J.T. Clerc, *^{13}C Data Bank*, Vol. 1, Copyright Bruker-Physik, Karlsruhe 1976; A.R. Tarpley and J.H. Goldstein, *J. Am. Chem. Soc.* **76**, 515 (1972).
† Solvent not specified.

(Table 4.16), five-membered heterocycles (Table 4.17), six-membered heterocycles (Table 4.18), and polycyclic heteroaromatics (Table 4.19).

4.1.3.5 Carbonyl Carbons

Carbonyl carbon atoms are strongly deshielded on account of the polarization

$$\text{>C=O} \longrightarrow \text{>}\overset{+}{\text{C}}-\overset{-}{\text{O}}$$

Typical values of chemical shifts for various carbonyl compounds such as ketones, aldehydes, acids, esters, anhydrides, etc., are given in Table 4.20. Carbonyl carbons of aliphatic ketones appear farthest downfield as compared to other carbonyl groups. Alkyl substitution at the α-carbon atoms causes downfield shifts of 2 to 3 ppm in the carbonyl resonances of both cyclic and acyclic ketones. Solvents capable of hydrogen bonding cause deshielding of the carbonyl carbons of ketones, the shifts being usually small (0 to 4 ppm), but in acidic media the downfield shifts may be up to 40 ppm. The chemical shifts of alkyl or aryl carbon atoms are also markedly affected by the introduction of carbonyl-bearing substituents. This is shown in Table 4.21. The resonances of carbonyl carbon atoms of carboxylic acids normally appear between 165 and 185 ppm, whereas the corresponding resonances of the anions are shifted downfield by about 5 ppm. Ester carbonyl carbons resonate between 160 and 180 ppm, and their chemical shift is affected more by the nature of the substituent at the α-carbon atom than by the nature of the alcohol residue

Table 4.16. ^{13}C chemical shifts of polynuclear aromatics[a]

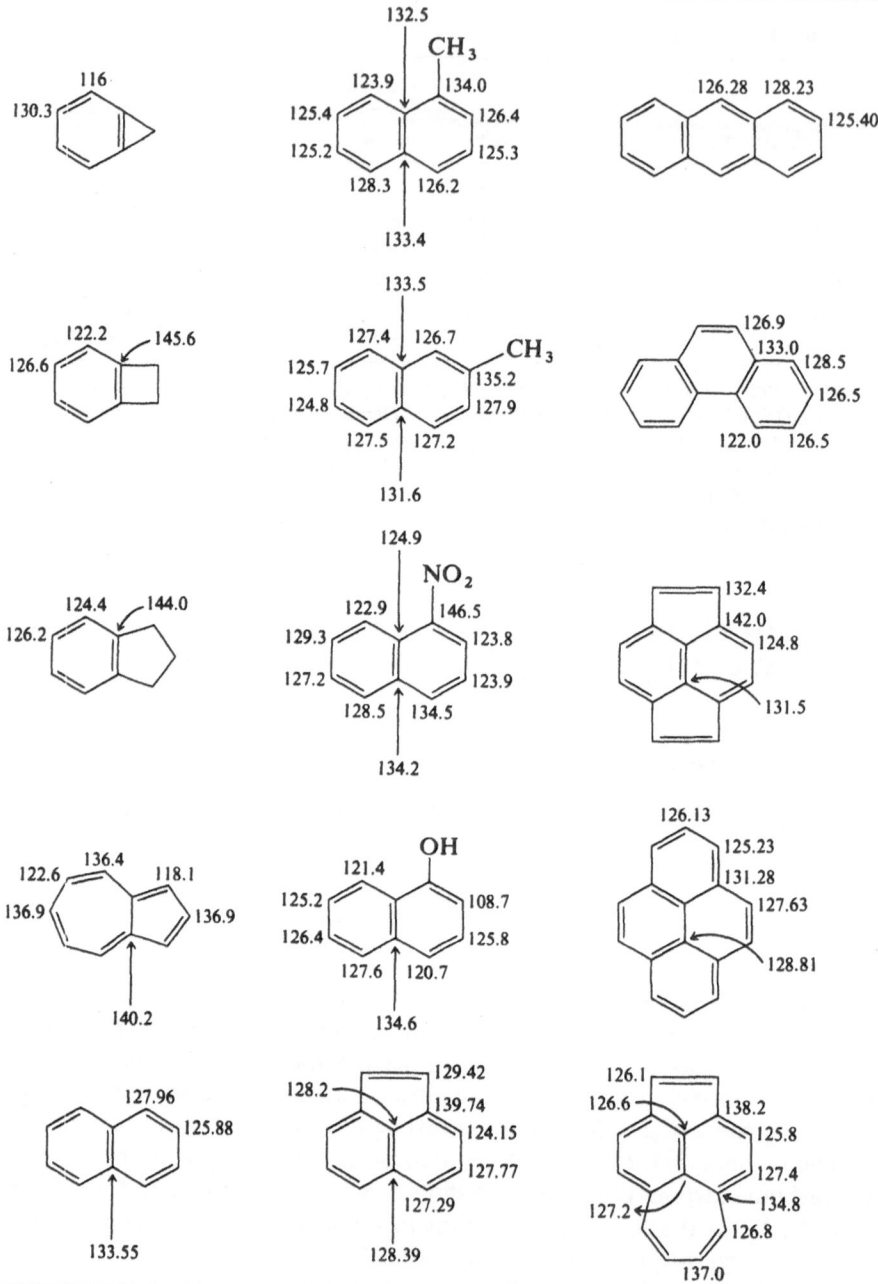

[a] T.D. Alger, D.M. Grant, and E.G. Paul, *J. Am. Chem. Soc.* **88**, 5397 (1966); P.E. Hanson, *Org. Magn. Resonance* **12**, 109 (1979); B.M. Trost and W.B. Herdle, *J. Am. Chem. Soc.* **98**, 4080 (1976); S. Gronointz, I. Johnson, and A.-B. Hornfeldt, *Chem. Scr.* **7**, 76 (1975); *ibid.* **7**, 211 (1975); M.T.W. Hearn, *Aust. J. Chem.* **29**, 107 (1976); F. Fringuelli, S. Gronowitz, A.-B. Hornfeldt, I. Johnson, and A. Taticchi, *Acta Chem. Scand. B* **28**, 125 (1974).

Table 4.17. ^{13}C shifts of five-membered heterocycles^a

Furan: 110.4, 143.6 (O)

2-Methylfuran: 111.1, 106.3, 141.7, 152.7, CH₃ 13.3 (O)

2-Methoxyfuran: 111.8, 79.9, 133.8, 163.1, OCH₃ 57.5 (O)

2-Nitrofuran: 112.6, 114.5, 146.9, 153.7, NO₂ (O)

3-Methylfuran: 112.9, 120.5, CH₃, 143.5, 140.2 (O)

3-Methoxyfuran: 58.3 OCH₃, 104.9, 151.5, 143.2, 124.2 (O)

3-Nitrofuran: NO₂, 106.6, 144.8, 146.0 (O)

Pyrrole: 108.2, 118.5 (N–H)

1-Methylpyrrole: 108.0, 121.6, 35.4 CH₃ (N)

2-Methylpyrrole: 108.1, 105.9, 116.7, 127.2, CH₃ 12.4 (N–H)

Pyrazole: 105.8, 134.6 (N–N, H)

1-Methylpyrazole: 105.4, 139.0, 129.8, 38.2 CH₃ (N–N)

Imidazole: 122.8, 136.3 (N, N–H)

1-Methylimidazole: 129.6, 120.3, 138.3, 32.6 CH₃ (N)

1,2,3-Triazole: 130.6 (N, N–NH)

2-Methyltriazole: 134.7, N–N–CH₃ 41.5

1,2,4-Triazole: HN, 147.9 (N)

Tetrazole: 143.3 (N–N, N–N, H)

2-Methyltetrazole: 144.2, 33.7 CH₃ (N–N, N)

1-Methyltetrazole: 153.4, N–N–CH₃ 38.8

1-Methyl-1,2,3-triazole: 134.3, 125.5, 35.7 CH₃ (N, N)

4-Methyl-1,2,4-triazole: 30.7 CH₃, 144.1 (N–N, N)

1-Methyl-1,2,4-triazole: 36.0 H₃C–N, 144.7, 152.6 (N)

Thiophene: 127.3, 125.6 (S)

2-Methylthiophene: 127.5, 125.7, 123.7, 139.8, CH₃ 14.9 (S)

2-Methoxythiophene: 125.5, 104.2, 112.4, 167.4, OCH₃ 60.5 (S)

2-Nitrothiophene: 128.4, 129.9, 134.7, 151.2, NO₂ (S)

3-Methylthiophene: 15.5 CH₃, 130.1, 138.2, 126.1, 121.3 (S)

3-Methoxythiophene: 119.9, 159.8 OCH₃, 125.6, 97.3 (S)

3-Nitrothiophene: 123.1, 149.7 NO₂, 128.9, 129.2 (S)

Selenophene: 129.5, 131.4 (Se)

Isoxazole: 103.7, 157.9, 149.1 (N, O)

Isothiazole: 123.4, 147.8, 157 (N, S)

3-Methylisothiazole: 123.9, 148.1, 166.7 H₃C (N, S)

^{a} V. Formacek, L. Desnoyer, H.P. Kellerhals, and J.T. Clerc, ^{13}C *Data Bank*, Vol. 1, Copyright Bruker-Physik, Karlsruhe (1976); F. Fringuelli, S. Gronowitz, A.-B. Hornfeldt, I. Johnson, and A. Taticchi, *Acta Chem. Scand.* **28B**, 175 (1974); R.E. Wasylishen, T.R. Clem, and E.D. Becker, *Can. J. Chem.* **53**, 596 (1975); R. Faure, A. Assaf, E.-J. Vincent, and J.-P. Aune, *J. Chim. Phys.* **75**, 727 (1978); N. Plavac, I.W.J. Still, M.S. Chauhan, and D.M. McKinnon, *Can. J. Chem.* **53**, 836 (1975); J. Gainer, G.A. Howarth, W. Hoyle, and W.M. Roberts, *Org. Magn. Resonance* **8**, 226 (1976); M. Garreau, G.J. Martin, M.C. Martin, M. Morel, and C. Paulmier, *Org. Magn. Resonance* **6**, 648 (1974); J. Elguero, C. Marzin, and J.D. Roberts, *J. Org. Chem.* **39**, 357 (1974); S. Gronowitz, I. Johnson, and A.-B. Hornfeldt, *Chem. Scr.* **7**, 76 (1975); *ibid.* **7**, 211 (1975); M.T.W. Hearn, *Aust. J. Chem.* **29**, 107 (1976); F. Fringuelli, S. Gronowitz, A.-B. Hornfeldt, I. Johnson, and A. Taticchi, *Acta Chem. Scand. B* **28**, 125 (1974).

Table 4.18. ^{13}C chemical shifts of six-membered heterocycles[a]

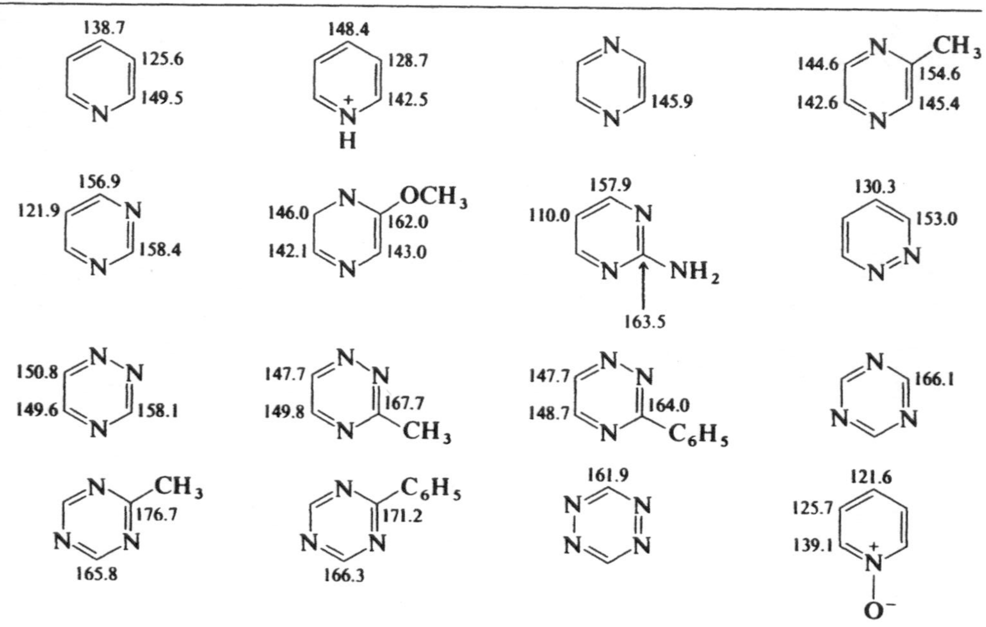

138.7 125.6 149.5

148.4 128.7 142.5

145.9

144.6 142.6 144.6 CH$_3$ 154.6 145.4

156.9 121.9 158.4

146.0 142.1 OCH$_3$ 162.0 143.0

157.9 110.0 NH$_2$ 163.5

130.3 153.0

150.8 149.6 158.1

147.7 149.8 CH$_3$ 167.7

147.7 148.7 C$_6$H$_5$ 164.0

166.1

176.7 CH$_3$ 165.8

171.2 C$_6$H$_5$ 166.3

161.9

125.7 139.1 121.6 N$^+$—O$^-$

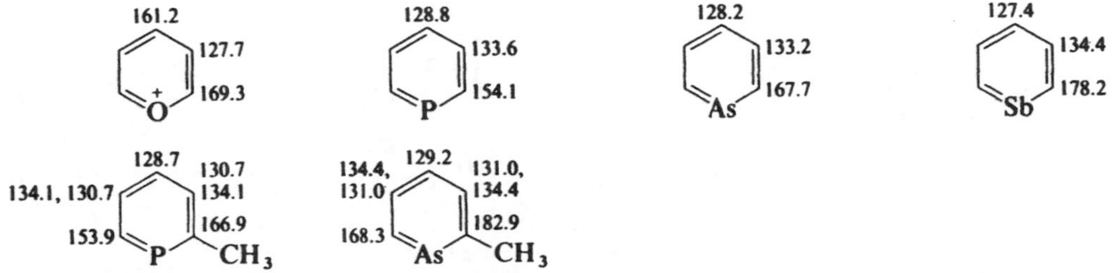

127.4
134.4
178.2
Sb

128.2
133.2
167.7
As

128.8
133.6
154.1
P

129.2
131.0
134.4
182.9
As—CH$_3$
168.3

134.4, 131.0

161.2
127.7
169.3
O$^+$

128.7
130.7
134.1
166.9
P—CH$_3$
153.9
134.1, 130.7

^{a} V. Formacek, L. Desnoyer, H.P. Kellerhals, and J.T. Clerc, ^{13}C Data Bank, Vol. 1, Copyright Bruker-Physik, Karlsruhe (1976); C.J. Turner and G.W.H. Cheeseman, Org. Magn. Resonance 6, 663 (1974); A.T. Balaban and V. Wray, Org. Magn. Resonance 9, 16(1977); J. Riaud, M. Th. Chenon, and N. Lumbroso-Bader, J. Am. Chem. Soc. 99, 6838 (1977); A.J. Ashe, III; R.R. Sharp, and J.W. Tolan, J. Am. Chem. Soc. 98, 5451(1976); J. Riaud, M. Th. Chenon, and N. Lumbroso-Bader, Tetrahedron Lett. 3123 (1974); C.J. Turner and G.W.H. Cheesman, Org. Magn. Resonance 8, 357 (1976); R.J. Pugmire and D.M. Grant, J. Am. Chem. Soc. 90, 697 (1968); W. Adam, A. Grimison, and G. Rodriguez, J. Chem. Phys. 50, 645 (1969); P. Van de Weijer and C. Mohan, Org. Magn. Resonance 9, 53 (1977); S. Braun and G. Frey, Org. Magn. Resonance 7, 194 (1975).

Table 4.19. ^{13}C chemical shifts of polycyclic heterocycles[a]

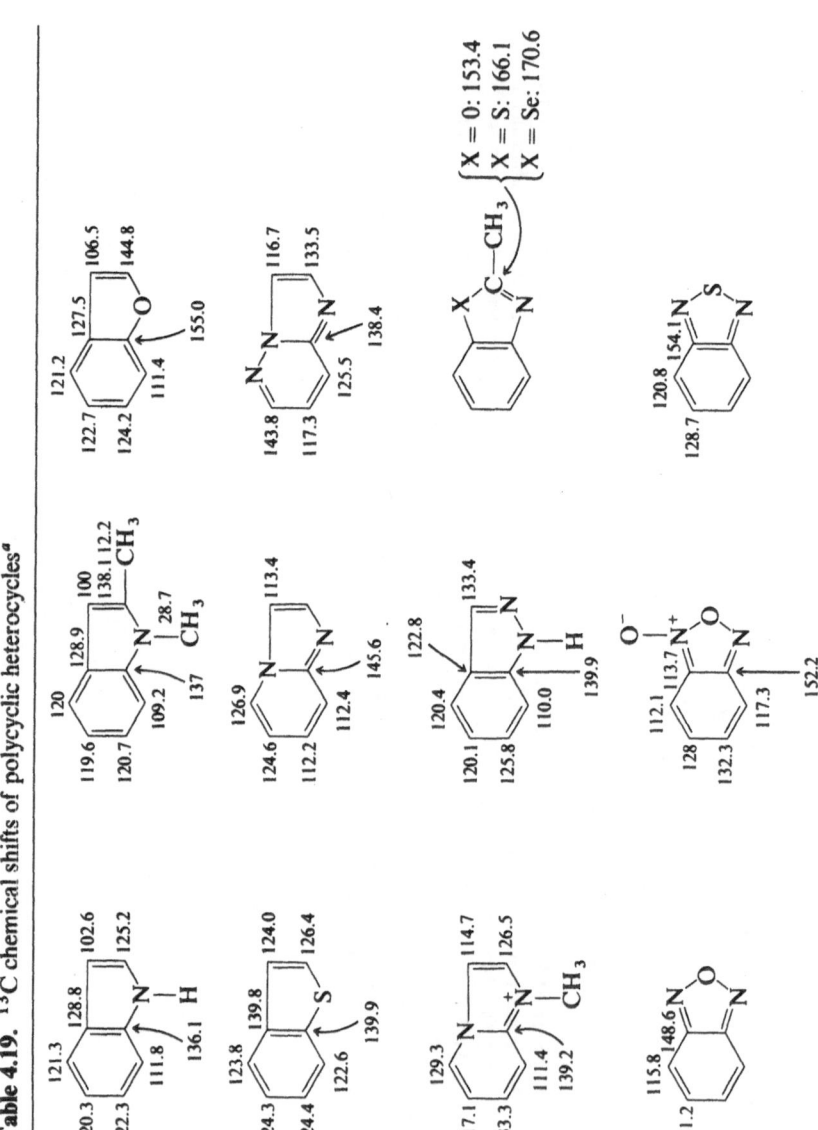

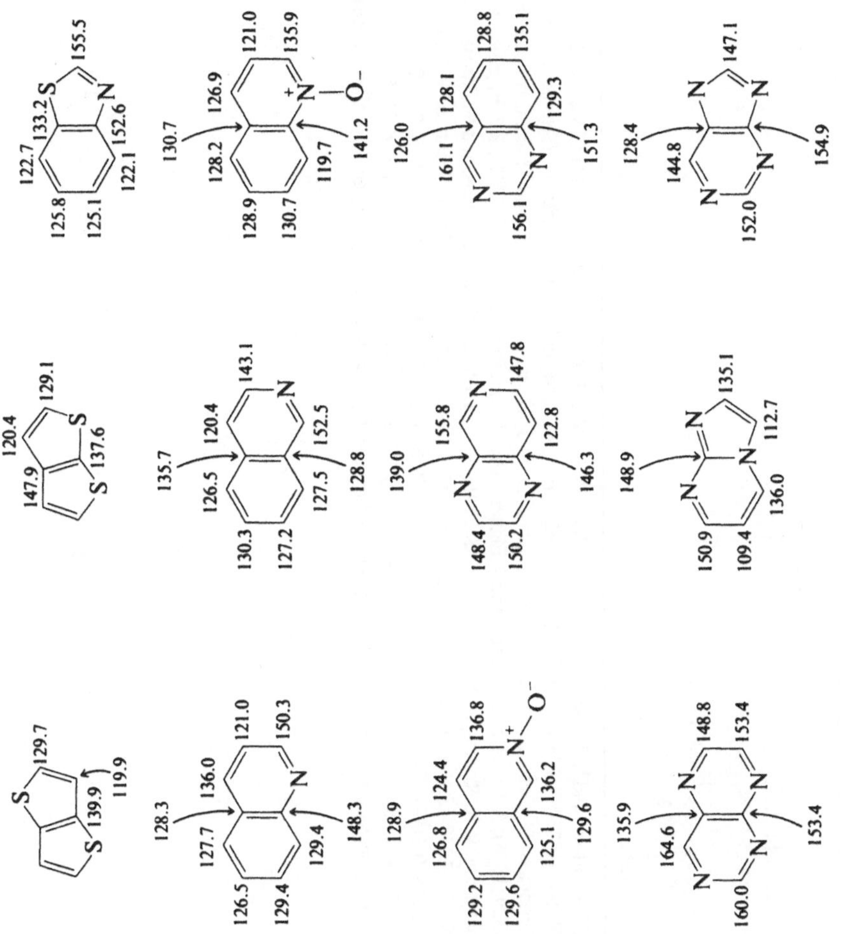
S 129.7 139.9 S 119.9
147.9 120.4 129.1 S 137.6 S
122.7 133.2 S 155.5 125.8 125.1 122.1 152.6 N
128.3 127.7 136.0 121.0 150.3 N 126.5 129.4 129.4 148.3
135.7 126.5 120.4 143.1 N 152.5 130.3 127.2 127.5 128.8
130.7 128.2 126.9 121.0 135.9 N O 128.9 130.7 119.7 141.2
128.9 126.8 124.4 136.8 N O 136.2 129.2 129.6 125.1 129.6
139.0 155.8 N 147.8 N 122.8 148.4 150.2 N 146.3
126.0 161.1 128.1 128.8 135.1 129.3 N 156.1 N 151.3
135.9 164.6 N 148.8 153.4 N N 160.0 N 153.4
148.9 N 135.1 N 112.7 150.9 109.4 136.0 N
128.4 144.8 N 147.1 N N 152.0 N 154.9

Table 4.19. (*Continued*)

[a] U. Ewers, H. Gunther, and L. Jaenicke, *Angew. Chem.* **87**, 356 (1975); J. P. Geerts, A. Nagel, and H.C. Van der Plas, *Org. Magn. Resonance* **8**, 607 (1976); R.J. Pugmire, J.C. Smith, D.M. Grant, B. Stanovnik, and M. Tisler, *J. Heterocycl. Chem.* **13**, 1057 (1976); E. Dradi and G. Gatti, *J. Am. Chem. Soc.* **97**, 5472 (1975). J. Elguero, A. Fruchier, and M. del Carmen Pardo, *Can. J. Chem.* **54**, 1329 (1976); A. Fruchier, E. Alcade, and J. Elguero, *Org. Magn. Resonance* **9**, 235 (1977); R.J. Pugmire, D.M. Grant, L.B. Townsend, and R.K. Robins, *J. Am. Chem. Soc.* **95**, 2791 (1973); S. Florea, W. Kimpenhaus, and V. Farcasan, *Org. Magn. Resonance* **9**, 133 (1977); S.N. Sawhney and D.W. Boykin, *J. Org. Chem.* **44**, 1136 (1974); D.M. Grant and R.J. Pugmire, *J. Am. Chem. Soc.* **93**, 1880, (1971); D.M. Grant, R.J. Pugmire, M.J. Robins, and R.K. Robins, *J. Am. Chem. Soc.*, **93**, 1887 (1971); **92**, 2386 (1970); J.A. Su, E. Siew, E.V. Brown, and S.L. Smith, *Org. Magn. Resonance* **10**, 122 (1977); *ibid.* **11**, 565 (1978); I. Yavari and F.A.L. Anet, *Org. Magn. Resonance*, **8**, 158 (1976); S. Gronowitz, I. Johnson, and A. Bugge, *Acta Chem. Scand.* **30B**, 417 (1976).

Table 4.20. Carbonyl chemical shifts[a]

(a) Ketones

Table 4.20. (Continued)

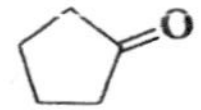

213.9

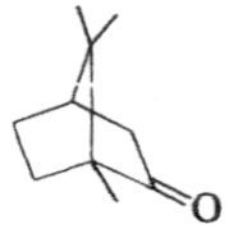

196.5

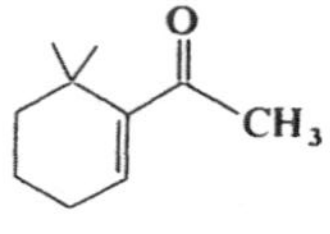

202.6

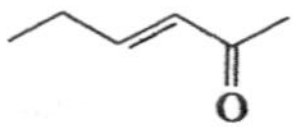

(b) Aldehydes

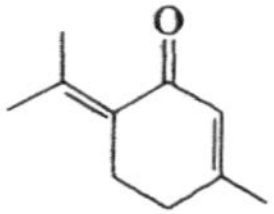

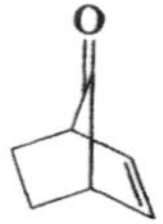

200.5

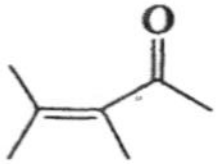

202.7

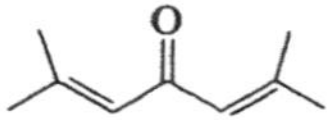

204.9

202.7

216.7

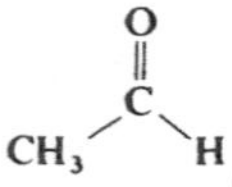

190.2

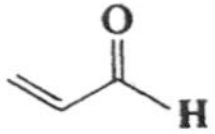

188.8

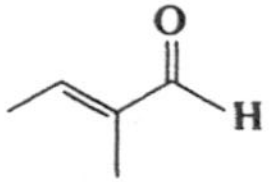

193.3

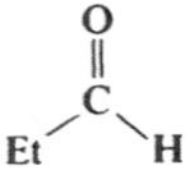

192.3

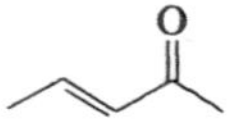

192.7

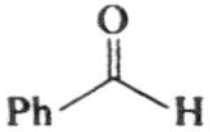

206.4

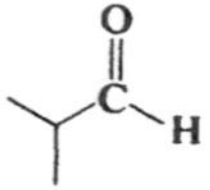

203.3

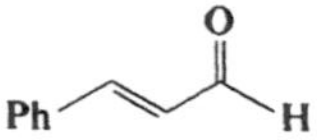

193.7

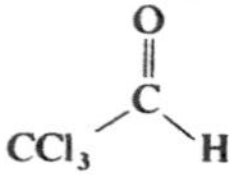

190.7

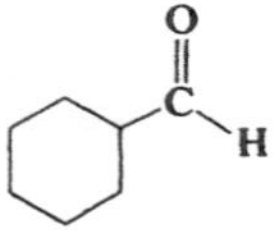

175.9

168.0   CCl$_3$—COOH

166.0   CF$_3$—COOH

181.4   (norbornane)—COOH

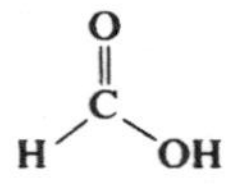

180.5   (norbornane)—COOH

161.6   CH$_3$—CO—CO—OEt

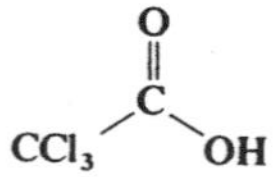

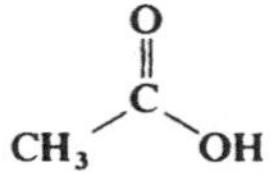

167.1

166.6

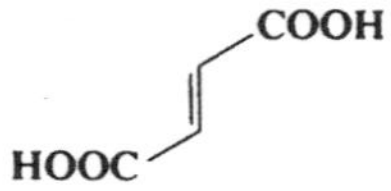

173.5

174.7

170.6

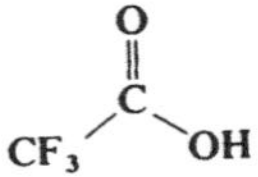

167.0

(c) Carboxylic acids

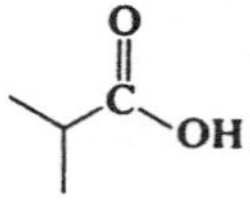

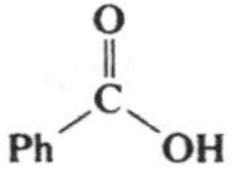

166.7   (cis) HOOC—CH=CH—COOH

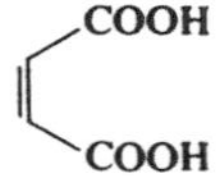

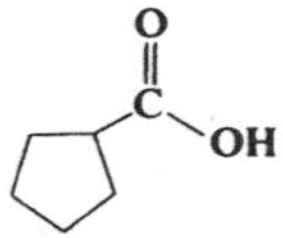

177.3   HOOC—CH=CH—COOH

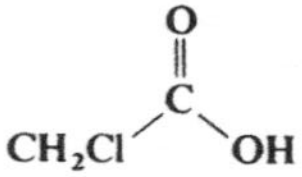

184.8   Ph—COOH

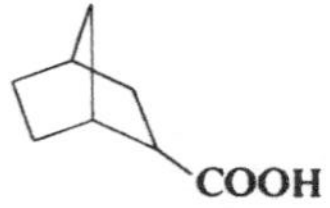

180.7   CH$_2$Cl—COOH

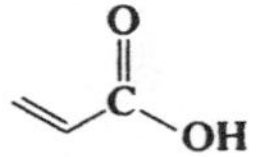

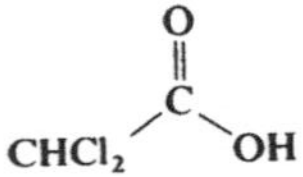

173.2   CHCl$_2$—COOH

171.0   Ph—CO—OCH$_3$

H—COOH

CH$_3$—COOH

(CH$_3$)$_2$CH—COOH

(cyclopentyl)—COOH

CH$_2$=CH—COOH

(d) Carboxylic esters and lactones

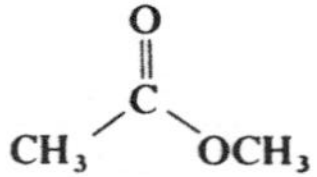

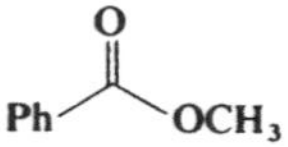

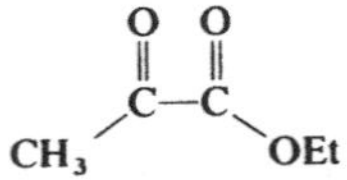

CH$_3$—CO—OCH$_3$

Table 4.20. (Continued)

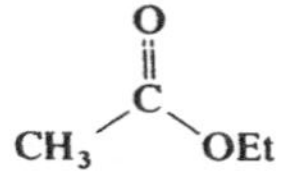

167.7 $O=C(OEt)-CH_2-C(=O)-CH_3$ (diethyl/ethyl acetoacetate type)

173.9 $O=C(OEt)-CH_2-CH_2-C(=O)-CH_3$

171.2 β-lactone (four-membered)

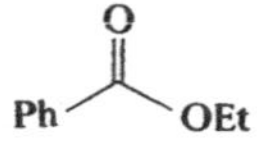

178.0 γ-butyrolactone (five-membered)

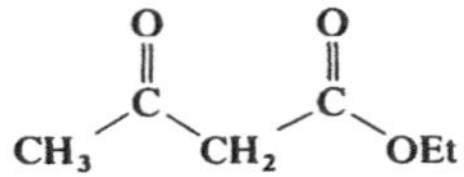

175.2 δ-valerolactone (six-membered)

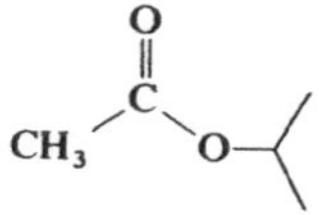

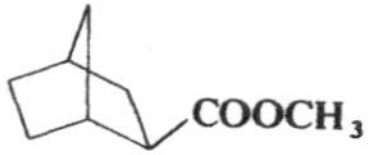

169.8 H_3C ... CH_3 substituted six-membered lactone

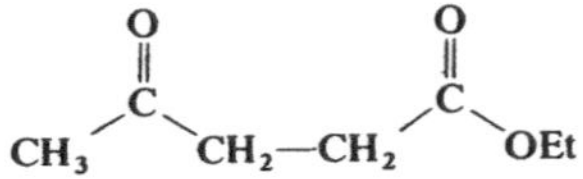

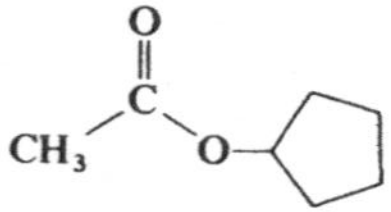

164.9 $O=C(OEt)-Ph$ (ethyl benzoate)

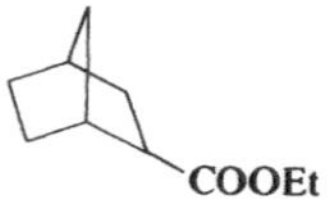

175.7 norbornane-$COOCH_3$

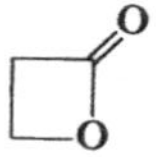

174.6 norbornane-$COOEt$

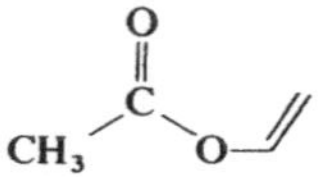

163.3 EtOOC–CH=CH–COOEt (trans, fumarate)

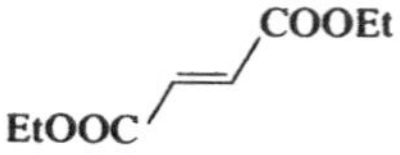

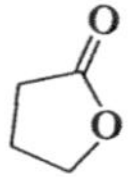

164.8 EtOOC–CH=CH–COOEt (cis, maleate)

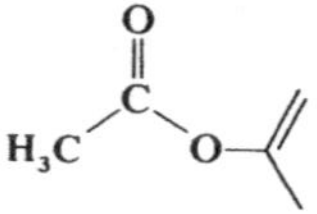

170.8 $EtOOCCH_2–CH_2–COOEt$

160.7 $O=C(OEt)-H$ (ethyl formate)

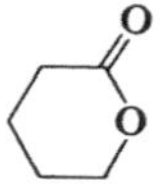

170.3 $O=C(OEt)-CH_3$ (ethyl acetate)

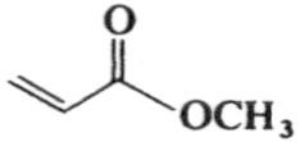

169.8 $O=C(O-iPr)-CH_3$ (isopropyl acetate)

169.9 $O=C(O-cyclopentyl)-CH_3$

167.7 $O=C(O-CH=CH_2)-CH_3$ (vinyl acetate)

168.9 $O=C(O-C(=CH_2))-CH_3$ (isopropenyl acetate)

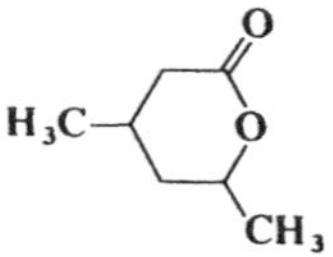

164.5 $O=C(OCH_3)-CH=CH_2$ (methyl acrylate)

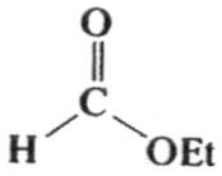

174.0   175.3   172.4   171.7   163.6   162.8

169.6   151.5   172.9   165.9   158.6   167.3
165.1

167.3   170.8   173.1   174.0   172.8   168.2

(e) Carboxylic acid anhydrides

Table 4.20. (Continued)

(f) Carboxylic acid halides

Compound	δ	Compound	δ	Compound	δ
CH_3COCl	170.5	CH_3CH_2COCl	174.5	$ClCH_2COCl$	169.7
CH_3COBr	166.5	$(CH_2CH_2COCl)_2$	174.3	$Cl_2CHCOCl$	166.2
CH_3COI	159.8	$PhCOCl$	168.7	Cl_3CCOCl	163.5

(g) Amides

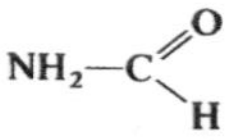

Compound	δ	Compound	δ	Compound	δ
NH_2–CHO	165.5	$CH_2{=}C(CH_3)$–CO–NH_2	170.3	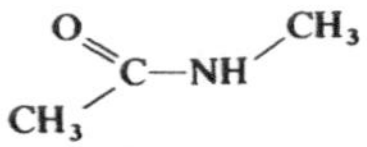 CH_3–CO–NH–CH_3	171.6
CH_3–NH–CHO	163.4	3-pyrrolin-2-one	163.4	C_6H_5, N–C_6H_5 (azetidinone)	166.7
H–N(CH_3)–CHO	166.7		175.9	CH_3, N(CH_3)–CO–Ph	162.1
CH_3–N(CH_3)–CHO	162.7	CH_3–CO–NH_2	162.7	Ph–N(CH_3)–CO–CH_3	170.8
nBu–N(nBu)–CHO	162.3	CH_3–CO–N(CH_3)$_2$	162.3	Ph–NH–CO–CH_3	168.3
				O_2N–C$_6$H$_4$–NH–CO–CH_3	169.2

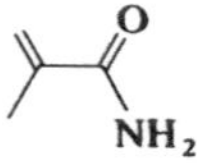

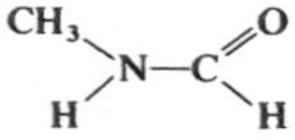

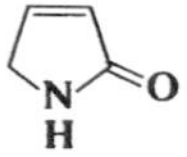

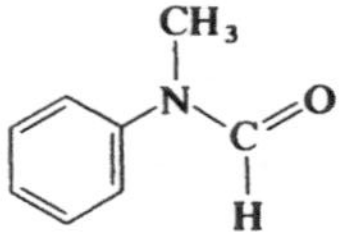

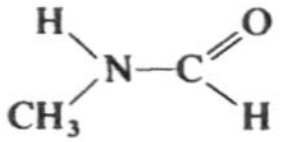

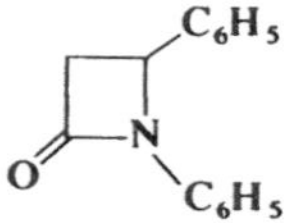

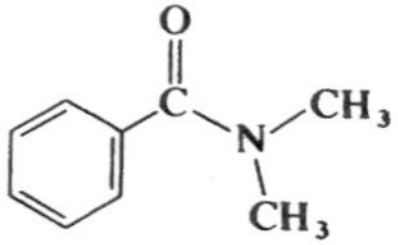

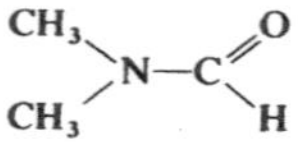

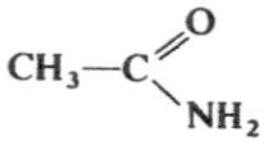

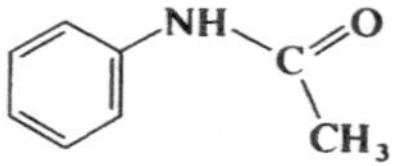

Additional values listed: 175.9, 172.7, 170.7

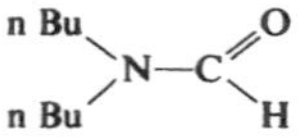

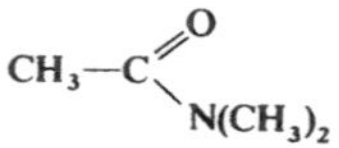

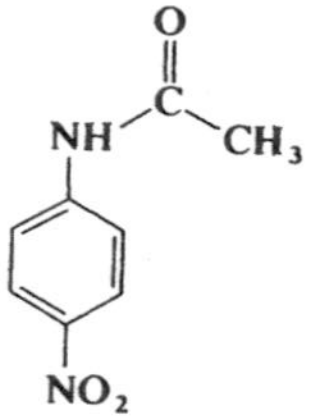

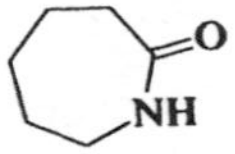	178.6	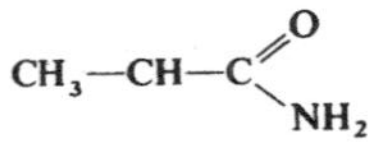	177.2	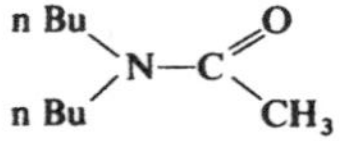	169.1

(*h*) *Imides*

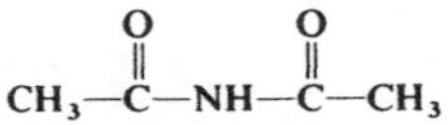	171.8	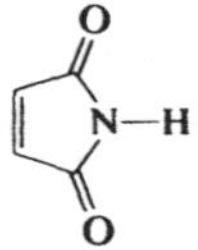	172.5	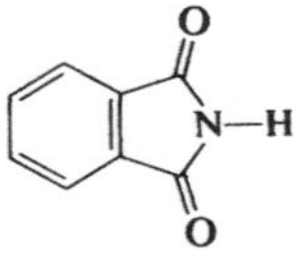	169.7
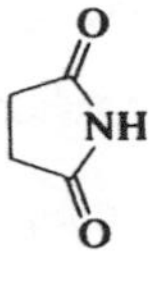	179.1	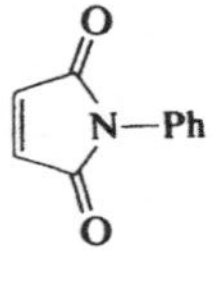	170.0	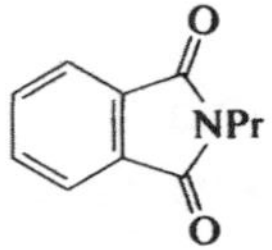	168.5
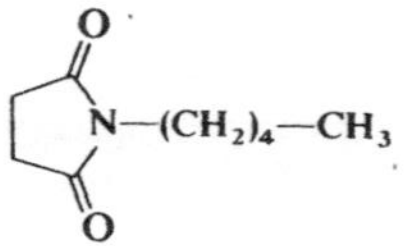	177.5	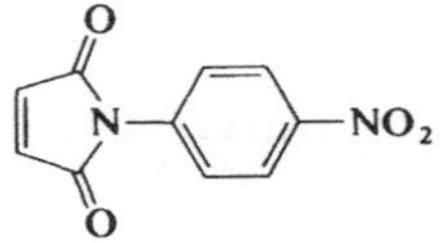	170.0	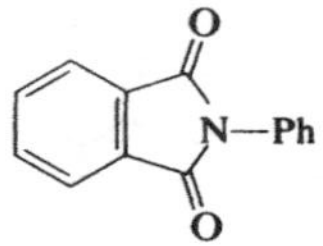	167.7
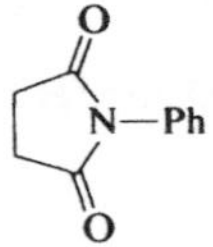	177.9				

(*i*) *Ureas*

$H_2N{-}C(=O){-}NH_2$	161.3	$(CH_3)_2NCON(CH_3)_2$	164.8	$PhNHC(=O)NHPh$	152.6

Table 4.20. (*Continued*)

CH$_3$NHCNH$_2$ ‖O 159.8	PhNHCNH$_2$ ‖O 156.6	H$_2$N—C—NH—C—NH$_2$ (‖O ‖O) (biuret) 155.5
CH$_3$NHCONHCH$_3$ 159.5		

(j) Carbonates

Dimethyl carbonate 156.6	Dibutyl carbonate 155.5	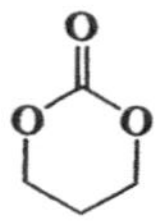 (1,3-dioxan-2-one) 155.2
Diethyl carbonate 155.8	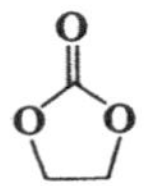 (ethylene carbonate) 156.7	CH$_3$CH$_2$CHOCOCH$_3$ / CH$_3$ 155.6
Diphenyl carbonate 153.0		

(k) Quinonoid compounds

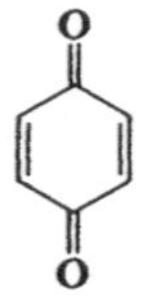

(p-benzoquinone) 187.0	(o-benzoquinone) 180.2	(1,4-naphthoquinone) 184.6
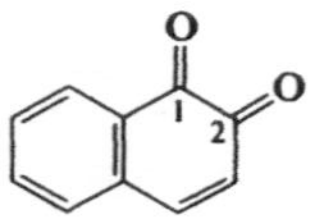 (anthraquinone) 179.8(C$_1$) 180.9(C$_2$) 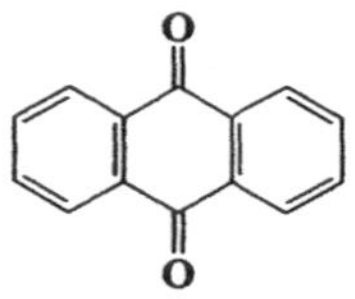	(1,2-naphthoquinone) 183.0	

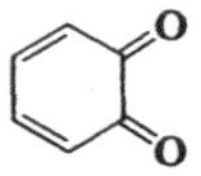

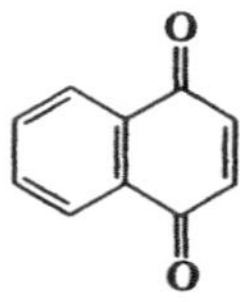

[a] L.M. Jackman and D.P. Kelley, *J. Chem. Soc.*, *(B)*, 102, (1970); E.V. Dehmlow, R. Zeisberg, and S.S. Dehmlow, *Org. Magn. Resonance* **7**, 418 (1975); D.H. Marr and J.B. Stothers, *Can. J. Chem.* **43**, 596 (1965); F.J. Weigert and J.D. Roberts, *J. Am. Chem. Soc.* **92**, 1338, 1347 (1970); E. Lippmaa, T. Pehk, J. Paasivirta, N. Belikova, and A. Plate, *Org. Magn. Resonance* **2**, 581 (1970); J.B. Stothers and P.C. Lauterbur, *Can. J. Chem.* **42**, 1563 (1964); J.B. Grutzner, M. Jautelat, J.B. Dence, R.A. Smith, and J.D. Roberts, *J. Am. Chem. Soc.* **92**, 7107; J.B. Stothers and C.T. Tan, *Can. J. Chem.* **53**, 581 (1975); R. Bicker, H. Kessler, A. Steigel, and G. Zimmermann *Chem. Ber.* **111**, 3215, (1978); F.J. Koer, A.J. de Hoog, and C. Altona, *Recl. Trav. Chim. Pays-Bas* **94**, 75 (1975); F.J. Koer and C. Altona, *ibid.* **75**, 127 (1975); S. Berger and A. Rieker, *Tetrahedron* **28**, 3123 (1972); *Chem. Ber.* **109**, 3252 (1976); G. Hofle, *Tetrahedron* **32**, 1431 (1976); M. Christl, H.J. Reich, and J.D. Roberts, *J. Am. Chem. Soc.* **93**, 3453 (1971); S.W. Pelletier, Z. Djarmati, and C. Pape, *Tetrahedron* **32**, 995 (1976); P.A. Couperus, A.D.H. Clague, and J.P.C.M. Van Dongen, *Org. Magn. Resonance* **11**, 590 (1978); I.A. McDonald, T.J. Simpson, and A.F. Sierakowski, *Aust. J. Chem.* **30**, 1727 (1977); A. Arnone, G. Fronza, R. Mondelli, and J. St. Pyrek, *J. Magn. Resonance* **28**, 69 (1977); W. Stadeli, R. Hollenstein, and W. Von Philipsborn, *Helv. Chim. Acta* **50**, 948 (1977); G.C. Levy and G.L. Nelson, *J. Am. Chem. Soc.* **94**, 4897 (1972); R. Buchi and E. Pretsch, *Helv. Chim. Acta* **58**, 1573 (1975); G. Severin-Ricca, P. Manitto, D. Monti, and E.W. Randall, *Gazz. Chim. Ital.* **105**, 1273 (1975); A.K. Bose and P.R. Srinivasan, *Org. Magn. Resonance* **12**, 34 (1979); D.B. Bigley, C. Brown, and R.H. Weatherhead, *J. Chem. Soc. Perkin. Trans.* II, 701, (1976); G.L. Nelson, unpublished results; S. Combrisson, J.-P. Lautie, and M. Olomuczki, *Bull. Soc. Chim. Fr.* 2769, (1975); L.F. Johnson and W.C. Jankowski, *Carbon-13 Spectra*, Wiley-Interscience, New York (1972); H.O. Kalinowski and H. Kessler, *Org. Magn. Resonance* **6**, 305 (1974); M.P. Sibi and R.L. Lichter, *J. Org. Chem.* **44**, 3017 (1979); A.R. Tarpley, Jr., and J.H. Goldstein, *J. Am. Chem. Soc.* **93**, 3573 (1971); J.W. Triplett, G.A. Digenis, W.J. Layton, and S.L. Smith, *Spectrosc. Lett.* **10**, 141 (1977); G.C. Levy, R.L. Lichthler, and G.L. Nelson, *Carbon-13 NMR Spectroscopy*, John Wiley and Sons, New York (1980).

Table 4.21. Effect of carbonyl and phenyl groups on alkyl chemical shifts[a]

R	α		β		γ
	st[b]	br[c]	st[b]	br[c]	
Ph	23	17	9	7	−2
COOH	21	16	3	2	−2
COOR	20	17	3	2	−2
COCl	33	28	—	2	—
CHO	31	—	0	—	−2
COR	30	24	1	1	−2

[a] Values in table are to be added to the normal alkyl shift for that carbon.

[b] Straight chain: R—CH$_2$—CH$_2$—CH$_2$.
 $\quad\quad\quad\quad\quad\quad\;\alpha\quad\;\;\beta\quad\;\;\gamma$

[c] Branched chain:
$$\begin{array}{c} \text{R} \\ | \\ \text{C—C—C—C—C.} \end{array}$$
$\quad\quad\quad\quad\quad\quad\gamma\;\;\beta\;\;\alpha\;\;\beta\;\;\gamma$

from which the ester has been formed. Anhydride carbonyl carbons resonate ~ 10 ppm upfield from the resonance positions of carbonyl carbon atoms of the corresponding acids. Carbonyl carbon resonances of acid halides occur 4 to 8 ppm upfield from their positions in the corresponding acids. The carbonyl carbon resonances of amides are found in the region between 160 and 180 ppm, and are affected by the conformation of the substituent on the nitrogen relative to the carbonyl group. Imide carbonyl carbons fall in the same region as amides while in carbonates they resonate between 150 and 160 ppm.

4.1.3.6   Other Functional Groups

Chemical shifts of nitrile carbon atoms appear in the region between 112 and 126 ppm (Table 4.22). In isocyanides, the resonances of the carbon atoms occur between 155 and 160 ppm, whereas in aldoximes and ketoximes they appear between 145 and 163 ppm. The isocyanate carbon atoms resonate between 118 and 132 ppm (Table 4.23). The chemical shift ranges of imines, iminoethers, amidines, oximes, semicarbazones, and phenylhydrazones are given in Table 4.24.

The resonances of aliphatic carbocations occur at very low fields. The chemical shifts of carbocations are given in Table 4.4. In charged aromatic systems, an upfield shift is observed with increasing π-electron density (Table 4.25).

Table 4.22. ^{13}C chemical shifts of nitrile carbons[a]

Structure	Shift	Structure	Shift	Structure	Shift
CH_3CN	117.7	EtO—CN	118.2	(bicyclo)—CN	122.6
=CN	117.5	Cl—CH=CH—CN	114.4		
=CN	117.6	Ph—CN	118.7	PhO—CH_2—CH_2—CN	118.4
NC—CN	119.3	p-F—PhCN	114.1	HO—C(CH$_3$)$_2$—CN	123.5
(CH$_3$)$_2$CH—CN	123.7	NC—C(CH$_3$)$_2$—CN	125.1	NC—C(CH$_3$)$_2$—CN	123.5
Ph—CH_2—CN	118.0	(cyclohexyl)—CN	121.5	N(CH$_2$CN)$_3$ (nitrilotriacetonitrile)	115.1
(bicyclo)—CN	120.8	(bicyclo)—CN	123.4		

[a] J.B. Grutzner, M. Jautelat, J.B. Dence, R.A. Smith, and J.D. Roberts. *J. Am. Chem. Soc.* **92**, 7107 (1970); L.F. Johnson and W.C. Jankowski, *Carbon-13 Spectra*, Wiley-Interscience, New York (1972); G.E. Maciel and D.A. Beatty, *J. Phys. Chem.* **59**, 3920 (1965); G.C. Levy, R.L. Lichthler, and G.L. Nelson, *Carbon-13 NMR Spectroscopy*, John Wiley and Sons, New York (1980).

Table 4.23. ^{13}C chemical shifts of oximes and isocyanates[a]

Structure	Shift	Structure	Shift
(butanone oxime)	159.2	(butanone oxime, anti)	159.2
benzaldoxime	149.6	benzaldoxime	—
salicylaldoxime	152.4	cyclohexanone oxime	152.4
CH_3NCO	158.7 / 121.4	CH_3NCO	—
EtNCO	146.4 / 122.5	EtNCO	—
cyclohexyl NCO	159.4 / 123.6		

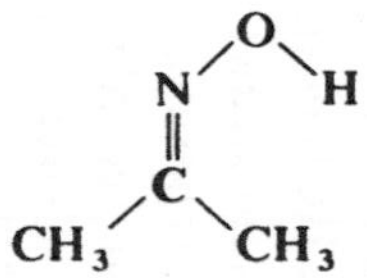	154.3	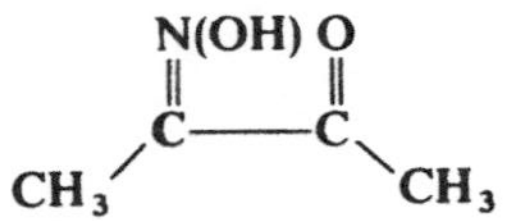	156.6
PhNCO	129.5		
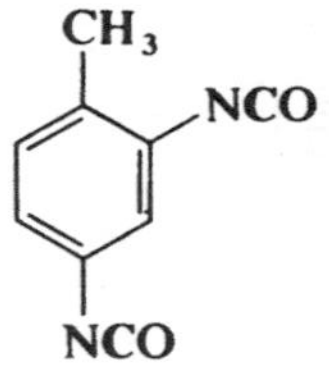	125.0		
$CCl_3-C(=O)-NCO$	130.2		
$(\overset{+}{K})NCO^-$	129.7		

[a] N. Naulet, M.L. Filleux, G.J. Martin, and J. Pornet, *Org. Magn. Resonance* **7** (7), 326 (1975); S. Combrisson, J.-P. Lautie, and M. Olomuczki, *Bull. S. Chim. Fr.* 2769 (1975); G.C. Levy and G.L. Nelson, *J. Am. Chem. Soc.* **94**, 4897 (1972); L.F. Johnson and W.C. Jankowski, *Carbon-13 Spectra*, Wiley-Interscience, New York (1972); G.E. Maciel and D.A. Beatty, *J. Phys. Chem.* **59**, 3920 (1965); G.C. Levy, R.L. Lichther, and G.L. Nelson, *Carbon-13 NMR Spectroscopy*, John Wiley and Sons, New York (1980).

Table 4.24. Range of ^{13}C chemical shifts for imine functional groups[a]

Imines (aliphatic)	168–175	Oximes (aliphatic)	154–159
Imines (aromatic)	157–163	Oximes (aromatic)	148–155
Imino ethers (aromatic)	153–156	Semicarbazones (aliphatic)	158–160
Amidines (aliphatic)	149–154	Phenylhydrazones (aliphatic)	145–149
Amidines (aromatic)	149–151	Phenylhydrazones (aromatic)	145–146

[a] N. Naulet, M.L. Filleux, G.J. Martin, and J. Pornet, *Org. Magn. Resonance* **7** (7), 326 (1975).

Table 4.25. Chemical shifts of charged aromatic compounds[a]

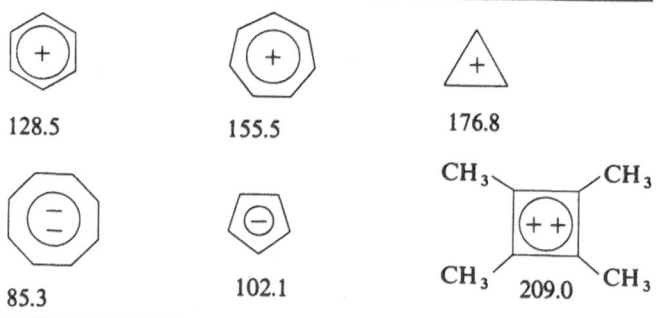

[a] G.A. Olah and G.D. Mateescu, *J. Am. Chem. Soc.* **92**, 1430 (1970).

4.2 COUPLINGS IN ^{13}C-NMR SPECTROSCOPY

4.2.1 Carbon–Proton and Carbon–Carbon Coupling

4.2.1.1 One-Bond Couplings

4.2.1.1.1 $^{1}J_{CH}$ Direct carbon–proton couplings, $^{1}J_{CH}$, can be readily obtained by observing the ^{13}C satellite peaks in proton spectra. Since the abundance of ^{13}C is only 1%, the ^{1}H spectrum of pure CHCl$_3$ would contain a large singlet (Figure 4.9) due to ^{12}CHCl$_3$, corresponding to 99% of molecules in the sample, and two "satellite" peaks (one on each side of the ^{12}CHCl$_3$ peak with a separation from one another of 210 Hz) due to the ^{13}CHCl$_3$ molecules. The ^{13}C spectrum will contain an identical doublet with the same separation when measured in Hz (but not in ppm; in a ^{13}C-NMR spectrum, recorded at 25.2 MHz, the separation of the peaks would be 8.33 ppm whereas in the ^{1}H-NMR spectrum recorded at 100 MHz, a separation of 209 Hz is observed, which is equal to 2.09 ppm).

In simple hydrocarbons $^{1}J_{CH}$ couplings are directly proportional to the extent of s-character (ρ) of the CH bond:

$$^{1}J_{CH} = 500\rho$$

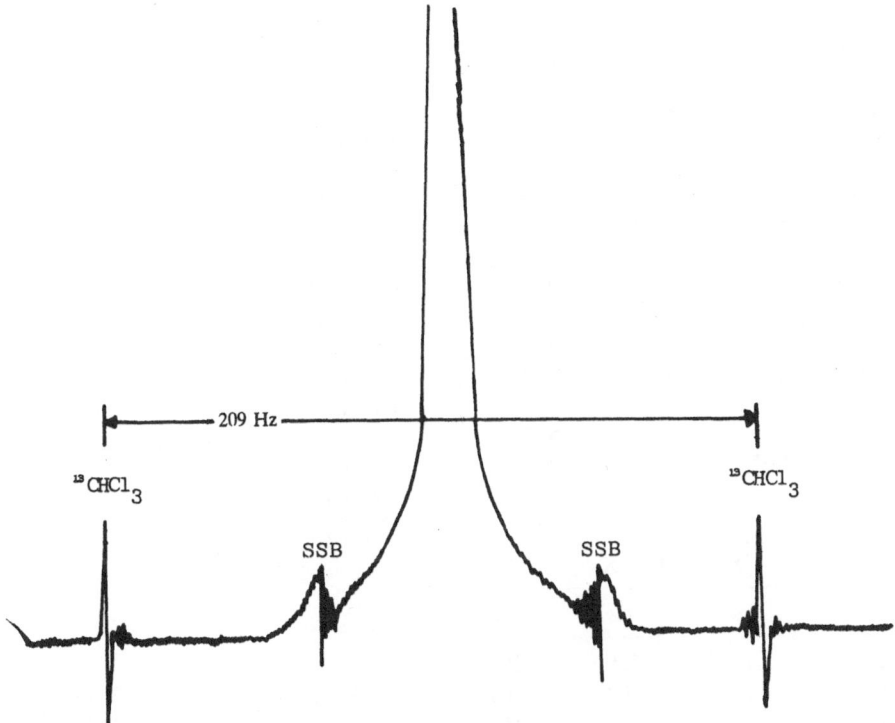

Figure 4.9. The 100-MHz ^{1}H spectrum of pure CHCl$_3$, showing the spinning side bands (SSB) and ^{13}C satellites.

Thus in methane (sp^3-hybridized carbons; s-character, 0.25) $^1J_{CH}$ = 125 Hz, in ethylene (sp^2-hybridized carbons; s-character, 0.33) $^1J_{CH}$ = 167 Hz while in acetylene (sp-hybridized carbons; s-character, 0.5) $^1J_{CH}$ = 250 Hz. The values of $^1J_{CH}$ couplings thus allow one to probe the extent of s-character in a given bond. For example, in cyclopropane $^1J_{CH}$ = 162 Hz, which is evidence of sp^2 hybridization in this molecule.

The magnitudes of $^1J_{CH}$ couplings vary considerably with the nature of the substituents, and the effects of substituents are largely additive. Thus $^1J_{CH}$ is 125 Hz for CH$_4$, 150 Hz for CH$_3$Cl, 178 Hz for CH$_2$Cl$_2$, and 209 Hz for CHCl$_3$ (the additivity being less accurate for trisubstituted methanes).

The effect of electronegative substituents is even more marked when they are attached to sp^2-hybridized carbons. Thus $^1J_{CH}$ in methyl cyanide is 136 Hz but in vinyl cyanide it is 177 Hz. Some $^1J_{CH}$ values are presented in Table 4.26. The additivity of substituent effects on aliphatic carbons is given by the equation

$$^1J_{CHZ_1Z_2Z_3} = 125.0 + \sum_i Z_i$$

where Z_1, Z_2, and Z_3 are the substituents attached to a given carbon atom and Z_i is the increment factor for each group. The increments Z_i for various functional groups are presented in Table 4.27.

Table 4.26. Some characteristic $^1J_{CH}$ couplings[a]

Compound	Coupling (Hz)	Compound	Coupling (Hz)
C_2H_6	125	$CH_3.CO.CH_3$	127
Cyclohexane	123	$CH_3.CO_2H$	130
C_2H_4	156	$C\underline{H}_3.C\text{:}CH$	132
Cyclopropane	162	CH_3NH_2	133
Cyclobutane	134	$CH_3.CN$	136
Cyclopentane	128	$CH_3.OH$	141
Benzene	159	$CH_3.\overset{+}{N}H_3$	145
Acetylene	248	$CH_3.NO_2$	147
$CH_3C\underline{H}O$	173	CH_3F	149
$CCl_3C\underline{H}O$	207	CH_3Cl	150
$\underline{H}.CO_2H$	222	CH_3Br	152
$Me_2N.C\underline{H}O$	191	CH_3I	151
$Me_2\overset{+}{C}H$ SbF_5Cl	168	$CH_2(OEt)_2$	161
CH_4	125	CH_2Cl_2	178
$(CH_3)_4Si$	118	$CHCl_3$	209

$$CH_2{=}CH{-}X$$

X	α	cis	trans
F	200	159	162
Cl	195	163	161
CHO	162	157	162
CN	177	163	165

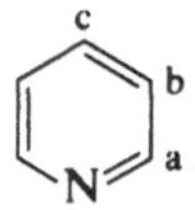

a, 170
b, 163
c, 152

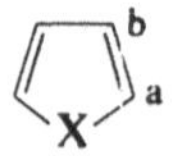

X	a	b
O	201	175
NH	184	170
S	185	167
CH_2	170	170

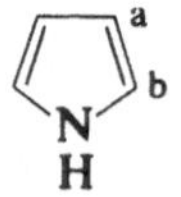

a, 170
b, 182

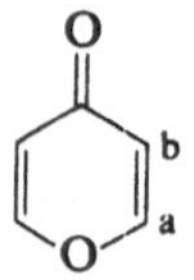

a, 200
b, 169

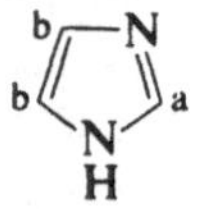

a, 208
b, 199

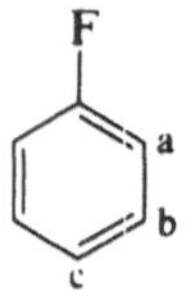

a, 155
b, 163
c, 161

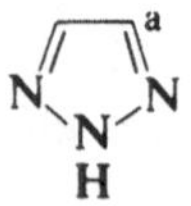

a, 205

Table 4.26. (*Continued*)

Compound	Coupling (Hz)	Compound	Coupling (Hz)
	a, 167 b, 198		a, 208
	a, 190 b, 178		

[a] G.E. Maciel, J.W. Mclver, Jr., N.S. Ostlund, and J.A. Pople, *J. Am. Chem. Soc.* **92**, I and II (1970) and references therein; F.J. Weigert and J.D. Roberts, *J. Am. Chem. Soc.* **90**, 3543 (1968); *ibid.* **91**, 4940 (1969); G.S. Handler and J.H. Anderson, *Tetrahedron* **2**, 345 (1958); J.J. Burker and P.C. Lauterbur, *J. Am. Chem. Soc.* **86**, 1870 (1964); G.A. Olah and A.M. White, *J. Am. Chem. Soc.* **89**, 7072 (1967); C.H. Yoder, R.H. Tuck, and R.E. Hess, *J. Am. Chem. Soc.* **91**, 539 (1969); E.F. Mooney and P.H. Winson, in: *Annual Review of NMR Spectroscopy*, Vol. 2, E.F. Mooney, Ed., Academic Press, New York, London (1969), p. 153.

Table 4.27. Additivity rules for $^1J_{CH}$ coupling constants in aliphatic carbons[a]

$$J_{CH\,Z_1Z_2Z_3} = 125.0 + \sum_i Z_i$$

Substituent	Increase Z_i^b	Substituent	Increase Z_i^b
—H	0.0		
—CH$_3$	1.0	—OH	18.0
—C(CH$_3$)$_3$	−3.0	—OCH$_3$	15.0
—CH=CH$_2$	−3.0	—O—Phenyl	18.0
—CH$_2$Cl	3.0	—NH$_2$	8.0
—CH$_2$Br	3.0	—NHCH$_3$	7.0
—CH$_2$I	7.0	—N(CH$_3$)$_2$	6.0
—CHCl$_2$	6.0	—NO$_2$	22.0
—CCl$_3$	9.0	—SCH$_3$	13.0
—C≡CH	7.0	—SOCH$_3$	13.0
—Phenyl	1.0	—CHO	2.0
—F	24.0	—COCH$_3$	−1.0
—Cl	27.0	—COOH	5.5
—Br	27.0	—COOCH$_3$	5.0
—I	26.0	—CN	11.0

[a] E. Pretsch, J. Seibl, W. Simon, and T. Clerc, *Strukturaufklarung Organicher Verbindungen*, Springer-Verlag, Berlin, New York (1981); N.J. Boboken and E.R. Malinowski, *J. Am. Chem. Soc.* **83**, 1479 (1961).
[b] Example: To calculate the coupling constant for CHCl$_3$: $J_{CH} = 125.0 + (3 \times 27.0) = 206.0$ Hz. This is quite close to the experimentally determined value of 209.0 Hz.

4.2.1.1.2 $^1J_{CC}$. One-bond carbon–carbon couplings, $^1J_{CC}$, vary over a large range, though they are generally smaller than $^1J_{CH}$ couplings because of the smaller magnetic moment of carbon in comparison to hydrogen. They are affected by the extent of s-character of the carbon atoms. Some typical values are presented in Table 4.28. In many compounds, it has been found that there is a relationship between $^1J_{CC}$ and $^1J_{CH}$ which in the case of a methyl group attached to a carbon atom may be expressed as

$$^1J_{C-CH_3} = (0.27)^1J_{C-H}$$

4.2.1.2 Two-Bond Couplings

4.2.1.2.1 $^2J_{CH}$ C—C   Couplings between carbon and hydrogen atoms separated by two bonds (e.g., C—C—H) are small, usually between -6 and -4 Hz in aliphatic systems. Introduction of electronegative substituents causes them to increase (i.e., their negative value decreases) and they can then occur between -5 and $+2$ Hz, the increase being maximum when the substituent is in the same plane as the C—C—H moiety. A useful feature of $^2J_{CH}$ couplings is that they are often 60–70% of the value of $^2J_{HH}$ in a similar geometrical environment. Increasing the s-character of the C—C or C—H bonds causes an increase in the value of $^2J_{CH}$. Some values of $^2J_{CH}$ are presented in Table 4.29.

4.2.1.2.2 $^2J_{CC}$ C—C   Two-bond carbon–carbon couplings, $^2J_{CC}$, are usually less than 3 Hz, particularly in saturated aliphatic systems. However, they can be larger if the central carbon atom is a carbonyl group, a cyano group, or acetylenic. Typical $^2J_{CC}$ values are given in Table 4.30.

4.2.1.3 Three-Bond Couplings

4.2.1.3.1 $^3J_{CH}$ C—C   Three-bond carbon–proton couplings, $^3J_{CH}$, depend on the dihedral angle ϕ between the planes containing the H—C—C and the C—C—C bonds, as is the case for vicinal proton couplings ($^3J_{HH}$).

The couplings are maximum when the dihedral angle is 180°, a little lower when it is 0°, and minimum when it is 90°. In olefins $^3J_{CH(cis)}$ is smaller than $^3J_{CH(trans)}$, a fact which may be used to ascertain the geometry of the olefins. Some values are presented in Table 4.31.

4.2.1.3.2 $^3J_{CC}$ C—C   Three-bond carbon–carbon couplings, $^3J_{CC}$, exhibit a dihedral angle dependence as do $^3J_{CH}$ couplings. Some $^3J_{CC}$ values are presented in Table 4.30.

Table 4.28. One-bond carbon–carbon coupling constants $^1J_{CC}{}^a$

Compound	$^1J_{CC}$(Hz)
Ethane	34.6
2-Methylpropane	36.9
Ethylbenzene	34.0 (aliphatic)
Propionitrile	33.0 (α, β)
1-Propanol	34.2 (2, 3)
	37.8 (1, 2)
Ethanol	37.7
2-Propanol	38.4
tert-Butylamine	37.1
tert-Butanol	39.5
2-Butanone	38.4 (1, 2)
Acetaldehyde	39.4
Acetone	40.1
3-Pentanone	35.7 (1, 2)
	39.7 (2, 3)
Acetophenone	43.3 (aliphatic)
Acetate anion (aq.)	51.6
N,N-Dimethylacetamide	52.2
Acetic acid	56.7
Ethyl acetate	58.8 (CH₃—C—) with O double bonded to C
tert-Butyl cyanide	52.0 (—C—CN)
Isopropyl cyanide	54.8 (CH—CN)
Propionitrile	55.2 (—CH₂—CN)
Acetonitrile	56.5
Propyne	67.4 (ĊH₃—Ċ≡C—)
Ethylene	67.6
Acrylic acid	70.4 (CH₂=CH—)
Acrylonitrile	70.6 (CH₂=CH—)
Styrene	70.0 ± 3 (aliphatic)
Benzene	57.0
Nitrobenzene	55.4(1–2)
	56.3(2–3)
	55.8(3–4)
Iodobenzene	60.4(1–2)
	53.4(2–3)
	58.0(3–4)
Anisole	58.2(2–3)
	56.0(3–4)
Aniline	61.3(1–2)
	58.1(2–3)
	56.6(3–4)

Table 4.28. (*Continued*)

Compound	$^1J_{CC}$(Hz)
Pyridine	53.8(2–3)
	56.2(3–4)
Thiophene	64.2
Pyrrole	65.9
Furan	69.1
Benzonitrile	80.3 (N≡C—C⟨)
1,1-Dimethylallene	99.5 (CH$_2$=C=)
Phenylethynyl cyanide	155.8 (—C≡C—)
Acetylene	171.5
Phenylacetylene	175.9 (H—C≡C—)
Methylcyclopropane	44.0(1–α)
Dicyclopropylketone	54.0(1–α)
	10.2(1–2)
Cyclopropanecarboxylic acid	72.5(1–α)
	10.0(1–2)
Cyclopropyl cyanide	77.9(1–α)
	10.9(1–2)
Cyclopropyl iodide	12.9(1–2)
Cyclopropyl bromide	13.3(1–2)
Cyclopropyl chloride	13.9(1–2)

X	
H	60.3
OH	68.8
OAc	74.0

X	
OH	66.8
OAc	72.0
CH$_3$	60.1
CHO	60.1
CN	62.8

[a] G.E. Maciel, J.W. McIver, Jr., N.S. Ostlund, and J.A. Pople, *J. Am. Chem. Soc.* **92**, 1, 11(1970) and references therein; F.J. Weigert and J.D. Roberts, *J. Am. Chem. Soc.* **94**, 6021 (1972); S. Berger and K.P. Zeller, *Org. Magn. Resonance* **11**, 303 (1978); P.E. Hansen, *Org. Magn. Resonance* **11**, 215 (1978); P.E. Hansen, O.K. Poulsen, and A. Berg, *Org. Magn. Resonance* **7**, 475 (1975); R.E. Wasylishen, *Annu. Rep. NMR Spectroscopy* **7**, 245 (1977).

Table 4.29. Two-bond coupling constants $^2J_{CH}$ of organic compounds[a]

Compound	$^2J_{CH}$(Hz)
Ethane	-4.5
1,2-Dichloroethane	-3.4
1,1,2,2-Tetrachloroethane	1.2
Ethylene	-2.4
1,2-Dichloroethylene, *trans*	0.8
1,2-Dichloroethylene, *cis*	16.0
Methylenecyclopentane	4.2 $\overset{*}{H_2C}=\overset{*}{C}<$
Methylenecyclohexane	5.2 $\overset{*}{H_2C}=\overset{*}{C}<$
Methylenecycloheptane	5.5 $\overset{*}{H_2C}=\overset{*}{C}<$
Acetone	5.5
3-Aminoacrolein	6.0 $OH\overset{*}{C}-\overset{*}{CH}=$
Acetaldehyde	26.7 $\overset{*}{CH_3}-\overset{*}{CHO}$
2-Ethylbutyraldehyde	22.1 $>\overset{*}{CH}-\overset{*}{CHO}$
Acrolein	26.9 $=\overset{*}{CH}-\overset{*}{CHO}$
3-Aminoacrolein	20.0 $=\overset{*}{CH}-\overset{*}{CHO}$
Propynal	33.2 $\equiv\overset{*}{C}-\overset{*}{CHO}$
Chloroacetaldehyde	32.5 $Cl\overset{*}{CH_2}-\overset{*}{CHO}$
Dichloroacetaldehyde	35.3 $Cl_2\overset{*}{CH}-\overset{*}{CHO}$
Trichloroacetaldehyde	46.3 $Cl_3\overset{*}{C}-\overset{*}{CHO}$
Acetylene	49.3 $H-\overset{*}{C}\equiv C-\overset{*}{H}$
Phenoxyacetylene	61.0 $-\overset{*}{C}\equiv C-\overset{*}{H}$
1-Phenoxy-1-propyne	10.8 $-\overset{*}{C}\equiv\overset{*}{C}-CH_3$

Chlorobenzene

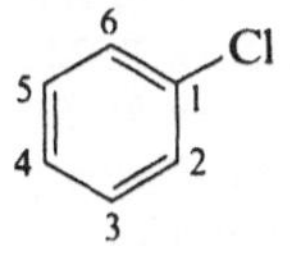

C_1H_2,	-3.4
C_2H_3,	$+1.4$
C_3H_2,	$+0.3$
C_3H_4,	$+1.6$
C_4H_3,	$+0.9$

o-Dichlorobenzene

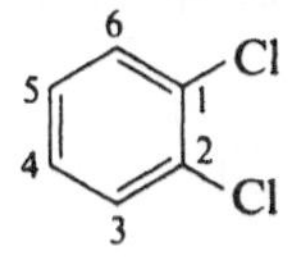

C_2H_3,	-3.5
C_3H_4,	$+1.9$
C_4H_3,	-0.02
C_4H_5,	$+1.1$

Table 4.29. *(Continued)*

Compound	$^2J_{CH}(Hz)$
Pyridine	

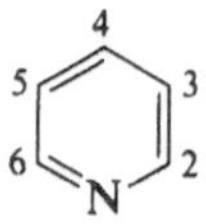

C$_2$H$_3$, +3.1
C$_3$H$_2$, +8.5
C$_3$H$_4$, +0.8
C$_4$H$_3$, +0.7

3-Cyanopyridine

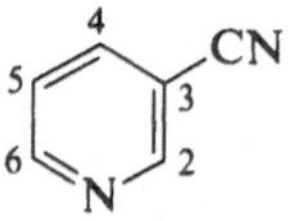

C$_3$H$_2$, 8
C$_3$H$_4$, <1
C$_4$H$_5$, <1
C$_5$H$_4$, <1
C$_5$H$_6$, 9
C$_6$H$_5$, 4
C$_2$H$_3$, 7.6
C$_3$H$_2$, 7.8
C$_3$H$_4$, 4.6

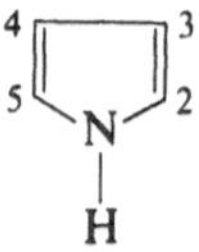

Pyrrole

Furan

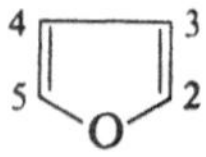

C$_2$H$_3$, +11.0
C$_3$H$_2$, +13.8
C$_3$H$_4$, +4.1

Thiophene

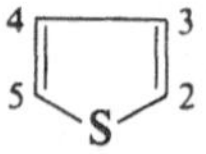

C$_2$H$_3$, 7.4
C$_3$H$_2$, 4.7
C$_3$H$_4$, 5.9

[a] G. Gray, P.D. Ellis, D.D. Traficante, and G.E. Maciel, *J. Magn. Resonance* 1, 41 (1969); F.J. Weigert and J.D. Roberts, *J. Am. Chem. Soc.* **90**, 3543 (1968); M. Hansen, R.S. Hansen, and H.J. Jakobsen, *J. Mag. Resonance* **13**, 386 (1974); Y. Takeuchi and N. Dennis, *J. Am. Chem. Soc.* **96**, 3657 (1974); M. Hansen and H.J. Jakobsen, *J. Mag. Resonance* **10**, 74 (1973); A.R. Tarpley and J.H. Goldsetin, *J. Phys. Chem.* **76**, 515 (1972); C.H. Yoder, R.H. Tuck, and R.E. Hess: *J. Am. Chem. Soc.* **91**, 539 (1969); E.F. Mooney and P.H. Winson, in: *Annual Review of NMR Spectroscopy*, Vol. 2, E.F. Mooney, Ed., Academic Press, New York, London (1969), p. 153; G.A. Olah and A.M. White, *J. Am. Chem. Soc.* **89**, 7072 (1967); E. Breitmaier and W. Voelter, ^{13}C *NMR Spectroscopy* Verlag Chemie, Weinheim, New York (1978).

Table 4.30. Longer-range carbon–carbon coupling constants $^2J_{CC}$ and $^3J_{CC}$[a]

Compound	$^2J_{CC}$ (Hz)	Compound	$^3J_{CC}$ (Hz)
*CH₃—C≡*C—H	11.8	Pyridine ($J_{2,5}$)	13.95
CH₃—C(O)—*CH₂—CH₃	9.5	Aniline ($J_{2,5}$)	7.9
(cyclobutanone, *C's marked)	9.0	Iodobenzene ($J_{2,5}$)	8.6
CH₃*CH₂C(OH)(Me)*CH₃	2.4	Nitrobenzene ($J_{2,5}$)	7.6
CH₃*CH₂CH₂*CO₂H	1.8	*CH₃CH₂C(OH)(Me)*CH₃	1.9
(cyclohexyl)*CH₃	1.7	*CH₃CH₂CH₂*CO₂H	3.6
(cyclohexyl)*CO₂H	ca.0	(cyclohexane OH, CH₃ structure)	3.2
*CH₃CO*CH₃	16	(cyclohexane CH₃, OH structure)	<0.5
		(cyclohexyl)*CO₂H	4.0
*CH₃CH₂*CN	33	*CH₃CH₂*CH*CH₃	b

[a] F.J. Weigert and J.D. Roberts, *J. Am. Chem. Soc.* **94**, 6021 (1972); R.J. Abraham and P. Loftus, *Proton and Carbon-13 NMR Spectroscopy*, Heyden and Son Ltd., London (1979).
[b] Calculated couplings: $\Phi = 0$, $J = 5.8$; $\Phi = 30$, $J = 4.0$; $\Phi = 60$, $J = 1.9$; $\Phi = 90$, $J = 0.6$; $\Phi = 120$, $J = 1.5$; $\Phi = 150$, $J = 3.3$; $\Phi = 180$, $J = 4.6$.

Table 4.31. Typical $^3J_{CH}$ values in aromatic systems[a]

Compound	$^3J_{CH}$ (Hz)
Chlorobenzene	
(benzene ring with Cl, positions 1-6)	C_1H_3, + 11.1
	C_2H_4, + 7.9
	C_2H_6, + 5.0
	C_3H_5, + 8.2
	C_4H_2, + 7.4
o-Dichlorobenzene	
(benzene ring with 2 Cl, positions 1-6)	C_2H_4, + 11.6
	C_2H_6, + 7.9
	C_3H_5, + 8.4
	C_4H_6, + 8.6

Table 4.31. (*Continued*)

Compound	$^3J_{CH}$ (Hz)
Pyridine	

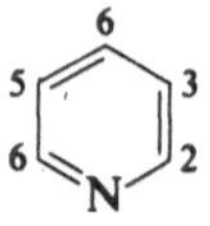

	$C_2H_4, +6.9$
	$C_2H_6, +11.2$
	$C_3H_5, +6.6$
	$C_4H_2, +6.3$

3-Cyanopyridine

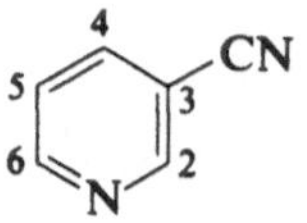

	$C_2H_4, 5$
	$C_2H_6, 13$
	$C_3H_5, 5$
	$C_4H_2, 4$
	$C_4H_6, 6$
	$C_6H_2, 12$
	$C_6H_4, 8$

Pyrrole

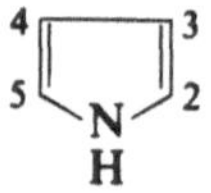

	$C_2H_4, 7.6$
	$C_2H_5, 7.6$
	$C_3H_5, 7.8$

Furan

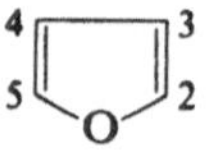

	$C_2H_4, +7.0$
	$C_2H_5, +6.9$
	$C_3H_5, +6.0$

Thiophene

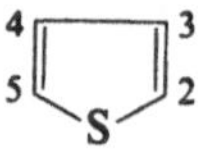

	$C_2H_4, 10.0$
	$C_2H_5, 5.2$
	$C_3H_5, 9.5$

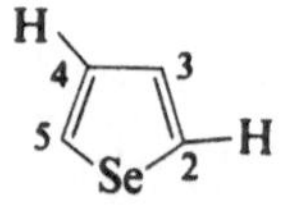	$C_2H_4, 6.0$

[a] F.J. Weigert and J.D. Roberts, *J. Am. Chem. Soc.* **90**, 3543 (1968); *ibid*, **91**, 4940 (1969); E. Breitmaier and G. Bauer, *Pharm. Unserer Zeit* **5**, 113 (1976); F. J. Weigert and J.D. Roberts, *J. Am. Chem. Soc.* **90**, 3543 (1974); M. Hansen, R.S. Hansen, and H.J. Jakobsen, *J. Magn. Resonance* **13**, 386 (1974); Y. Takuchi and N. Dennis, *J. Am. Chem. Soc.* **96**, 3657 (1974); M. Hansen and H.J. Jakobsen, *J. Magn. Resonance* **10**, 74 (1973); A.R. Tarpley and J.H. Goldstein, *J. Phys. Chem.* **76**, 515 (1972).

4.3 HIGH-RESOLUTION NMR IN SOLIDS

Nuclear magnetic resonance spectroscopy operates at the low-energy end of the electromagnetic spectrum and allows small energy differences, corresponding to chemical shifts and coupling constants, to be measured. The 0.1-Hz resolution which can be routinely achieved corresponds to an energy difference of 10^{-13} Joule/mole. Such subtle differences could not be measured in solids in the past for three principal reasons.

Firstly, the resonance lines are broadened in solids to widths in the KHz range due to *anisotropic dipole–dipole* (DD) and *quadrupole field gradient* (QF) interactions. The magnetic field that a nucleus experiences when placed in an external magnetic field is dependent not only on the magnitude B_0 of the external field but also on the local fields B_{loc} due to the effects of neighboring nuclei. Figure 4.10 shows two nuclei ^{13}C and ^1H bound to one another. The local field experienced by the nuclei is given by the equation

$$B_{loc} = \pm \frac{\mu}{r^3}(3\cos^2\phi - 1)$$

where r is the distance between the two nuclei, and ϕ is the angle which the vector connecting the two nuclei makes with the external field. The + and − signs in this equation reflect the spin alignment of the neighboring nucleus (parallel or antiparallel, respectively) with respect to the applied field B_0. In solids a given nucleus is affected by contributions from a number of different local fields from various close-lying nuclei. These fields are several gauss in

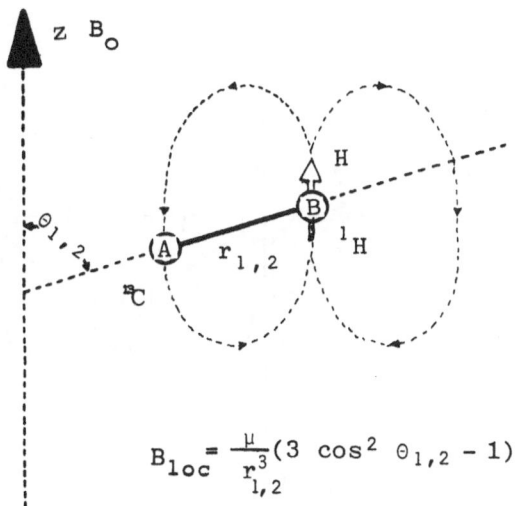

Figure 4.10. Local magnetic field B_{loc} at nucleus A produced by the nuclear magnetic dipole moment μ of nucleus B at a distance $r_{1,2}$.

magnitude and they can result in the broadening of lines by several kilohertz. Different internal electric field gradients may also broaden the resonance lines of those nuclei which have spin quantum number $I > \frac{1}{2}$ and thus possess a quadrupole moment.

Different orientations of the axis of a bond in an external field result in different shielding or deshielding influences with respect to the applied field. This is shown in Figure 4.11 in which the magnetic moment μ of nucleus A in bond A—B has been (a) aligned with the z-axis, (b) aligned with the y-axis, and (c) aligned with the x-axis. It can be seen that in the first two orientations the induced field at B augments the applied field but in the last orientation it opposes the applied field at B. In solid or powdered samples, the molecules may be oriented in all possible directions with respect to the applied B_0, and this *chemical shift anisotropy* will result in a spread of chemical shifts, which is again proportional to $3\cos^2\phi - 1$.

Another factor which has posed problems in the past in recording the NMR spectra of solids has been the very long spin–lattice relaxation times in solids. Since the delays between successive pulses have to be $5T_1$ in order to allow the nuclei to relax fully to their original equilibrium state of magnetization, the long relaxation times in solids make the intervals between pulses in FT spectroscopy so great that only a few scans could be accumulated, resulting in a reduction in sensitivity.

The three major problems in the recording of high-resolution NMR of solids mentioned above have been solved by (a) high-power decoupling, (b) magic angle spinning (MAS), and (c) cross-polarization (CP).

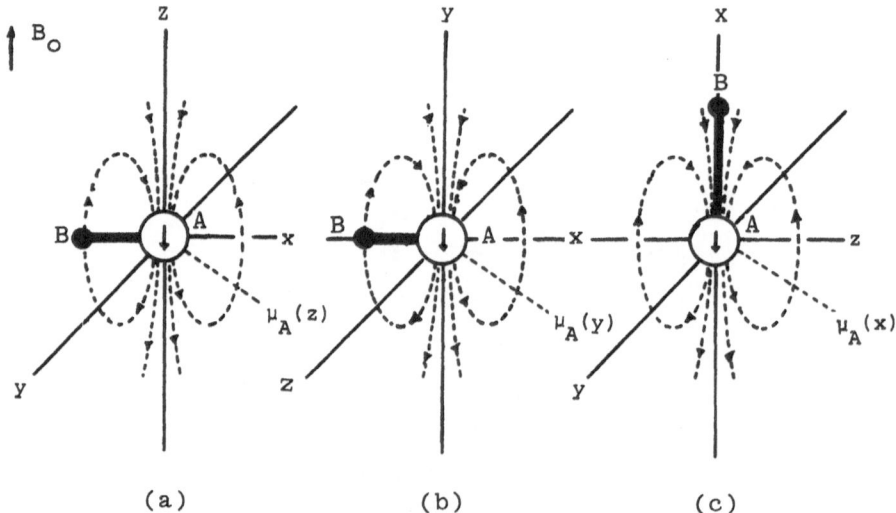

Figure 4.11. Chemical shift anisotropy: different orientations of the axis of the bond between two nuclei, A and B, in applied field B_0 result in nucleus B experiencing different shielding or deshielding effects.

High-power decoupling has been employed to nullify heteronuclear interactions. Thus when recording ^{13}C spectra of solids, one can apply an RF field
to the protons which causes their spins to rotate steadily. As a result, the local
field at the ^{13}C nucleus due to the neighboring protons becomes zero (Figure
4.12). In contrast to the case of liquids, the dipolar couplings in solids may be
several thousand hertz in width, and to overcome these a very high-power
decoupler (100 watts or more) has to be employed. Homonuclear dipolar
interactions remain unaffected by high-power decoupling.

Magic angle spinning (MAS) has been used to overcome peak broadening in
solids due to homonuclear dipolar interactions, chemical shift anisotropy
(CSA) effects, and quadrupole field (QF) gradient interactions. All these
interactions depend on the magnitude of $3\cos^2\theta - 1$, but this term is zero
when θ has a value of $54°44'$. This is the so-called "magic angle," i.e., if the
material is spun about an axis which is tilted with respect to the applied field by
an angle of $54°44'$, then the homonuclear dipolar interactions, CSA, and QF
effects are all reduced to zero values (Figure 4.13). The solid-state ^{13}C spectra
of poly(butylene terephthalate) shown in Figure 4.14(a)–(c) illustrate how
high-power decoupling [Figure 4.14(b)] and magic angle spinning [Figure
4.14(c)] can increase signal-to-noise ratio and sharpen lines.

The problem of long relaxation times of ^{13}C nuclei in solids has been
overcome by utilizing *cross-polarization* (CP) and *proton enhanced* (PE)
spectroscopy. By this process the ^{13}C spin system transfers its energy to the ^{1}H
spin system. By relaxation of the protons, the energy of the ^{13}C nuclei can then
be transferred to the lattice. Since the ^{1}H spin–lattice relaxation times are

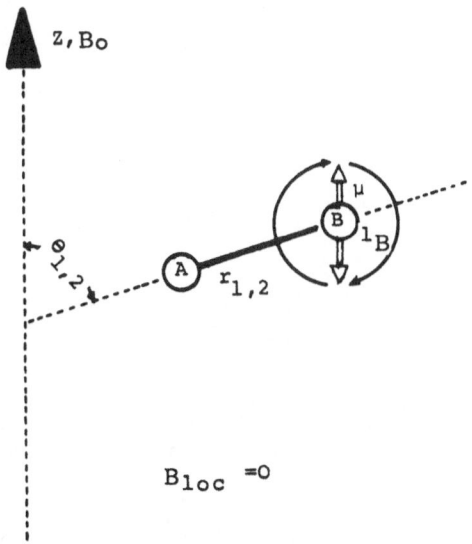

Figure 4.12. High-power decoupling of protons (nucleus A) results in steady rotation
of their spins, resulting in zeroing of the effect due to local fields on ^{13}C (nucleus B).

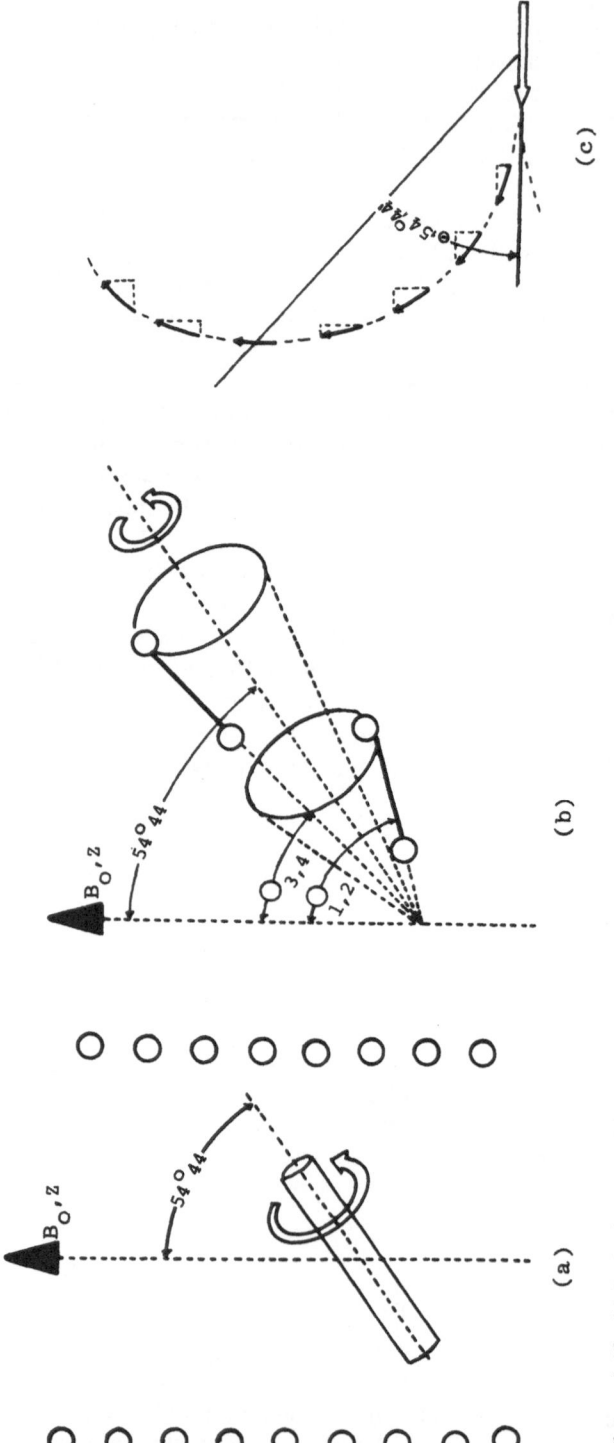

Figure 4.13. (a) "Magic Angle Spinning" of a specimen in a field at the magic angle of 54°44'. (b) Effect of magic angle spinning: Rotation about the magic angle causes the time-averaged value of all vectors to become 54°44', e.g., $\langle \theta_{1,2} \rangle = \langle \theta_{3,4} \rangle = 54°44'$ although $\theta_{1,2} \neq \theta_{3,4}$. (c) Magnetic field line of a magnetic dipole showing positive and negative z-components. At $\theta = 54°44'$, the z-component becomes zero.

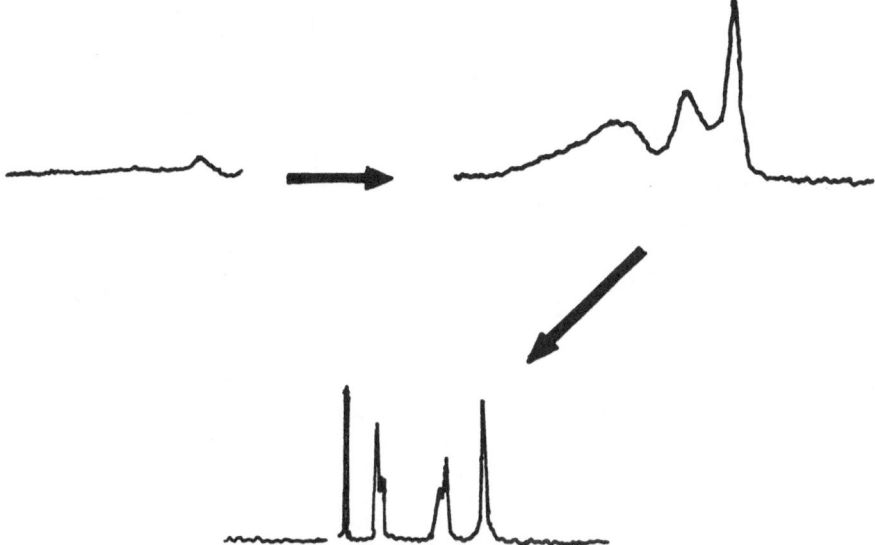

Figure 4.14. (a) Solid-state ^{13}C-NMR spectrum of poly(butylene terephthalate) with low-power decoupling. (b) Improved signal-to-noise ratio with high-power decoupling which removes proton and *J*-coupling. (c) High-power decoupling and magic angle spinning gives sharper isotropic lines. (Part (a) reprinted with permission from *Chem. Eng. News*, **62**. Copyright 1984, American Chemical Society.

much smaller than those of ^{13}C nuclei, the pulses can be applied at smaller delay intervals, and many more scans can be accumulated in a given time. In order to allow cross-polarization to occur, one has to bring the ^{1}H and ^{13}C nuclei in thermal contact. This is done by adjusting the value of the applied field so that ^{1}H and ^{13}C nuclei have the same Larmor frequency in the rotating frame, i.e., $\omega_{1H} = \omega_{1C}$. Any energy released by the flipping of ^{13}C nuclei can be absorbed by ^{1}H nuclei (Figure 4.15). There is a substantial reduction in spin temperature of the ^{13}C nuclei which is accompanied by an increased polarization of the ^{13}C spins, but there is only a much smaller increase in the spin temperature (and a correspondingly small decrease in the polarization) of the ^{1}H nuclei due to their much higher heat capacity.

CP-MAS has allowed the study of conformations and cross-linkages in natural and synthetic polymers, and of structures of proteins, nucleic acids, phospholipid membranes, etc. The field is still in its infancy and its applications will undoubtedly grow. The CP-MAS solid state ^{13}C-NMR spectrum of 19S-vindolinine, has been studied by Atta-ur-Rahman and coworkers at H.E.J. Research Institute of Chemistry, and it is shown in Figure 4.16. It represents the first example of a CP-MAS spectrum of a solid *Aspidosperma* alkaloid ever recorded.

One of the problems encountered in recording NMR spectra of solids is that at lower spinning speeds the ^{13}C signals break up into a number of

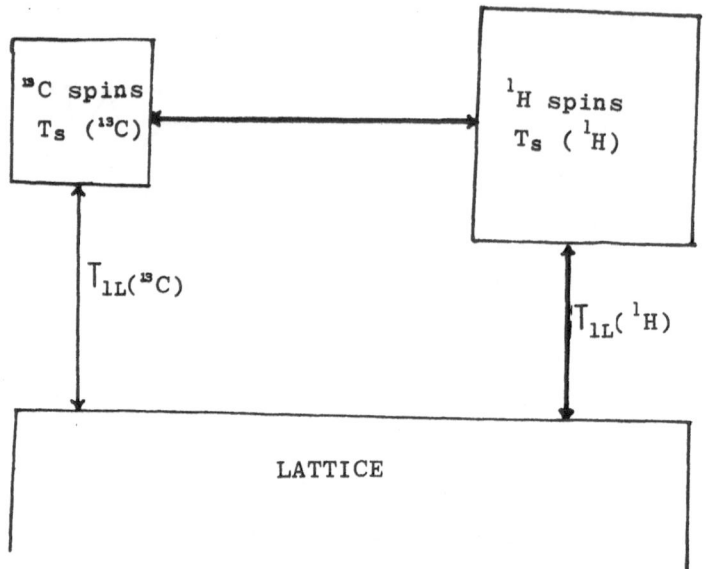

Figure 4.15. ^{13}C–^{1}H cross-polarization experiment (spin temperature concept; for more details, see text).

rotational side bands which severely reduces signal intensity. A number of pulse sequences have recently been developed to overcome this problem, involving "scaling" of shift anisotropies, use of four π-pulse sequences to suppress side bands (TOSS), or a combination of the two methods.

Recently two-dimensional solid-state heteronuclear chemical shift correlation experiments involving magic angle sample spinning (MASS) have been described.* This technique results in an increase in the resolution of solid-state proton spectra by a factor of 5–10. Solid-state NMR spectroscopy has hitherto been confined to ^{13}C-NMR spectra since the residual line widths in the spectra of solid samples are around 2 ppm, so that only four or five nonequivalent protons can be distinguished in the normal proton spectral region of 10–15 ppm. The increase in resolution attained by the 2D solid-state heteronuclear COSY experiment should promote further studies of the ^{1}H-NMR spectra of solid samples.

4.3.1 NMR Imaging

An exciting development in recent years has been that of NMR zeugmatography or "spin mapping." By employing alternating field gradients, it is possible to produce NMR images of the points where these gradients overlap. Computer-assisted scanning can lead to image construction. Similarly, by

---

* J.E. Roberts, S. Vega, and R.G. Griffin, *J. Am. Chem. Soc.* **106,** 2506 (1984) and references therein.

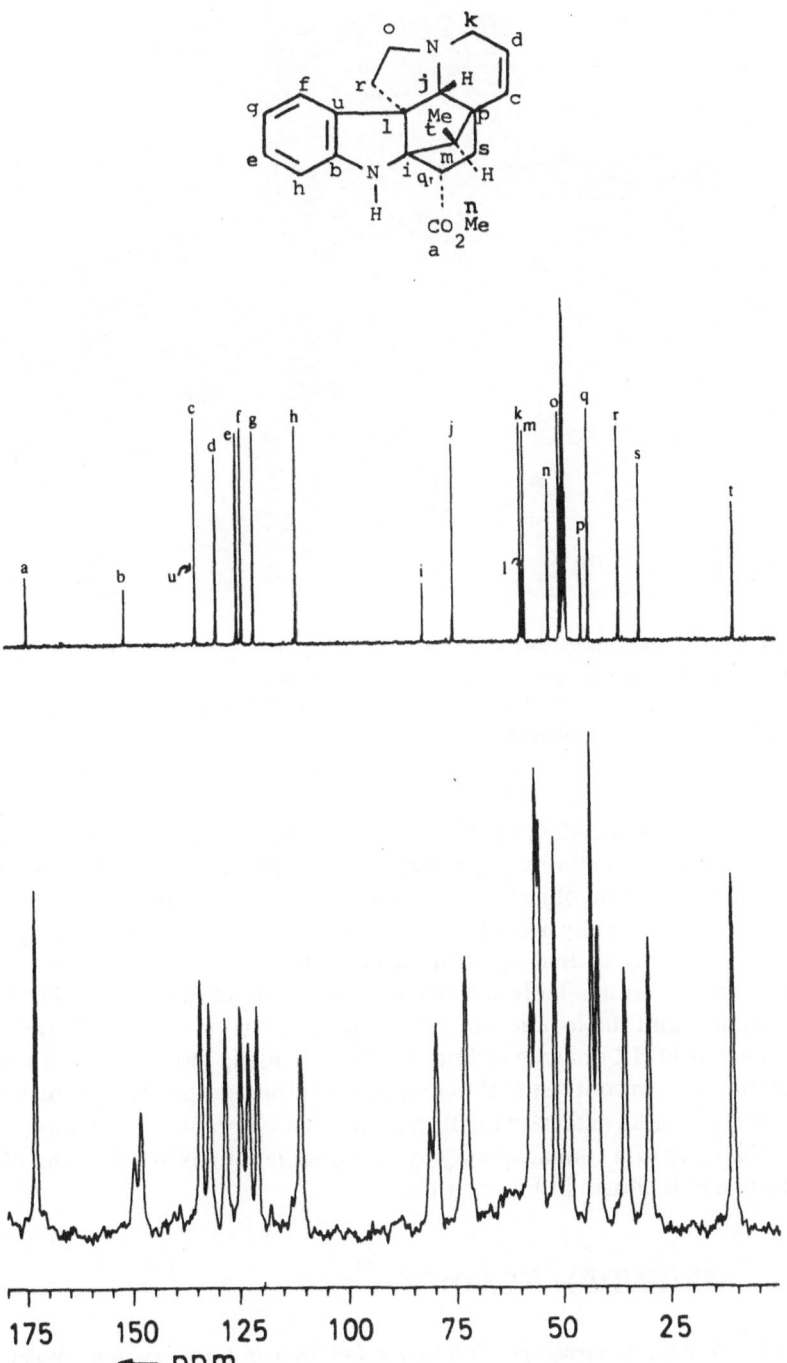

Figure 4.16. *Above*: Broad-band ^{13}C-NMR of 19S-vindolinine (solution spectrum); *below*: solid-state ^{13}C-NMR spectrum of the same compound.

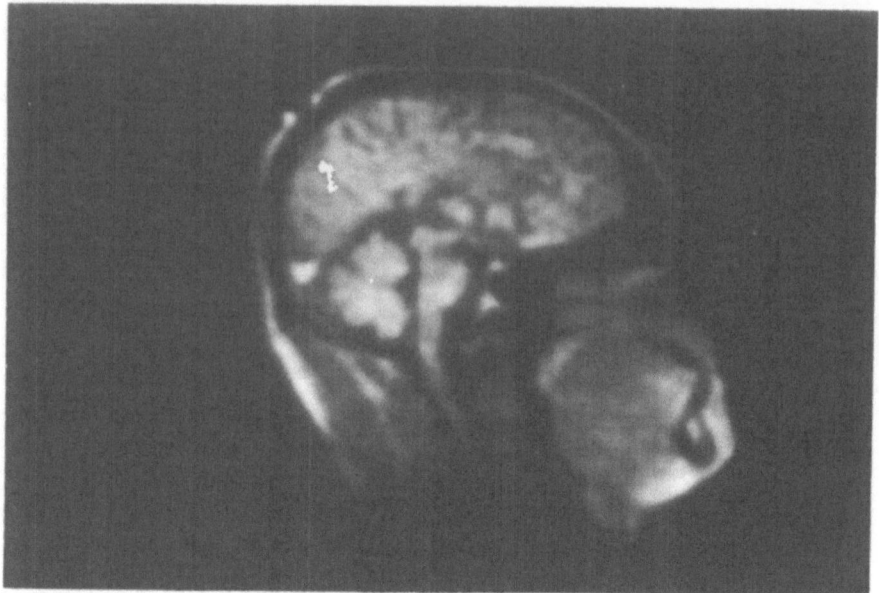

Figure 4.17. A two-dimensional NMR image of a cross section of the human head. Different cross sections provide valuable information about sizes and locations of tumors, etc. Courtesy of Dr. Klaus Albert, Institute of Organic Chemistry, University of Tübingen, Tübingen, West Germany.

using two time-dependent gradients one can create an NMR-sensitive line, and by using FT methods combined with computer processing techniques, NMR images of solid objects can be built up. Since magnetic and RF fields do not appear to cause any harm to living cells, the technique of NMR imaging is already beginning to find application in medicine. Thus NMR systems are now commercially available in which the whole patient can be placed between the magnets and the location of cancerous growths or other malformations examined by NMR imaging techniques. Since ionizing radiation is not used in NMR imaging, in contrast to the computerized tomography (CT) scan, it does not carry the risks associated with radiation-based scanning techniques, but affords images of a similar quality. A slice obtained by NMR imaging of the human head is shown in Figure 4.17.

RECOMMENDED READING

1. J.W. Cooper, *Spectroscopic Techniques for Organic Chemists*, John Wiley and Sons, New York (1980).
2. D. Shaw, *Pulsed Fourier Transform NMR Spectroscopy*, Elsevier Scientific Publishing Co., Amsterdam (1979).

3. F.W. Wehrli and T. Wirthlin, *Interpretation of Carbon-13 NMR Spectra*, Heyden and Son Ltd., London (1978).
4. G.C. Levy, Ed., *Topics in Carbon-13 NMR Spectroscopy*, Vols. 1–3, John Wiley and Sons, New York (1979).
5. E. Breitmaier and W. Voelter, *^{13}C-NMR Spectroscopy*, Verlag Chemie, Weinheim (1978).
6. C. Brevard and P. Graner, *Handbook of High Resolution Multinuclear NMR*, John Wiley and Sons, New York (1981).
7. G.C. Levy, R.L. Lichter, and G.L. Nelson, *Carbon-13 Nuclear Magnetic Resonance Spectroscopy*, John Wiley and Sons, New York (1980).
8. H. Günther, *NMR Spectroscopy*, John Wiley and Sons, New York (1980).
9. L.M. Jackman and S. Sternhell, *Applications of Nuclear Magnetic Resonance Spectroscopy in Organic Chemistry*, International Series in Organic Chemistry, Vol. 10, Pergamon Press, Oxford (1969).
10. F.A. Bovey, *Nuclear Magnetic Resonance Spectroscopy*, Academic Press, New York (1969).
11. E.D. Becker, *High Resolution NMR*, Academic Press, New York (1980).
12. M.L. Martin and G.J. Martin, *Practical NMR Spectroscopy*, Heyden and Son Ltd., London (1980).
13. T. Clerc and E. Pretsch, *Kernresonanz Spektroskopie*, Akademische Verlagsgesell-schaft, Frankfurt (1973).
14. A. Carrington and A.D. McLachlan, *Introduction to Magnetic Resonance*, Chapman and Hall, London (1979).
15. H. Saito, G. Izumi, T. Mamizuka, S. Suzuki, and T.R. Tabeta, A ^{13}C cross polarisation–magic angle spinning (CP-MAS) NMR study of crystalline cyclohexa-amylose inclusion complexes. Conformation-dependent ^{13}C chemical shifts are related to the dihedral angles of glycosidic linkages. *J. Chem. Soc., Chem. Commun.*, 1386 (1982).

Chapter 5

Special Pulse Sequences and Two-Dimensional NMR Spectroscopy

One of the most dramatic advances in recent years has been the development of several new pulse sequences, made possible by the advent of dedicated microcomputers, which have allowed precise manipulations of pulse angles, introduction of delays between pulses, and rapid Fourier transformations. These have heralded the advent of a number of extremely powerful procedures among which may be mentioned: (a) new pulse sequences for the unambiguous assignments of methyl, methylene, methine, and quaternary carbon atoms, side-stepping the difficulties associated with the overlapping of multiplets in the normal off-resonances measurements. Examples are APT, DEPT, ADEPT, etc.; (b) two-dimensional NMR spectroscopy for observing couplings between protons, between protons and carbon atoms, and more recently between carbon atoms themselves; and (c) precise measurements of nuclear Overhauser enhancements by NOE difference measurements involving alternate recording of normal and NOE enhanced spectra, and automatic computer-assisted subtractions which result in measurements of even small enhancements, which were previously not possible.

Many of the new pulse sequences being developed rely on the fact that it is possible to alter at will the time between the excitation of the nuclei and their detection. The maximum intensity of the signal will of course be obtainable immediately after the application of the pulse, but if one introduces different time delay intervals between the application of the pulse and the detection of the signal, then during this delay period the nuclei interact with one another ("evolve"). This period of evolution provides a powerful procedure by which the nature of the nuclei can be probed, as will be seen later.

An attempt will be made here to describe the various pulse sequences and their effects on nuclei under observation without involving a complex mathematical treatment. However, before doing so it would be appropriate to discuss in depth the effect of a strong magnetic field B_0 and the radio frequency pulse on nuclei and their subsequent behavior. As stated earlier, when placed in a strong magnetic field, nuclei such as ^1H or ^{13}C adopt one of two different spin states, aligned with or against the applied magnetic field B_0. There is a slight Boltzmann excess of nuclei (one nucleus in a hundred thousand) aligned with the applied field B_0 and there is a corresponding net magnetization, M_0, relating to this excess [Figure 5.1(a)]. The nuclei behave like tiny gyroscopes and precess around the z-axis with an angular velocity $\omega = \gamma B_0$, where γ is the gyromagnetic ratio of the nucleus.

If a radio frequency pulse is now applied along the x-axis, it bends the magnetization M_0 away from the z-axis and causes it to precess around the x'-axis (or B_1) at a rate $\omega = \gamma B_1$ in the z,y-plane [Figure 5.1 (b), (c)]. The duration and power of the pulse determines the extent to which the magnetization is bent away from the z-axis. Thus at a field strength of 25 gauss, the ^{13}C nuclei would take 22 microseconds to complete one full circle and return to their original position (i.e., from $+z$ to $+y$, then to $-z$, $-y$, and finally back to $+z$). This would be equivalent to a 360° pulse. A 90° pulse [Figure 5.1(b), (c)] would be of one-fourth the duration, i.e., 5.5 μs, while a 180° pulse, often also called the π-pulse, would be of 11-μs duration and would result in inversion of the magnetization. The accurate determination of the 180° or 90° pulse for the system (by the procedure described in an earlier section of this text) allows one to bend or "flip" the magnetization by whatever angle one wishes by an appropriate choice of the time for which the radio frequency pulse is applied. The components of magnetization after tipping by 35° are shown in Figure 5.1(d).

The pulse serves to convert the longitudinal magnetization along the z axis into detectable transverse magnetization along the y axis. The rotation of the transverse magnetization in the x,y plane results in a continuously varying magnetic field which induces an alternating current in the receiver coil. After the pulse, the NMR signal declines due to relaxation processes and field inhomogeneity. This decline in signal strength is observable as a weak alternating current, and is recorded in the form of a free induction decay (FID). In organic compounds, many nuclei with different Larmor frequencies contribute to the FID, and Fourier transformation of the resulting "pulse interferogram" leads to the NMR spectrum in which the Larmor frequencies determine the positions of the signals while the life spans of the transverse magnetization determine their width. A slow decline in the NMR signal (or a long effective transverse relaxation time, T_2^*), results in a sharp signal while a rapid free induction decay affords a broad signal.

Since all the nuclei in a molecule are not chemically identical and are shielded to various degrees, they will not all precess at exactly the same angular velocity. After they have been allowed to precess for a time t, they would separate into a number of magnetic vectors with different relative angular velocities $\omega_1 t$, $\omega_2 t$, $\omega_3 t$, etc. [Figure 5.1(e)]. Thus the difference in angles between a faster-moving magnetic vector as compared to a slower-moving vector will grow with time to a maximum of 180° before it starts decreasing again as the faster-moving vector crosses the halfway mark around the circle and starts "catching up" with the slower-moving vector. In the above discussion, the x- and y-axes were considered to be stationary but (as

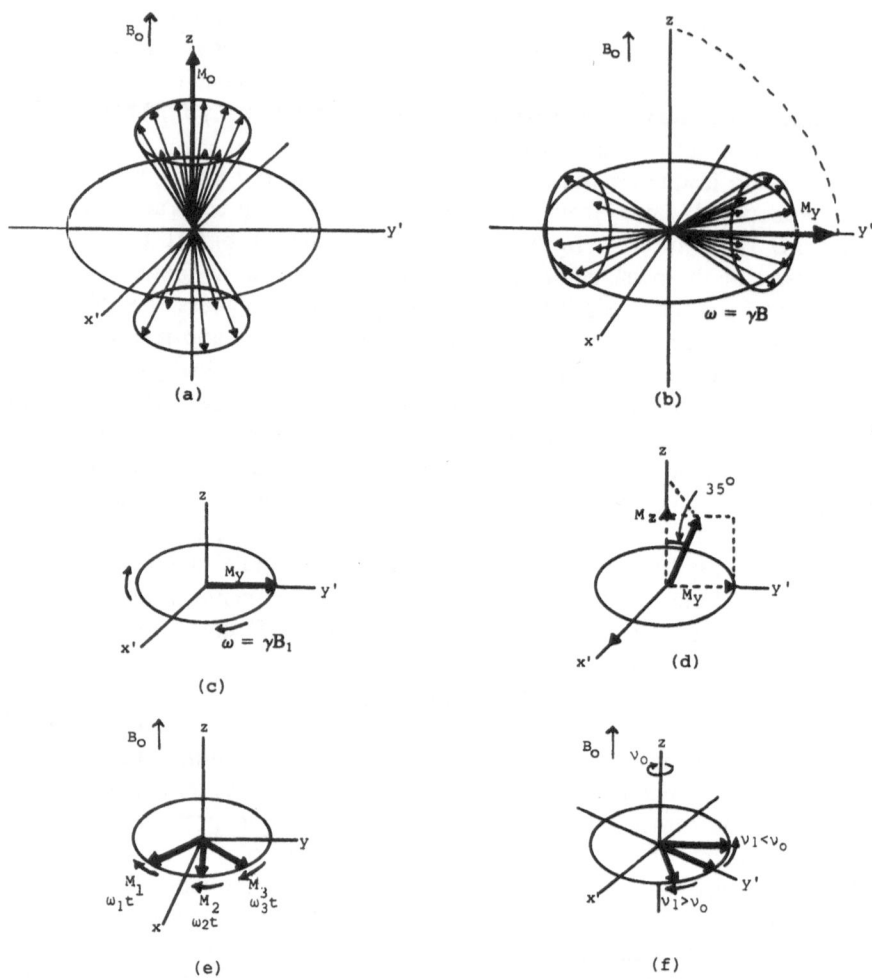

Figure 5.1. (a) Equilibrium magnetization state, M_0. (b) and (c) Application of 90°
pulse bends the magnetization by 90°. It is represented by M_y. The magnetization
precesses in the x,y plane around the z-axis with an angular velocity of $\omega = \gamma B$ in the
stationary frame of reference. (In the rotating frame of reference, the xy plane itself
rotates, so that M_y appears to be stationary). The effect that the pulse has on the
magnetization vector may be described by the "right hand rule" (straight thumb, partly
bent fingers). The thumb then represents the direction of the pulse, and the bent fingers
indicate the direction in which the magnetization vectors bend. (d) Application of 35°
pulse bends the magnetization M_0 by 35°. Components of the magnetization along the
y- and z-axes are then represented as M_y and M_z. (e) Magnetization vectors M_1, M_2,
and M_3 associated with three groups of nuclei with different chemical shifts are seen to
separate from the y-axis. The farther downfield the resonances are from TMS, the
greater will be the separation of the magnetization vectors from the TMS vector and
they precess away at different relative angular velocities, $\omega_1 t$, $\omega_2 t$, and $\omega_3 t$. (f) Same as
(e) but in a rotating frame of reference.

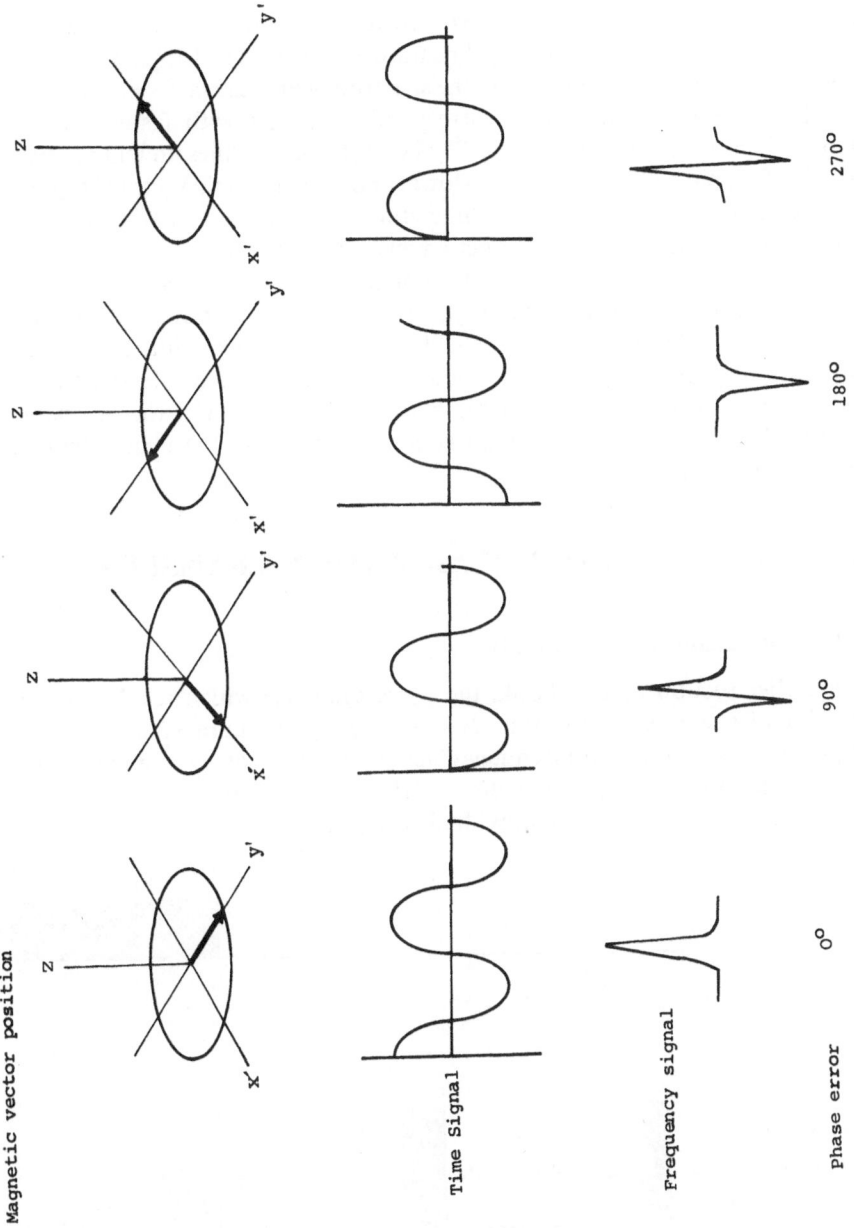

Figure 5.2. Relationship between position of magnetic vector, time signal, frequency signal, and phase error.

mentioned in an earlier section), if we rotate these two axes around the z-axis at a fixed frequency (say, the precession frequency of the protons in tetramethylsilane, TMS) [Figure 5.1(f)] and observe the movement of the magnetic vectors relative to the movement of the TMS vector, i.e., from this rotating frame of reference, the TMS vector will now appear to have a constant velocity along the y'-axis, while the other vectors will appear to move away much more slowly. The lines in the spectrum which are farther downfield than the TMS signal will then correspond to nuclei with faster-moving magnetic vectors as compared to the TMS vector whereas lines upfield relative to the TMS signal will correspond to slower-moving magnetic vectors [Figure 5.1(g)]. Chemical shifts can thus be understood in terms of relative variations in the angular velocities of the magnetic vectors for different types of nuclei in comparison to the velocity of the nuclei of the reference (TMS) vector. The relative movements of vectors are very slow in the rotating frame of reference and the effect of the static magnetic field B_0 along the z-axis thus essentially disappears, allowing the weaker interactions of the nuclei to be observed with much greater clarity. The relationship between the position of the magnetic vector in the x,y-plane, time signal, frequency signal, and the phase error is shown in Figure 5.2.

5.1 SPIN-ECHO AND POLARIZATION TRANSFER

5.1.1 Spin-Echo Measurements

In an earlier section of this book, the pulse sequence which results in the formation of a spin-echo after time 2τ was described [Figure 3.32(v); Figure 5.3A, B]. If, however, a radio frequency field is simultaneously applied in the ^{1}H region, it results in a *J-modulation of the echo magnitude*, which allows assignment of multiplicities to individual carbon atoms.

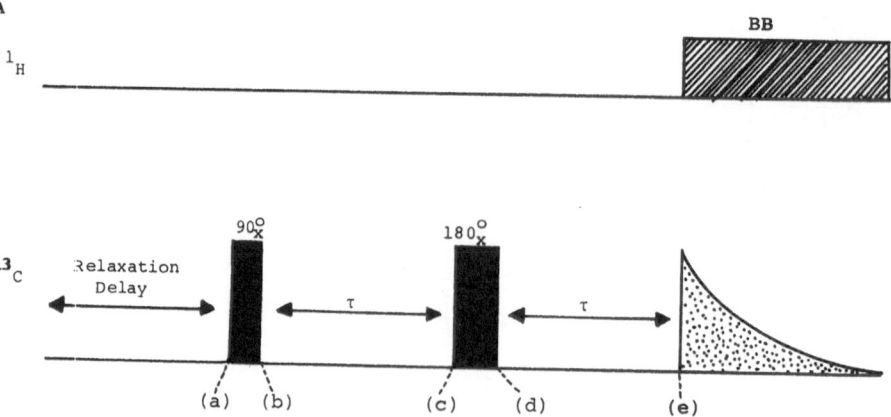

Figure 5.3A. ^{13}C Spin-echo pulse sequence.

B

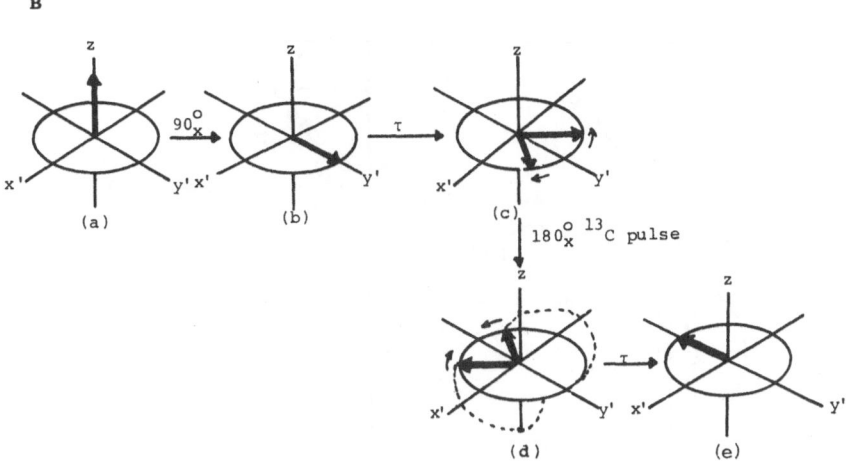

Figure 5.3B. (a) Longitudinal magnetization before the application of 90°_x pulse. (b) Application of 90°_x pulse bends the magnetization along x'- the y'-axes. (c) Fanned out magnetization vectors at time τ after the application of 90°_x pulse. (d) Application of 180° ^{13}C pulse causes inversion of magnetization so that the vectors occupy mirror image positions across the x'-axis. (e) Spin-echo formation occurs along the $-y'$-axis by focusing of magnetization vectors after a further delay τ.

To properly understand this phenomenon, one needs to consider how spin–spin coupling affects spin-echo experiments. Let us first consider a spin system AX containing two different nuclei A and X (a *heteronuclear* system) which are coupled to one another. The magnetization of A will experience the effects of the neighboring nucleus X, which is in two different states, α and β. The net magnetization of A can therefore be considered as being made up of two separate magnetization components, $M_A^{X\alpha}$ and $M_A^{X\beta}$. After the nucleus A is subjected to a 90° pulse, its component magnetizations, $M_A^{X\alpha}$ and $M_A^{X\beta}$, will precess about the applied field B_0 at different rates [Figure 5.4(a), presented in stationary frame of reference]. If J_{AX} has a positive value, $M_A^{X\beta}$ may be considered to be precessing at a slightly faster frequency, $v_A + \frac{1}{2}J_{AX}$, while $M_A^{X\alpha}$ will be precessing at a slightly slower frequency, $v_A - \frac{1}{2}J_{AX}$. The two magnetization vectors thus start to move apart and after a fixed time period τ, they may be shown as in Figure 5.4, diagram (b). If a 180°_A pulse is then applied along the y-axis, it causes the magnetization vectors of the A nuclei to adopt the mirror image positions across the y-axis indicated in Figure 5.4, diagram (c). Since the 180°_A pulse is selective for the A nuclei and does not affect the spin states of the X nuclei, the *direction* of movement of the A vectors is unchanged although their *position* changes; the faster-moving A vector, $M_A^{X\beta}$, now lies behind the slower-moving A vector $M_A^{X\alpha}$. The vectors thus draw closer and meet ($M_A^{X\beta}$ "catches up" with $M_A^{X\alpha}$) after time τ, thereby producing an echo (Figure 5.4, diagram (d)).

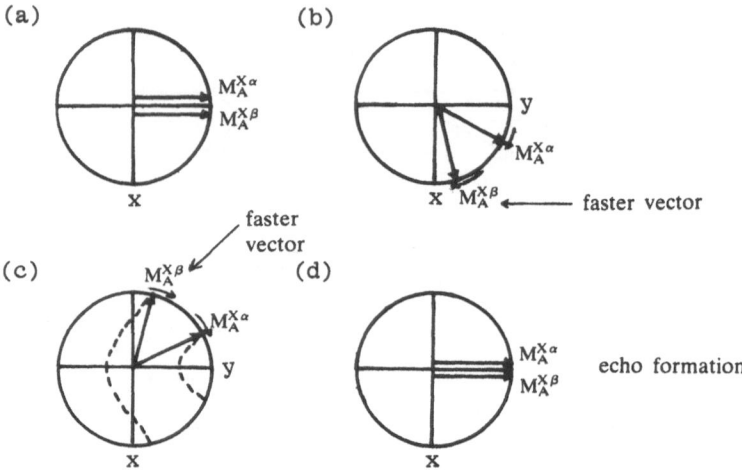

Figure 5.4. Evolution in the x,y-plane of magnetization vectors of nucleus A for a *heteronuclear* AX spin system during a spin-echo sequence (presented in a stationary frame of reference): (a) position of vectors immediately after 90°_x (A) pulse; (b) position of vectors after time τ; (c) position of vectors after 180°_y (A) pulse; (d) position of vectors after a further time τ.

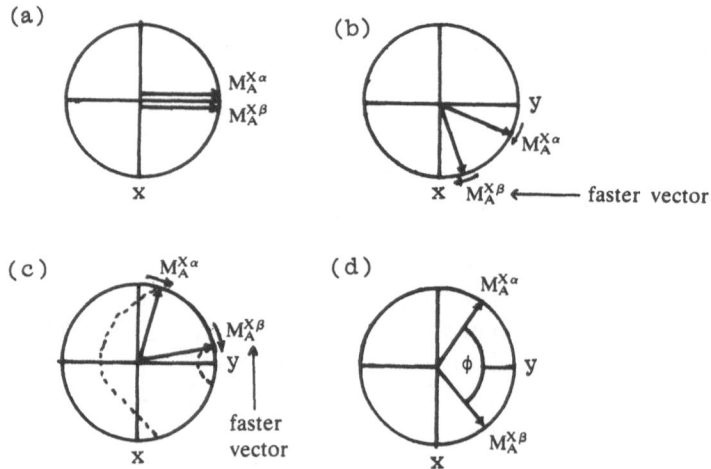

Figure 5.5. Evolution in the x,y-plane of the A magnetization vectors of a *homonuclear* AX spin system during a spin-echo experiment (presented in a stationary frame of reference) (a) immediately after the 90°_x (A) pulse; (b) after time τ later; (c) after nonselective 180°_y pulse [or 180°_y (A)/180°_y (X) pair of pulses], faster vector now lies *ahead* of the slower vector; (d) after further time τ later (showing the phase angle developed – no echo is produced).

The situation presented above is valid for a heteronuclear system. If both A and X nuclei are the same (i.e., a *homonuclear* system), then a different situation is encountered (Figure 5.5). When the A nuclei are subjected to a 90° pulse, the two A magnetization vectors, $M_A^{X\alpha}$ and $M_A^{X\beta}$, begin to move apart over a time period τ, as was the case in the heteronuclear case discussed above [Figure 5.5, diagrams (a) and (b)]. However, application of the 180° pulse to the A spins also causes the populations of X nuclei to invert in the homonuclear case, since the pulse is nonselective, both A and X nuclei being of the same type, e.g., both are protons. X_α therefore becomes X_β, and X_β becomes X_α. $M_A^{X\alpha}$ and $M_A^{X\beta}$ are therefore interchanged (compare diagram (c) in Figures 5.4 and 5.5). The faster-moving vector $M_A^{X\beta}$ now lies *ahead* of the slower vector $M_A^{X\alpha}$ and the two vectors move farther apart and do not refocus after the time period τ. This is shown in Figure 5.5, diagram (d). The difference in the angular frequency between the two components is $2\pi J_{AX}$ where J_{AX} is the coupling constant between the nuclei A and X. After a time period 2τ, the phase angle ϕ will be given by the equation $\phi = 4\pi J_{AX}\tau$.

Figure 5.6 portrays the formation of gated spin-echoes in heteronuclear and homonuclear systems in a rotating frame of reference. In the heteronuclear case, two variations can arise, depending on whether the 180_x° pulse is being applied to the A or X nuclei. The behavior of the magnetic vectors of the X nucleus after application of the 180_x° pulse to the X nucleus is shown in Figure 5.6A, whereas the behavior of the magnetic vectors of the X nucleus after application of the 180_x° pulse to the A nucleus is shown in Figure 5.6B. In the first case, the application of the 90_y° pulse results in the orientation of the magnetization of the X nuclei along the y'-axis [Figure 5.6A, (i)]. During the subsequent time period τ the two magnetization vectors of nucleus X arising due to the coupling with the neighboring nucleus A move apart and after time τ they may be represented as shown in Figure 5.6A, (ii). If a 180_x° pulse is now applied to the X nucleus (i.e., to the nucleus whose magnetic vectors are under discussion), the magnetic vectors will undergo a 180° inversion [Figure 5.6A, (iii)] and, after a further time period τ (i.e., time period 2τ after the original 90° pulse), they coalesce, resulting in the formation of an echo [Figure 5.6A, (iv)]. Since they come together along the $-y$-axis, a negative peak would be recorded [Figure 5.6A, (v)].

If, on the other hand, the 180_x° pulse is applied to the A nucleus and the behavior of the magnetization vectors of the X nucleus is observed, a different situation is seen to prevail [Figure 5.6B]. Since the 180_x° pulse is now applied to the A nuclei no inversion in the positions of the magnetization vectors of the X nuclei is observed. The 180_x° pulse causes a re-labeling of the spin states of the A nuclei, as mentioned earlier. The labels of the two vectors (M' and M'') are therefore interchanged but the direction of movement remains the same, resulting in their starting to come together. An echo results when they coalesce after a further time period τ [Figure 5.6B, (ii)–(iv)], giving a positive signal [Figure 5.6B, (v)]. A third possibility would be the simultaneous application of two 180_x° pulses to both the A and X nuclei. This would result in the

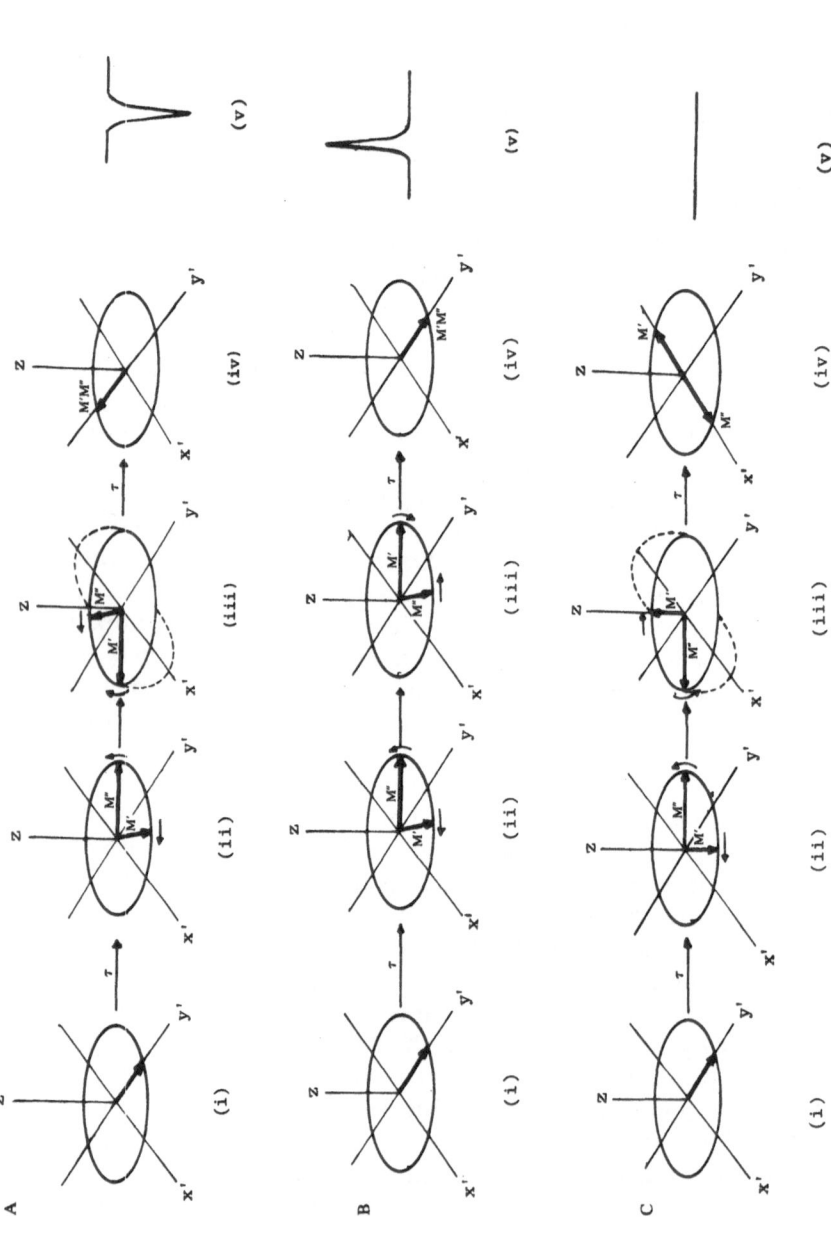

Figure 5.6. Spin-echo experiment in an AX system. Vector diagrams represent nucleus X. (A) Heteronuclear case. (B) Heteronuclear case, 180° pulse being applied to the X nucleus; (B) Heteronuclear case, 180ₓ pulse being applied to A nucleus; (C) homonuclear case (180° pulse affects both A and X nuclei); the heteronuclear case with 180ₓ pulses being applied simultaneously to *both* the *A* and *X* nuclei affords the same situation as the homonuclear case.

canceling out of the effect of the two magnetic vectors as they come to lie in opposite directions at time 2τ after the application of the 90°_y pulse, and no signal would be observed. The magnetic vector diagram would then be analogous to those shown in Figure 5.5 for the homonuclear system. In the homonuclear system, the 180°_x pulse is not selective and it effects both the A and X nuclei. The 180°_x pulse therefore interchanges the spin states of the neighboring nucleus as well as causing an inversion of the nucleus under observation [Figure 5.6C, (ii)–(iii)]. After a further time period τ, the magnetization vectors do not refocus but cancel one another [Figure 5.6C, (iv)] and no signal is observed [Figure 5.6C, (v)].

It may be noted that refocusing occurs only in the heteronuclear system, i.e., when a *selective* 180°_x pulse is applied. Field inhomogeneities or differences of chemical shift cause an additional fanning out of the magnetic vectors which can be eliminated by the application of the 180°_x pulse to the X nucleus. This is important for practical purposes since the Larmor frequency of the X nucleus is not normally identical to the carrier frequency ν_0, as assumed in the above discussion.

In heteronuclear systems the pulse sequence is modified so that 180° pulses are simultaneously applied to both types of nuclei (A and X). By using selective detection, i.e., A–{X} (observe A, decouple X), one obtains spectra in which the phases of the signals are dependent on time τ and the magnitude of the coupling constants.

The pulse sequence employed for a CH system is

$$[90^\circ(^1H)-\tau-180^\circ(^1H)/180^\circ(^{13}C)-\tau-]_n$$

This allows the FID of the echo sequence for A spins to be recorded with the X spins decoupled. It results in the refocusing of chemical shifts but does not refocus the coupling constants. To refocus the coupling constants without affecting the chemical shifts for a CH system, one uses the sequence

$$[90^\circ(^1H)-\tau-180^\circ(^{13}C)-\tau-]_n$$

The application of the 180° pulse to ^{13}C would then result in changing of the spin labels of $M_H^{C\alpha}$ and $M_H^{C\beta}$ and a refocusing of the doublet after a time τ.

5.1.2 Attached Proton Test by Gated Spin-Echo (GASPE) (or SEFT*) Measurements

The spin-echo experiments described above have been developed into a powerful method for differentiating between CH_3, CH_2, CH, and quaternary carbon atoms. The Attached Proton Test (APT) utilizes a *gated spin-echo* (GASPE) pulse presented in Figure 5.7. Since the proton decoupler is switched

---

* SEFT = *Spin-Echo Fourier Transform*.

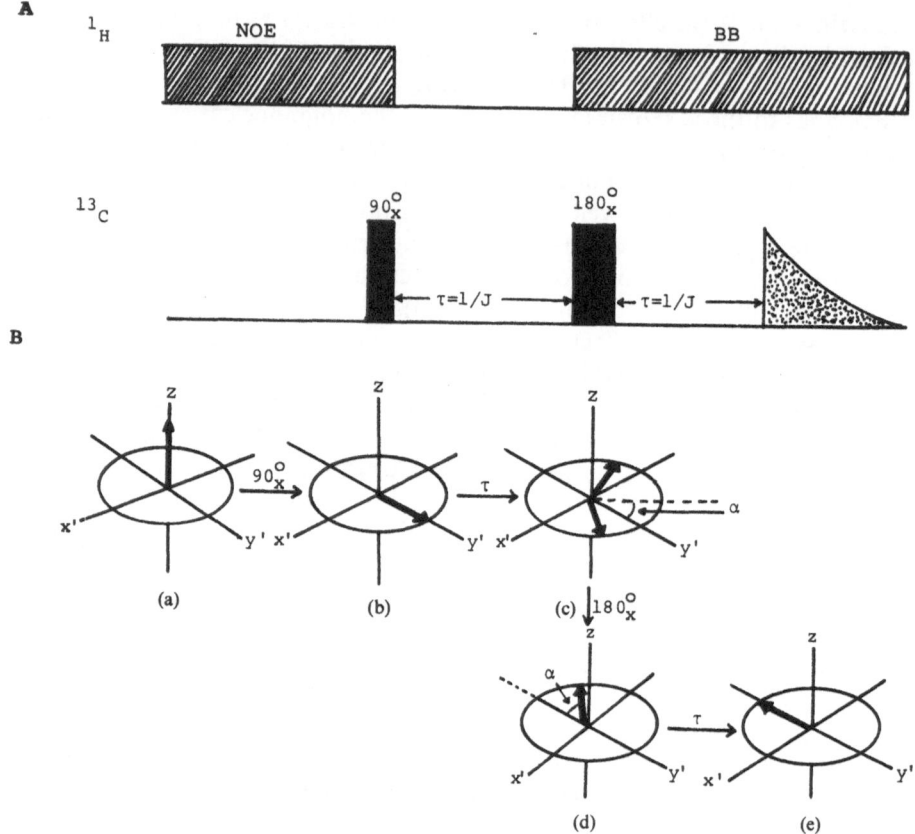

Figure 5.7. (A) The *gated spin-echo* (GASPE, or Attached Proton Test APT) pulse sequence. (B) Effects of a gated spin-echo sequence on a CH doublet. (a) At equilibrium; (b) after the 90° ^{13}C pulse; (c) after first period τ, angle $\alpha = (\Omega - \omega_0)\tau$ radians; (d) after the 180° ^{13}C pulse and ^{1}H broad band decoupling; (e) after final period τ.

off during the first half of the evolution period τ after the 90° pulse, "*J*-modulation" of the spin-echo amplitude occurs during this period due to spin–spin coupling. What this means is that the signal height (or echo amplitude) is seen to vary cosinusoidally as a function of the delay time τ. This variation does not occur in a synchronous manner for CH, CH_2, or CH_3 groups but is characteristically different for each, allowing a differentiation to be made between them. This difference is due to the fact that during the delay interval τ after the 90° pulse, the carbons of CH groups become split into two magnetization vectors which rotate with frequencies of $\pm J/2$; the carbons of CH_2 groups are split into three magnetization vectors of which the middle vector remains stationary (in the rotating frame) while the outer two vectors rotate with frequencies of $\pm J$; the carbons of CH_3 groups are split into four magnetization vectors—two slower vectors which rotate at frequencies of

$\pm J/2$ and two faster vectors which rotate with frequencies of $\pm 3/2J$. Since the quaternary carbon atoms are not directly coupled to any protons, they afford only one vector which remains stationary. The net resultant magnetization (or signal amplitude) detected by the receiver on the y'-axis will therefore vary in a characteristically different manner for CH_3, CH_2, and CH groups as a function of the delay interval τ (Figure 5.8). A series of measurements of echo heights affords spectra containing modulations produced by the corresponding couplings, and the method is therefore known as J-spectroscopy. Since the decoupler is off during the first delay (usually $1/J$) period, the coupling information is provided during this period. The chemical shift information is provided during both $1/J$ delays.

Let us consider the effect of the APT pulse sequence on CH and CH_2 groups. A simplifying assumption is first made that the chemical shift of a CH group being irradiated is identical to that of the reference TMS (which, of course, is not true). A 90° pulse would then flip the magnetization to the y'-axis and it will appear to lie along the y'-axis, and not precess away from it. Since there is a delay after the 90° pulse during which both the receiver and the proton decoupler are turned off, the CH magnetization no longer remains a single vector but becomes split into two lines, separated by J Hz, at $\omega_0 + (J/2)$ and $\omega_0 - (J/2)$. These two vectors correspond to half the molecules having a parallel (α) orientation with respect to the applied field and half the molecules having an antiparallel (β) orientation with respect to the applied field, and they represent the two component peaks of the doublet [Figure 5.8, diagram (b)(i)]. The receiver, which is aligned along the y-axis, detects the sum of these two vectors at any given time. One of two vectors moves in a clockwise direction, while the other moves in an anti-clockwise direction so that they precess away from each other. The net magnetization (which is the sum of these two vectors) therefore decreases as the angle between the two vectors increases, and it becomes zero at a time corresponding to $1/(2J)$ seconds when the two vectors are aligned at 180° to one another and therefore totally cancel each other [Figure 5.8, diagram (b)(ii)]. At the time $\tau = 1/J$ seconds the vectors are found to lie superimposed upon one another along the $-y$-axis [Figure 5.8, diagram (b)(iv)]. This results in the appearance of a negative signal for the CH lines. The delay after the 90° pulse thus provides information about the multiplicity of the peaks, there being two vectors if one proton is coupled to a carbon atom, three vectors if two protons are coupled to a carbon [Figure 5.8, diagram (c)], and four vectors when three protons are coupled to a carbon [Figure 5.8, diagram (d)].

In the case of a CH_2 group, the protons have three different spin states—$\alpha\alpha$, $\alpha\beta$ or $\beta\alpha$, and $\beta\beta$—which results in the ^{13}C resonances appearing as a 1:2:1 triplet. Again if we make the same simplifying assumption that the chemical shift of the CH_2 group corresponds exactly to that of the reference RF transmitter frequency, then the three magnetization vectors corresponding to the three different spin states of the methylene protons would all be aligned with the y-axis after a 90° pulse. The central line of the triplet, corresponding

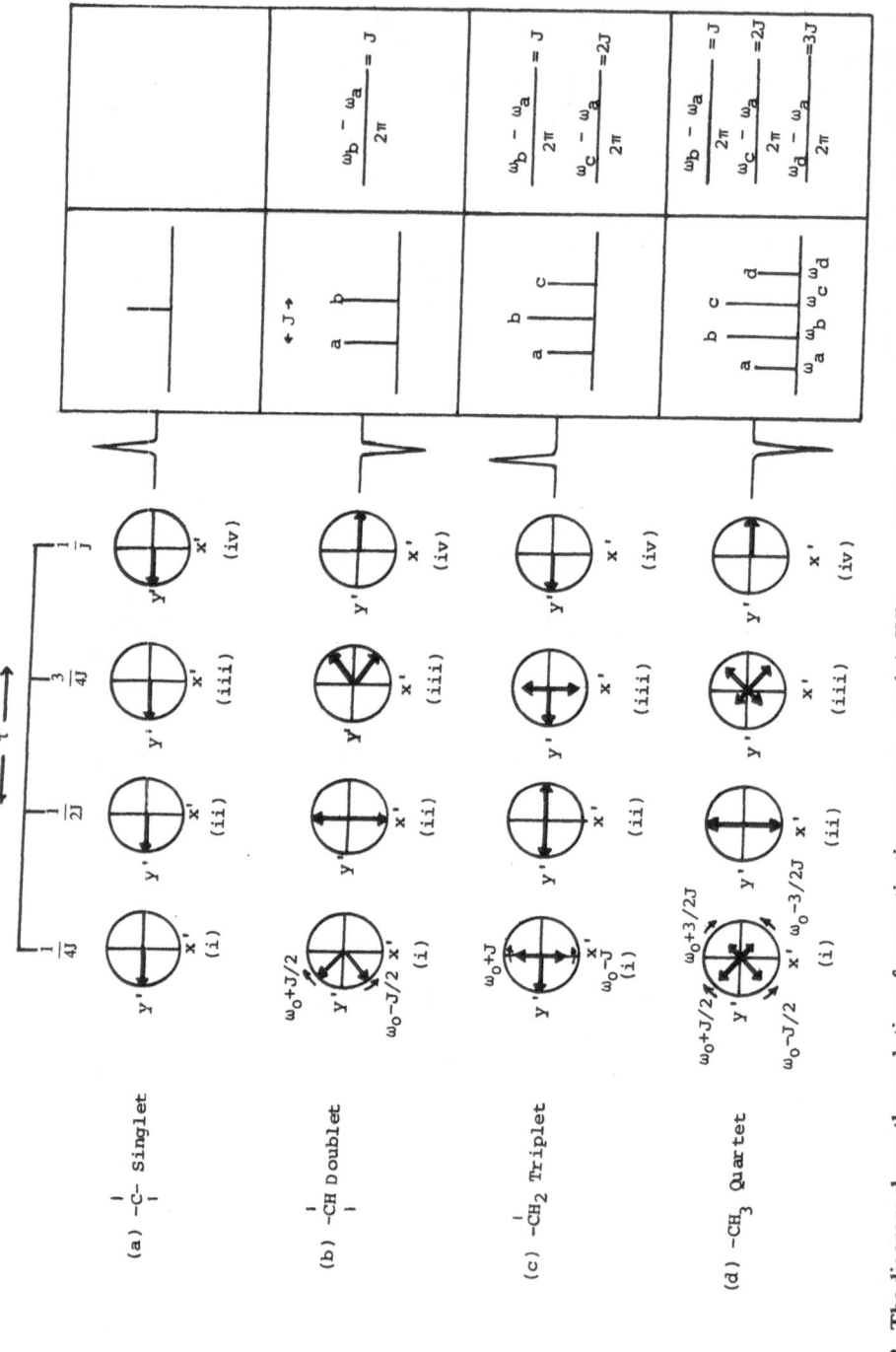

Figure 5.8. The diagram shows the evolution of magnetization vectors for C, CH, CH_2, and CH_3 when $\tau = 1/(4J)$, $1/(2J)$, $3/(4J)$, and $1/J$. When $\tau = 1/J$, the C and CH_2 carbons appear along the $+y$-axis (positive peaks) while CH and CH_3 appear along the $-y$-axis (negative peaks). When $\tau = 1/(2J)$ only quaternary carbons give signals. The vector positions are those in the second τ period, i.e., after the application of the 180°_x ^{13}C pulse. The signals represent the sum of the y components detected by the receiver.

to the $\alpha\beta$ and $\beta\alpha$ states, would then have the same angular velocity as the reference line and since the planes are also rotating at the same angular velocity, the central line would appear to be stationary in the rotating frame of reference. The outer and inner lines of the triplet, corresponding to the $\alpha\alpha$ and $\beta\beta$ states, would have vectors with angular velocities of $\omega_0 + J$ and $\omega_0 - J$ which would thus precess away from the central stationary vector, but after time $1/J$ they would complete one full circle and rejoin the central vector lying along the $+y$-axis [Figure 5.8, diagram (c)(iv); Figure 5.9(b)]. If the detector is switched on at time $1/J$, the CH_2 protons would appear as a single peak which would have the normal phase, unlike the CH peak which was $180°$ out of phase after time $1/J$ [compare Figure 5.8(b)(iv) with Figure 5.8(c)(iv)]. Similar considerations show that the CH_3 carbons would have a reverse phase (i.e., come to lie along the $-y$-axis) while quaternary carbon atoms would have normal phase (i.e., come to lie along the $+y$-axis) after time $1/J$. This therefore provides a powerful tool for differentiating the signals of CH_3 and CH carbon atoms from those of CH_2 and quaternary carbon atoms. The effects of echo modulation on the A nucleus for AX and AX_2 systems are shown in Figure 5.10.

In the above discussion we had made the simplifying assumption that the chemical shift of the CH_3, CH_2, CH, or quaternary carbon atoms was the same as that of the reference frequency situated on the $+y$-axis so that during the divergence of the magnetic vectors, the position of the *net* magnetization remained unchanged on the $+y$-axis, although its magnitude decreased with the divergence of the vectors (this is so because the net magnetization along the y-axis will be equal to the sum of the individual vectors). The magnitude of the magnetization was seen to reach a minimum when they came to lie along the $-y$-axis, and then started increasing again as the vectors moved towards the $+y$-axis. The situation in reality is different in that the chemical shifts of the CH_3, CH_2, CH, and quaternary carbon atoms are actually down-field from the reference TMS signal. The application of a $90°$ pulse will result

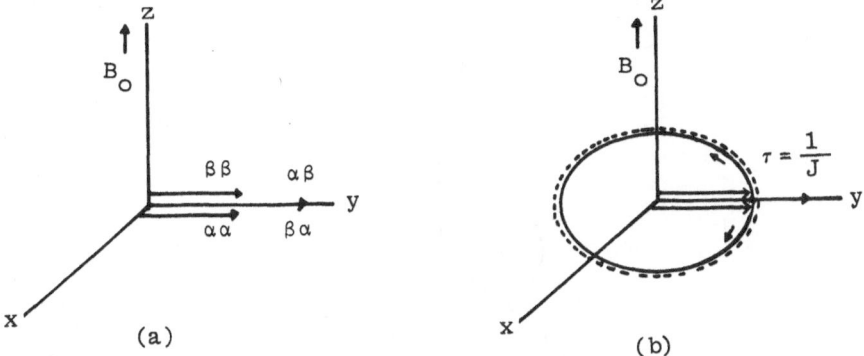

Figure 5.9. (a) Magnetization vectors of a CH_2 group after a $90°$ pulse. (b) Convergence of vectors with normal phase after $\tau = 1/J$.

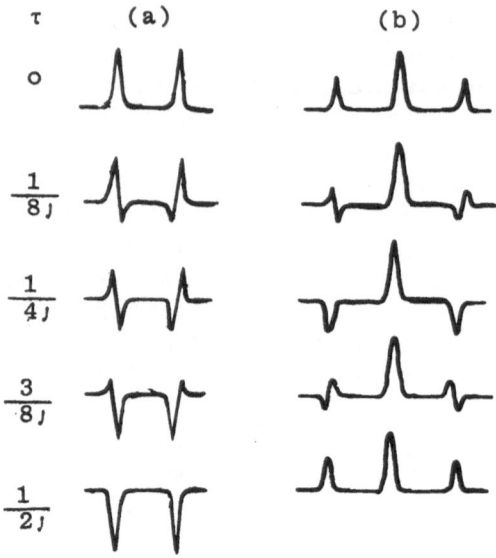

Figure 5.10. Echo modulation effects on the spectrum of the A nucleus for (a) an AX and (b) an AX_2 homonuclear spin system following a spin-echo pulse sequence, $90^\circ_x-\tau-180^\circ_y-\tau$–echo.

in their coming to lie on the y-axis but then their average position will not lie constant along the y-axis but will move clockwise with a phase shift of $\omega_0\,\delta/J$, causing a corresponding phase error. The greater the chemical shift of the nucleus, the greater will be the phase error (Figure 5.11). This results in a serious complication which needs to be corrected. The problem is solved by applying a 180° pulse by a coil along the y-axis. The refocusing effect of a 180° pulse has been mentioned earlier but it will be described again here. Let us imagine two vectors δ_1 and δ_2 which have diverged a certain distance from

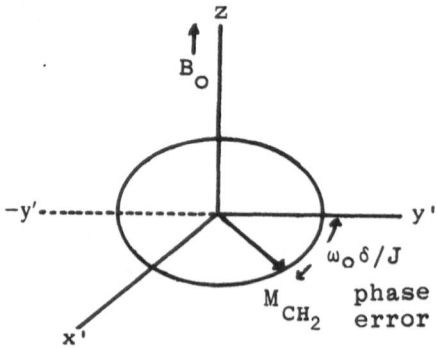

Figure 5.11. Phase error at time $\tau = 1/J$.

each other over a time period τ. The application of a 180° pulse results in the faster-moving vector coming to lie behind the slower vector and it takes a further time period τ for it to catch up with the slower vector, so that after a total time 2τ (counted from the application of the original 90° pulse) the vectors will refocus along the y-axis, resulting in the formation of a spin-echo. This 180° pulse therefore allows one to refocus various magnetization vectors resulting from differently shielded nuclei and correspondingly different rates of precession.

To illustrate how the spin-echo experiment can eliminate effects which result from different Larmor frequencies and chemical shifts, let us consider the ^{13}C resonances of chlorobenzene. Since the six aromatic carbon atoms of chlorobenzene are in four different environments, one would expect four signals. If after the application of the 90_x° excitation pulse a certain time delay τ, is incorporated before collection of data, then since the carbon atoms have four different Larmor frequencies, their magnetic vectors would have precessed to different degrees and the NMR signals obtained would have different phases [Figure 5.12(b)]. A 180_x° pulse would, however, invert the vector system across the x'-axis [Figure 5.12(c)], and a spin-echo would be created after a further time τ [Figure 5.12(d)]. The phase differences would now no longer exist and any field inhomogeneity effects would also have been compensated. An additional 180_x° pulse is finally applied so that an absorption spectrum in the correct phase can be recorded [Figure 5.12(e)].

Figure 5.13 shows the variation of the signal intensities of CH_3, CH_2, CH, and nonprotonated carbon atoms with time. If J_{CH} for sp^3-hybridized carbon atoms is assumed to be 125 Hz, then $1/J$ will correspond to $1/125 = 0.008$ s $= 8$ ms. If we set τ at 8 ms and carry out the above measurements, then CH_2 and quaternary carbon atoms will appear with positive signals (above the baseline) and CH_3 and CH carbon atoms with negative signals (below the baseline). We will then be observing along the vertical line (a) in Figure 5.13. If τ is adjusted to $1/(2J)$ (i.e., 4 ms, if $J = 125$ Hz), then it is apparent from Figure 5.13 that only quaternary carbon atoms will show stronger positive signals whereas the CH_3, CH_2, and CH carbon signals will be very weak, if visible at all [vertical line (c) being then the point of observation]. Adjustment of τ to $2/(3J)$ (5.3 ms for $J = 125$ Hz) results in CH_3 groups appearing with one-third of the negative intensity of CH groups since the point of observation of the signal amplitudes now lies along the vertical line (b) (Figure 5.13) where the CH groups have a stronger negative intensity than CH_3 groups. The position of magnetization vectors of C, CH, CH_2, and CH_3 as a function of the evolution time is given in Figure 5.8. The position of the vectors is shown after the 180_x° pulse. Only the sums of the y components of the vectors are sensed by the detector. The effect of choosing different τ values when recording APT spectra of cholesterol is shown in Figure 5.14. Thus APT spectra provide a superior alternative to off-resonance spectra for determining signal multiplicities since the overlap of peaks in close-lying multiplets is avoided. The angle α by which the magnetization rotates during the time

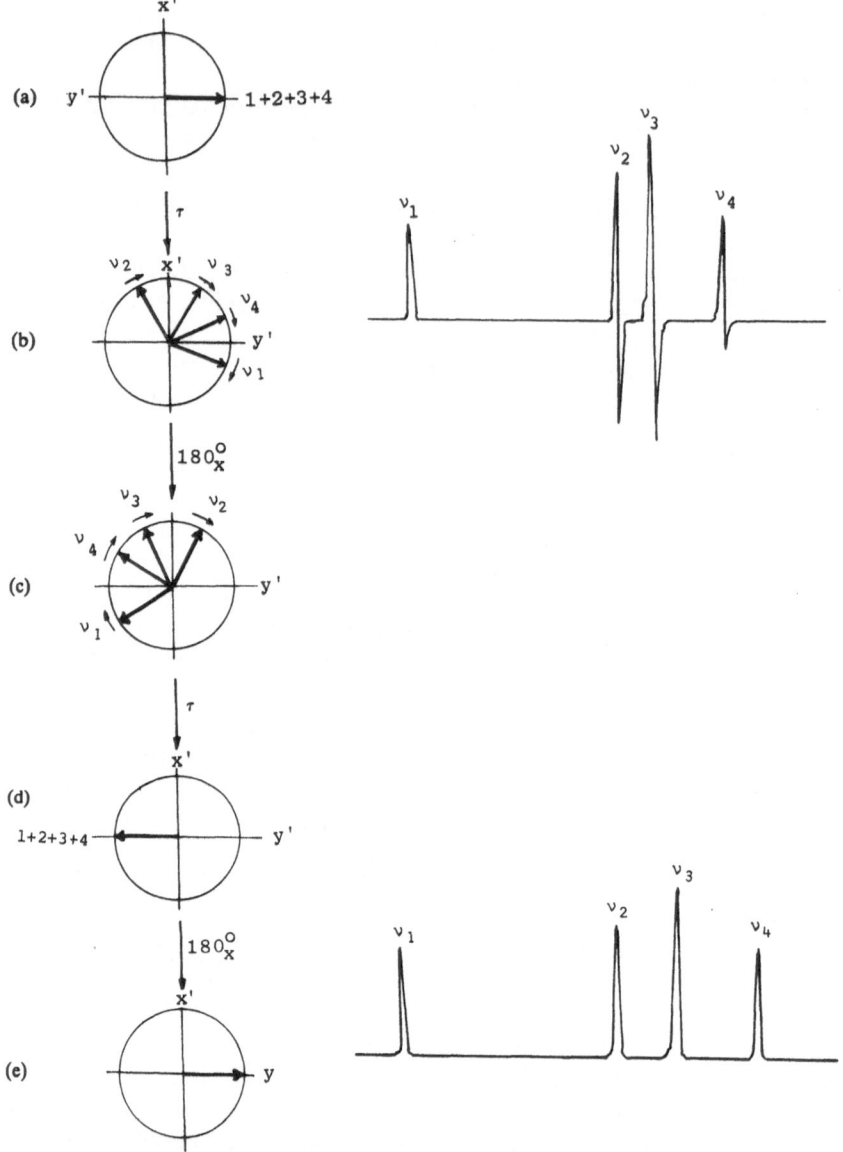

Figure 5.12. Spin-echo experiment. (a) Position of vectors after 90°_x pulse. (b) Position of vectors at time τ after 90° pulse. The corresponding spectrum shows some of the peaks as partly inverted on account of the phases of the corresponding magnetic vectors. (c) Application of 180°_x pulse brings the vectors to mirror-image positions across the x'-axis. (d) Spin-echo is created along the $-y'$-axis after further time τ. All peaks would appear inverted if recorded at this point (e) Application of 180°_x pulse corrects the phases, and allows the spectrum to be recorded correctly.

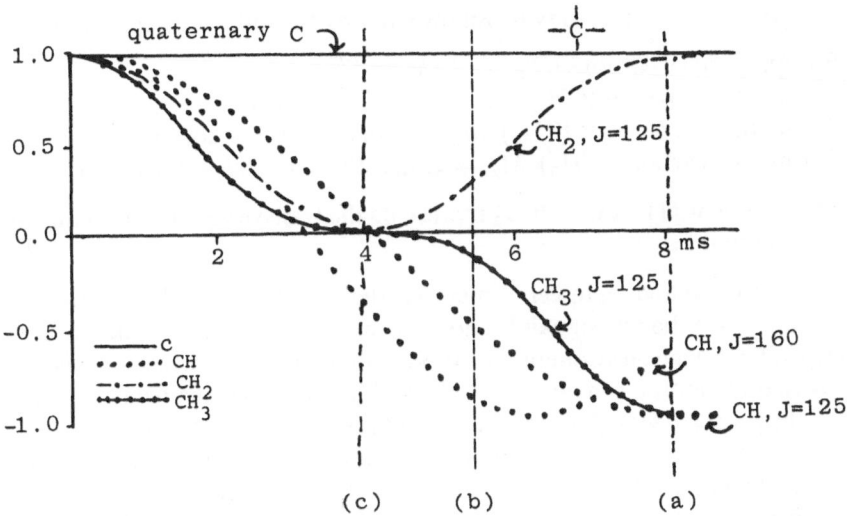

Figure 5.13. Dependence of signal intensities on time τ in the APT experiment.

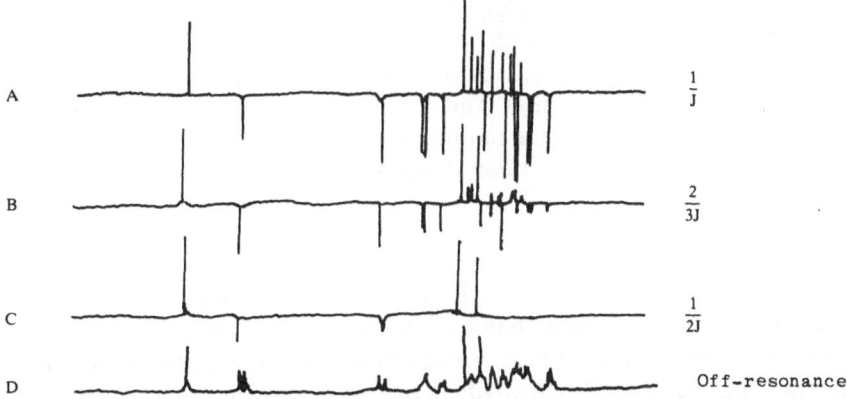

Figure 5.14. (A)–(C) APT spectra of cholesterol with $\tau = 1/J$, $2/(3J)$, and $1/(2J)$, respectively; (D) off-resonance decoupled spectrum of cholesterol.

period τ after the application of the 180° ^{13}C pulse is given by the equation:

$$\alpha = (\Omega - \omega_0)\tau \text{ radians}$$

where Ω is the average resonance frequency of the doublet and ω_0 is the observed frequency of rotation. It may be noted that after the final period τ the magnetization comes to lie along the y-axis, though its magnitude is reduced. All resonances thus appear with phases of 0° or 180°. The magnetization of

^{13}C multiplets at time 2τ may be represented by the following equations:

For quaternary carbons (C), $M_{2\tau} = M_0 e^{-2\tau/T_2}$

For methine carbons (CH), $M_{2\tau} = M_0 e^{-2\tau/T_2} \cos \pi\tau J$

For methylene carbons (CH$_2$), $M_{2\tau} = \frac{1}{2} M_0 e^{-2\tau/T_2}(1 + \cos 2\pi\tau J)$

For methyl carbons (CH$_3$), $M_{2\tau} = \frac{3}{4} M_0 e^{-2\tau/T_2}(\cos \pi\tau J + \frac{1}{3}\cos 3\pi\tau J)$

The exponential factor $-2\tau/T_2$ represents the transverse relaxation during the time period 2τ.

The J-modulated spin-echo method described here is of particular advantage when the ^{13}C signals lie close to one another, and determination of multiplicities by the off-resonance method is difficult. It finds wide use in NMR spectroscopy, but a disadvantage of the technique is that the delay τ during which the proton decoupler is off, has to suit the CH coupling constants. Since these can often vary considerably (alkyl groups $\simeq 125$ Hz, $\tau = 1/J = 8$ ms, alkenes, aromatic substances $\simeq 175$ Hz, $\tau = 1/J = 5.7$ ms etc.), the method works best when the coupling constants of the CH$_n$ multiplets lie close to one another. This can occasionally cause erroneous signals to appear which can be corrected by appropriate adjustments of the delay interval τ.

Another method for distinguishing quaternary carbons from others is by using low-power noise decoupling. Since the carbon atoms with directly bonded protons require a much higher amplitude to remove the larger couplings, the use of low power removes the long-range couplings but does not affect the large $^1J_{CH}$ splittings. This results in CH$_3$, CH$_2$, and CH carbons appearing as complex broadened multiplets whereas the quaternary carbon atoms appear as easily recognizable sharp singlets.

5.1.3 Cross-Polarization

A serious problem faced in NMR spectroscopy is that of sensitivity enhancement of nuclei present in low abundance. The difference in energy ΔE between the upper and lower energy levels of a precessing nucleus placed in a magnetic field, and hence its sensitivity to the NMR experiment, depends on its gyromagnetic ratio.

$$\Delta E = \frac{\gamma h B_0}{2\pi}$$

The populations N_a and N_g of the nuclei in the excited (a) and ground (g) states is determined by Boltzmann distribution

$$\frac{N_a}{N_g} = e^{-\Delta E/kT} = 1 - \Delta E/kT = \frac{1 - \gamma h B_0}{2\pi k T}$$

Nuclei such as ^{13}C and ^{15}N are much weaker as nuclear magnets in comparison to ^{1}H because of their low gyromagnetic ratios, and the population differences between their upper and lower states are correspondingly smaller:

$$\gamma_{1H} \cong 4\gamma_{13C}$$

If one could somehow enhance the intensity of the signals of the nuclei under observation by increasing the difference between the populations in their upper and lower states, then a substantial sensitivity enhancement would result. This can be achieved by a process of "population transfer," employing to advantage the fact that the gyromagnetic ratio of coupled protons (and hence their receptivity to the NMR experiment) is much higher than that of the carbon nuclei to which they are coupled. In practice, this is done in a CH system by applying a selective 180° pulse to the ^{1}H nuclei which not only inverts the population of their energy levels but also increases the population difference between the higher- and lower-energy states of the ^{13}C nuclei. This "cross-polarization" phenomenon therefore relies on the fact that when the population states of protons have been reoriented after the application of a 90° pulse, they can transfer their magnetization to the ^{13}C-nuclei; the process of cross-polarization is most effective when the nucleus with the low gyromagnetic ratio (e.g., ^{13}C) is being observed and the nucleus with the high gyromagnetic ratio (e.g., ^{1}H) is being irradiated.

Let us consider a proton bonded to a carbon atom in a CH group. Due to the 1% natural abundance of ^{13}C, 99% of the protons will be bonded to ^{12}C nuclei while only 1% of protons will be bonded to ^{13}C nuclei, and it is the latter with which we are concerned. The proton and carbon atoms will couple with each other so that doublets will be observed in the spectra of both. The energy level diagram of an H/C spin system is shown in Figure 5.15. The proton and carbon atoms can be oriented parallel to the applied field. The two nuclei are then conventionally designated to have $\alpha\alpha$ orientation and this is the lowest-energy orientation. There are two antiparallel orientations possible, one in which the proton is parallel and the carbon is antiparallel to the applied field B_0 ($\alpha\beta$), and the other in which the reverse situation exists ($\beta\alpha$). The fourth and highest-energy state is the one in which both nuclei have spins aligned against the applied field. The three vertical columns a, b, and c in Figure 5.15 show the energy levels without coupling (column a), with coupling with J positive (column b), and with coupling with J negative (column c). The proton transitions occur between levels 1 and 3, and between levels 2 and 4. Carbon transitions occur between levels 1 and 2, and between levels 3 and 4. When J = 0, both members of these pairs of transitions are of the same energy, and only one line is observed for each nucleus. However, when coupling is introduced with J having a positive value, the energy levels corresponding to parallel spin states (levels 1 and 4) are raised by $J/4$ while those corresponding to paired spin states (2 and 3) are lowered by $J/4$ (column b, Figure 5.15). The opposite situation prevails when J is negative (column c, Figure 5.15). The relative signs of the coupling constants cannot be ascertained from the appearance of the spectra since the magnitude of coupling (J) is the same

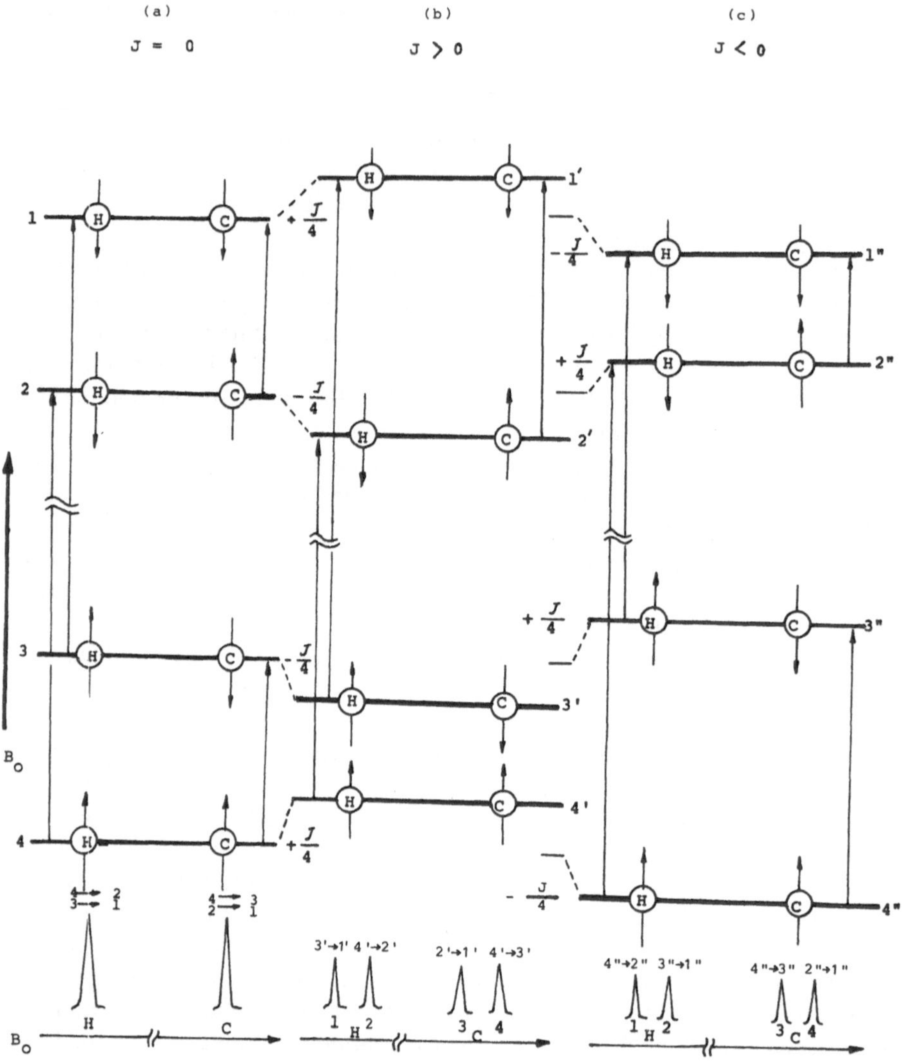

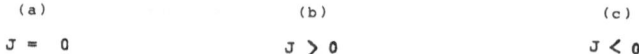

Figure 5.15. Energy level diagram of a CH spin system. (a) Without coupling. (b) Positive coupling constant results in raising of levels 1' and 4' and lowering of levels 2' and 3'. (c) Negative coupling constant raises levels 2" and 3" and lowers levels 1" and 4".

irrespective of whether J is positive or negative. In both cases (i.e., $J > 0$ or $J < 0$, shown in columns b and c, respectively) there are two transitions for ${}^1\text{H}$ nuclei, with the energy of each differing from that in the system without coupling by $\frac{1}{2}J$ (sum of two $\frac{1}{4}J$ differences) so that the total difference in energy levels remains equal to J. The same situation exists in the case of ${}^{13}\text{C}$ nuclei.

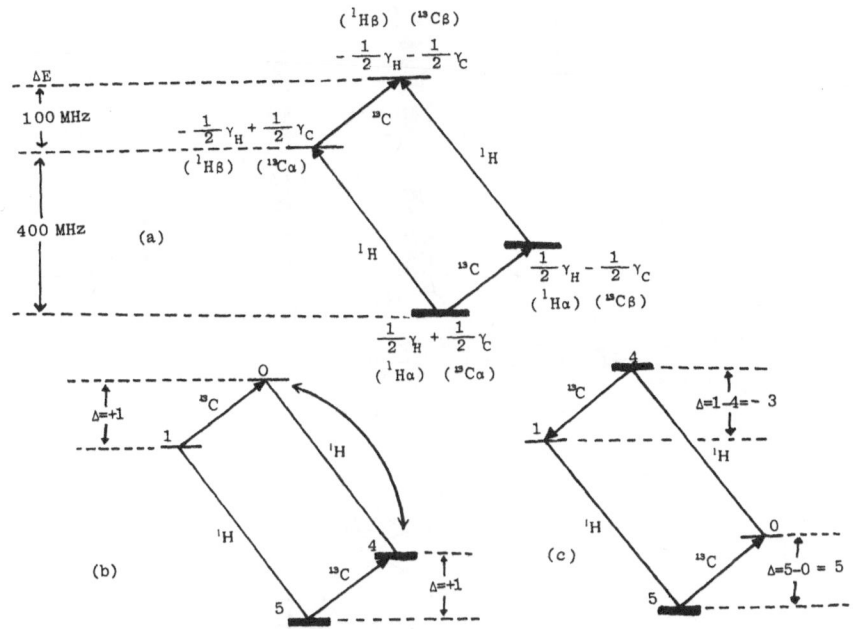

Figure 5.16. (a) Energy levels, populations, and single quantum transitions for an AX system. (b) → (c) Effect of population transfer on signal enhancement.

5.1.3.1 INEPT Spectra

Now let us consider the same energy diagram shown in column b of Figure 5.15 but drawn a little differently (Figure 5.16). The population of the lowest-energy state ($\alpha\alpha$) is given by $\frac{1}{2}\gamma_H + \frac{1}{2}\gamma_C$ while the other states are shown as $\frac{1}{2}\gamma_H - \frac{1}{2}\gamma_C$, $-\frac{1}{2}\gamma_H + \frac{1}{2}\gamma_C$, and $-\frac{1}{2}\gamma_H - \frac{1}{2}\gamma_C$, corresponding to the $\alpha\beta$, $\beta\alpha$, and $\beta\beta$ states, respectively. After a "preparation" period of five times the longest proton T_1, the proton states would reach a Boltzmann equilibrium distribution.

If, for the sake of clarity, we add a common factor $\frac{1}{2}\gamma_H + \frac{1}{2}\gamma_C$ to all the energy levels, then the four energy levels become $\gamma_H + \gamma_C$, γ_H, γ_C, and 0, respectively. If we assume that $\gamma_H = 4$ and $\gamma_C = 1$* then the populations of the $\alpha\alpha$, $\alpha\beta$, $\beta\alpha$, and $\beta\beta$ energy states may be represented as 5, 4, 1, and 0, respectively. The population difference for the two ^{13}C transitions before the application of the INEPT pulse sequence is given by the lower population minus the upper population, i.e., $1 - 0$ and $5 - 4 = 1$. [Figure 5.16(b)]. The INEPT pulse sequence causes an exchange of populations between the upper and lower energy states so that the population difference (i.e., the lower level

---

* These are arbitrary figures; the sensitivity of a nucleus to an NMR experiment depends on its magnetic moment μ which determines the energy difference as well as the population difference between the spin states. The magnetic moment μ of ^1H is 2.675 while for ^{13}C it is 0.70216.

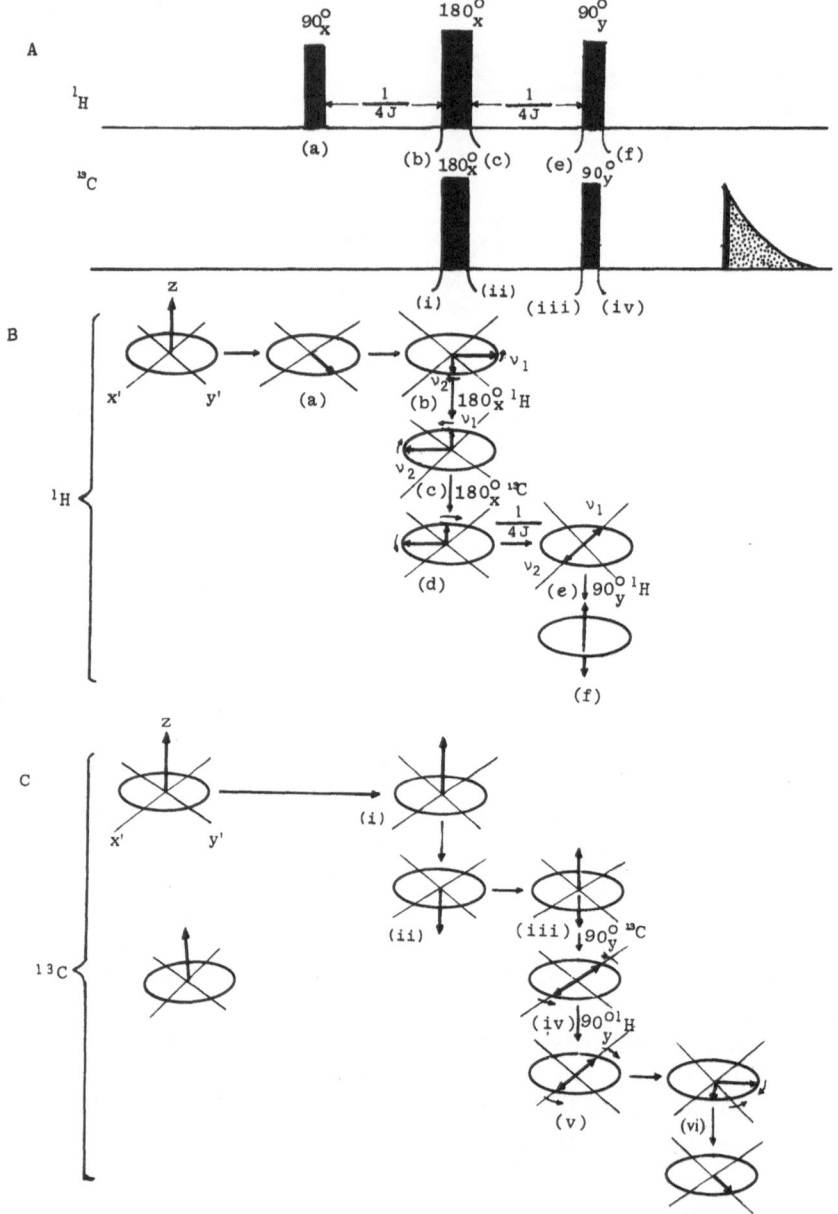

Figure 5.17. (A) INEPT pulse sequence. (B) Effect on ¹H magnetization. (C) Effect on ¹³C magnetization.

minus the upper level) becomes $+1 - 4$ and $+5 - 0 = -3$ and $+5$ (Figure 5.16(c)]. The spectra thus display an enhancement of γ_H/γ_C ($= \sim 4$).

To understand the sequence of events in INEPT* one needs to consider the fate of the magnetization vectors in the conventional rotating frame of reference. The pulse sequence applied to the ^1H and ^{13}C nuclei and the effect that this sequence has on ^1H and ^{13}C vectors are illustrated in Figure 5.17. Application of the 90_x° ^1H pulse aligns the ^1H magnetization along the y-axis [Figure 5.17B, (a)]. During the subsequent $\tau = 1/(4J)$ delay interval, coupling with the ^{13}C nuclei causes the vectors to split into two components, v_1 and v_2 [Figure 5.17B, (b)]. The 180_x° ^1H pulse causes an inversion of vectors across the x'-axis [Figure 5.17B, (c)]. Simultaneously, the 180_x° ^{13}C pulse interchanges the spin labels so that they now start to move apart [Figure 5.17B, (d)]. After a subsequent $\tau = 1/(4J)$ delay, they are found to be aligned along the x'-axis [Figure 5.17B, (e)]. A 90_y° ^1H pulse is then applied which rotates one vector to the equilibrium position on the $+z$-axis while the other vector is aligned along the $-z$-axis. It is this 90_y° ^1H pulse which inverts the magnetization of one proton line and results in *selective population inversion*. This is equivalent to the exchange of populations shown in Figure 5.16(c). The polarization of the spin system is detected by a 90_x° ^{13}C pulse, and the ^{13}C-NMR spectrum shows signals enhanced by a factor of γ_H/γ_C due to the increased population difference between lower and upper levels. The effect of the pulse sequence on the ^{13}C nuclei is shown in Figure 5.17C.

The two lines for the ^{13}C nuclei corresponding to intensities of $+5$ and -3 [Figure 5.16(c)] are at different phases and they would therefore cancel each other. A further delay is therefore introduced to bring them into phase. This delay time varies between CH_3, CH_2, and CH groups, as in the APT experiment. Once the two lines are in phase, the intensity of each line for the ^{13}C nucleus which is coupled to a proton will be increased fourfold. Normal proton-decoupled ^{13}C spectra show an enhancement by NOE of $1 + \frac{1}{2}(\gamma_H/\gamma_C)$ which equals 3, approximately. Thus the enhancement of ~ 4 obtained by INEPT is only marginally better than that obtained by NOE. The advantage is, however, much greater when nuclei with low gyromagnetic ratios are being studied. For instance, in the case of nitrogen the NOE enhancement is -4 while the INEPT enhancement is $+10$ so that considerable improvement in sensitivity results. Moreover, even in the case of carbon, since the spin–lattice relaxation times for protons are much shorter than those of ^{13}C, the delays between successive sequences can be fairly short, resulting in considerably improved spectra in a given time.

The multiplets encountered in INEPT spectra do not have the normal binomial intensity distribution of 1:1 for doublets, 1:2:1 for triplets, 1:2:2:1 for quartets, etc. Instead, the intensities obtained are shown in the form of a Pascal triangle (Figure 5.18) in which the outer peaks of the multiplets are generally more intense than the inner ones. Figure 5.19 shows the NOE-

---

* INEPT = *I*nsensitive *N*uclei *E*nhanced by *P*olarization *T*ransfer.

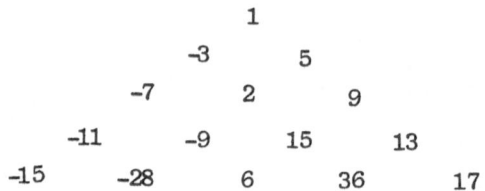

Figure 5.18. Relative intensities in an INEPT spectrum represented as a variation of the Pascal triangle.

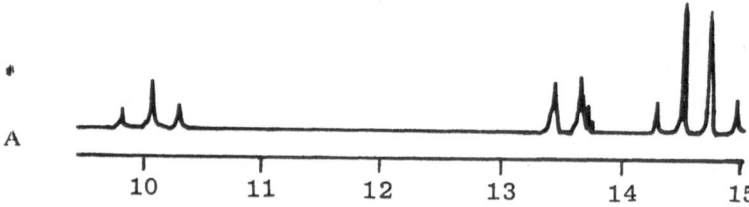

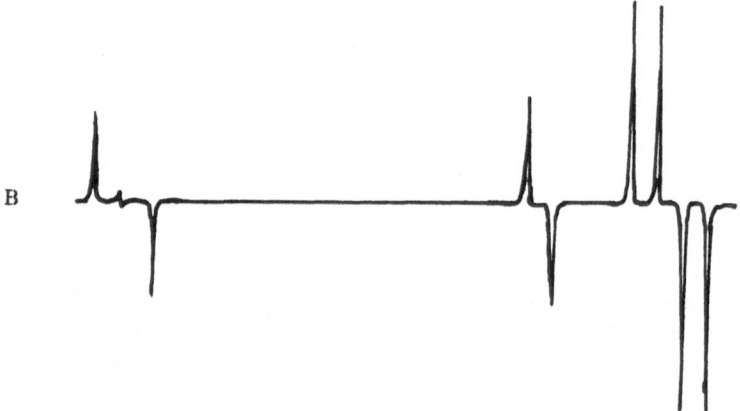

Figure 5.19. NOE enhanced carbon spectrum (A) and INEPT spectrum (B) of isobutyl alcohol.

enhanced carbon spectrum and the INEPT spectrum of isobutyl alcohol. The intensities predicted for CH, CH_2, and CH_3 groups in the Pascal triangle for INEPT spectra are strikingly apparent.

In order to decouple INEPT spectra and allow all the peaks to appear positive, another delay D_3 is introduced in the INEPT sequence and a $180°$ pulse applied to both protons and carbons at the center of the delay (Figure 5.20). The value of D_2 is kept at $1/(4J)$ while the value of D_3 may be adjusted

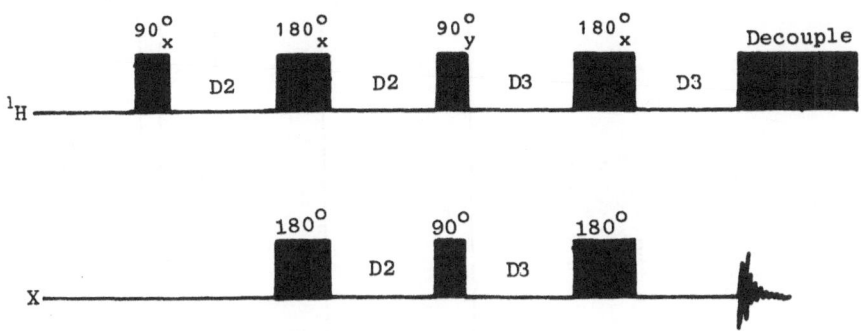

Figure 5.20. A modified INEPT pulse sequence.

depending on the multiplets to be observed. Thus for observing doublets D_3 should be set at $1/(4J)$, for triplets at $1/(8J)$, and at some intermediate value for observing quartets. Setting D_3 at $1/(6J)$ results in the observation of all signals irrespective of their multiplicities. When D_3 is $3/(8J)$, the triplets appear out of phase with respect to the doublets or quartets. Thus INEPT can compete with APT or off-resonance techniques for multiplicity assignments.

The spectrum of menthol shown in Figure 5.21 demonstrates the utility of INEPT in structural assignments. Figure 5.21D shows the proton-decoupled ^{13}C spectrum of menthol. Figure 5.21C is the off-resonance ^{13}C spectrum of menthol with the decoupler offset frequency adjusted to avoid overlapping of multiplets. Figure 5.21A is the INEPT spectrum, showing the carbons corresponding to doublets and quartets in phase (positive) but the carbons corresponding to CH_2 groups out of phase (negative peaks). In Figure 5.21B, the delay D_3 (see Figure 5.20) has been adjusted at $1/(4J)$ to allow only CH signals to appear.

5.1.3.2 DEPT Spectra

Another recently developed pulse sequence which also relies on the cross-polarization phenomenon for achieving sensitivity enhancement is known as "Distortionless Enhancement by Polarization Transfer" (DEPT). Let us consider the effect of the DEPT pulse sequence on a CH system. The pulse sequence employed on the ^1H and ^{13}C nuclei and the effects on the corresponding magnetization vectors are shown in Figure 5.22. After an equilibrium delay T_1, which allows a Boltzmann equilibrium of the protons to be reached, a 90°_x pulse is applied to the ^1H nuclei ("a" in the ^1H pulse sequence, Figure 5.22; note that a–f in the pulse sequence correlate with the vector diagrams a'–f', giving the corresponding effects on the magnetization vectors of the ^1H and ^{13}C nuclei) which causes the magnetization of the protons to be oriented along the y'-axis [Figure 5.22(a')]. In contrast to the INEPT sequence described earlier, a time delay of $\tau = 1/(2J)$ is now inserted during which the transverse magnetization of the ^1H nucleus is modulated through

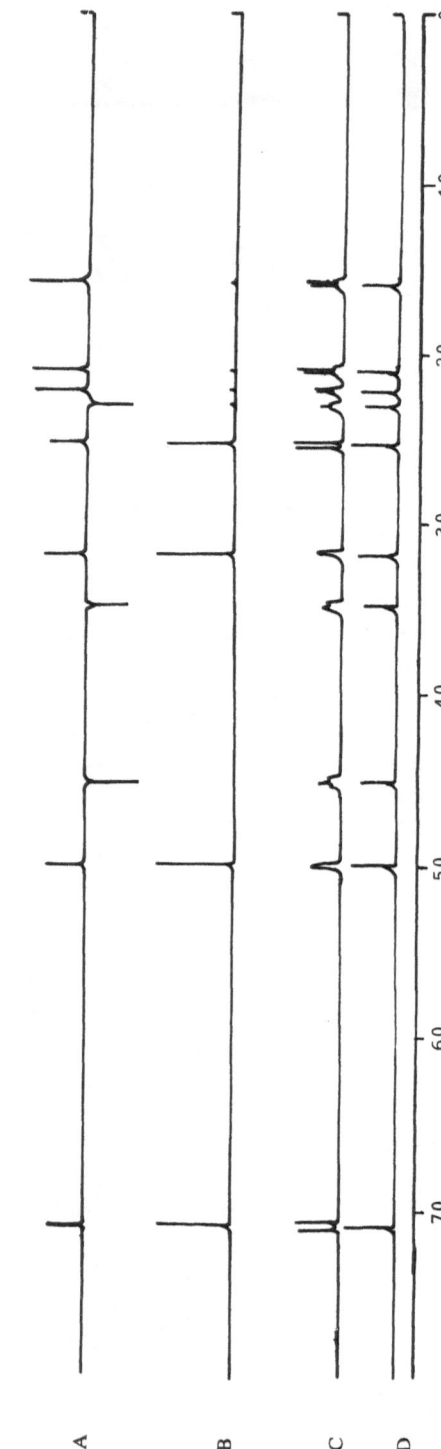

Figure 5.21. (A) INEPT spectrum of menthol with doublets and quartets in phase and triplets out of phase; (B) value of D_3 in modified INEPT sequence (shown in Figure 5.20) set at $1/(4J)$ to allow only doublet (CH) signals to appear; (C) off-resonance spectrum of menthol; (D) broad-band decoupled spectrum of menthol.

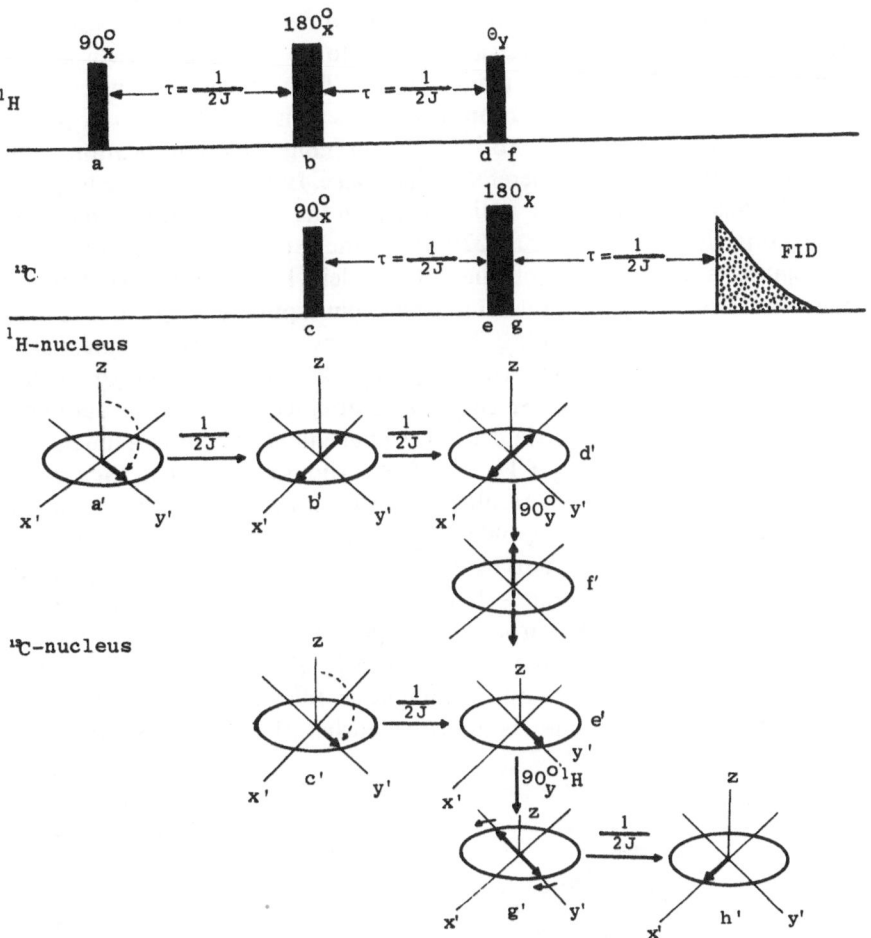

Figure 5.22. *Top*: DEPT pulse sequence. *bottom*: Vector diagrams for ^{1}H and ^{13}C nuclei. Note the effect of each pulse and delay on the corresponding ^{1}H and ^{13}C vectors drawn directly under the pulse for ease of viewing. The ^{13}C magnetizations may be detected as doublets or, if ^{1}H decoupling is applied, as singlets.

coupling with the ^{13}C nucleus. At the end of the $\tau = 1/(2J)$ time period, the two doublet vectors of the ^{1}H nuclei are found to have a phase difference of 180°, and they lie along the x'-axis [Figure 5.22(b')]. A 180°_x pulse is now applied to the protons in order to refocus any inhomogeneities. Simultaneously a 90°_x pulse is applied to the ^{13}C nuclei which creates transverse magnetization of these nuclei along the y'-axis [Figure 5.22(c')]. Since there is no magnetization of the ^{1}H or ^{13}C nuclei along the z-axis, the two are now decoupled, and during the following $\tau = 1/(2J)$ delay period, the vectors of both nuclei remain static in the rotating frames [see Figure 5.22(d'), (e')]. A proton pulse θ (the length of which is adjusted depending on the number of attached protons and the spectrum desired to be recorded) is then applied which serves to polarize the ^1H magnetization. The population transfer occurs at this stage, since the polarization of protons caused by the proton pulse θ also effects the ^{13}C nuclei due to $^{13}C-^1H$ coupling. An intensification of the ^{13}C magnetization vectors therefore takes place. In Figure 5.22 the length of this pulse has been kept at 90° which results in the ^1H magnetization coming to lie along the z-axis [Figure 5.22(f')]. At the same time this proton pulse also causes a polarization of the ^{13}C nuclei [Figure 5.22(g')]. Since z-magnetization now exists in the ^1H nucleus, spin–spin coupling can take place with the ^{13}C nucleus, resulting in a refocusing of the ^{13}C magnetization vectors during the last $1/(2J)$ period [Figure 5.22(h')]. The ^{13}C magnetization can now be detected at $t = 3\tau$ either as a doublet or, if ^1H decoupling is simultaneously applied, as a singlet. The signal enhancements as a function of the variable delay θ for CH_3, CH_2, and CH carbons are shown in Figure 5.23. An experiment with $\theta_y = 90°$ results in a spectrum containing peaks for CH carbon atoms only. A 135° θ_y pulse, on the other hand, affords a spectrum with CH_3 and CH carbon atoms bearing positive amplitudes and CH_2 carbon atoms with negative amplitudes. If the value of the θ_y pulse is kept at 45°, then CH_3, CH_2, and CH carbons all appear with positive amplitudes. A comparison of the spectrum resulting from θ_y at 135° with the spectrum derived from θ_y at 90° allows an unambiguous differentiation between CH_3, CH_2, and CH carbon atoms. As indicated earlier, the DEPT spectra can be

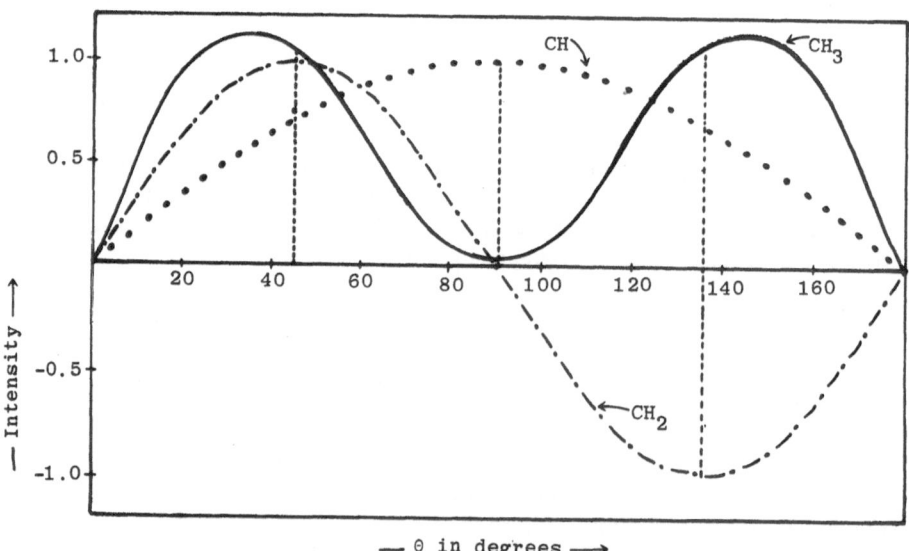

Figure 5.23. Signal enhancements for ^{13}C doublets, triplets, and quartets as a function of the theta pulse flip angle.

recorded with or without proton broad-band decoupling, to afford decoupled or coupled spectra, respectively.

In contrast to the INEPT spectra, DEPT spectra afford multiplets which have the same appearance as in the normal spectrum. Morever, the acquisition of INEPT spectra requires knowledge of J values for setting delays between pulses and variations in the settings would result in different enhancements in the decoupled INEPT spectra. DEPT spectra, on the other hand, depend on variation in the length of the θ pulse and are independent of the exact delay times $\frac{1}{2}J$ between the pulses so that an error of $\pm 20\%$ in J values would still afford good DEPT spectra. By appropriate computer-assisted subtraction and addition of different DEPT spectra obtained at various pulse delays, it is possible to selectively observe CH_3, CH_2, or CH signals and by subtracting the sum of the CH_3, CH_2, and CH signals from the normal broad-band decoupled spectrum, one can obtain the signals for the quaternary carbon atoms. This is illustrated in Figure 5.24. An automatic version of DEPT, known as ADEPT, automatically supplies ^{13}C spectra according to their multiplicities.

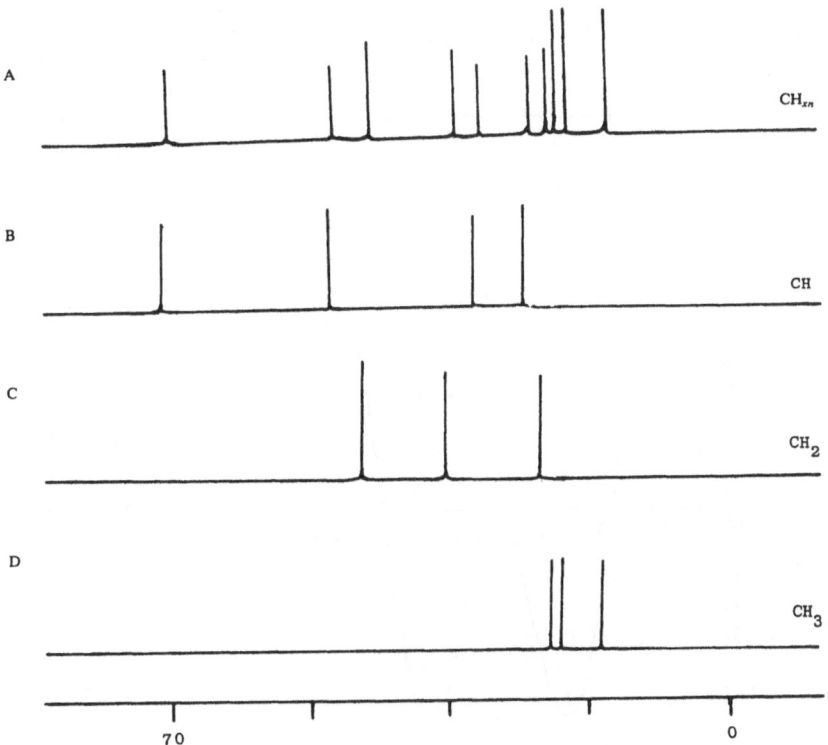

Figure 5.24. Edited ADEPT ^{13}C subspectra of menthol: (A) all protonated carbons; (B) methine carbons; (C) methylene carbons; and (D) methyl carbons.

$$\underset{P_1}{\frac{\pi}{2}[H,x]} - \frac{1}{2J} - \frac{\pi}{2}[C,x] \, \pi \, \underset{P_2}{[H,x,y]} - \frac{1}{2J} -$$

$$\underset{P_3}{\frac{\pi}{2}[H,x]} \underset{P_4}{\frac{\pi}{2}[H,\psi]} \, \pi \, [C] - \frac{1}{2J} - \text{acquire } ^{13}\text{C, decouple } ^{1}\text{H}$$

Figure 5.25. Pulse sequence for POMMIE

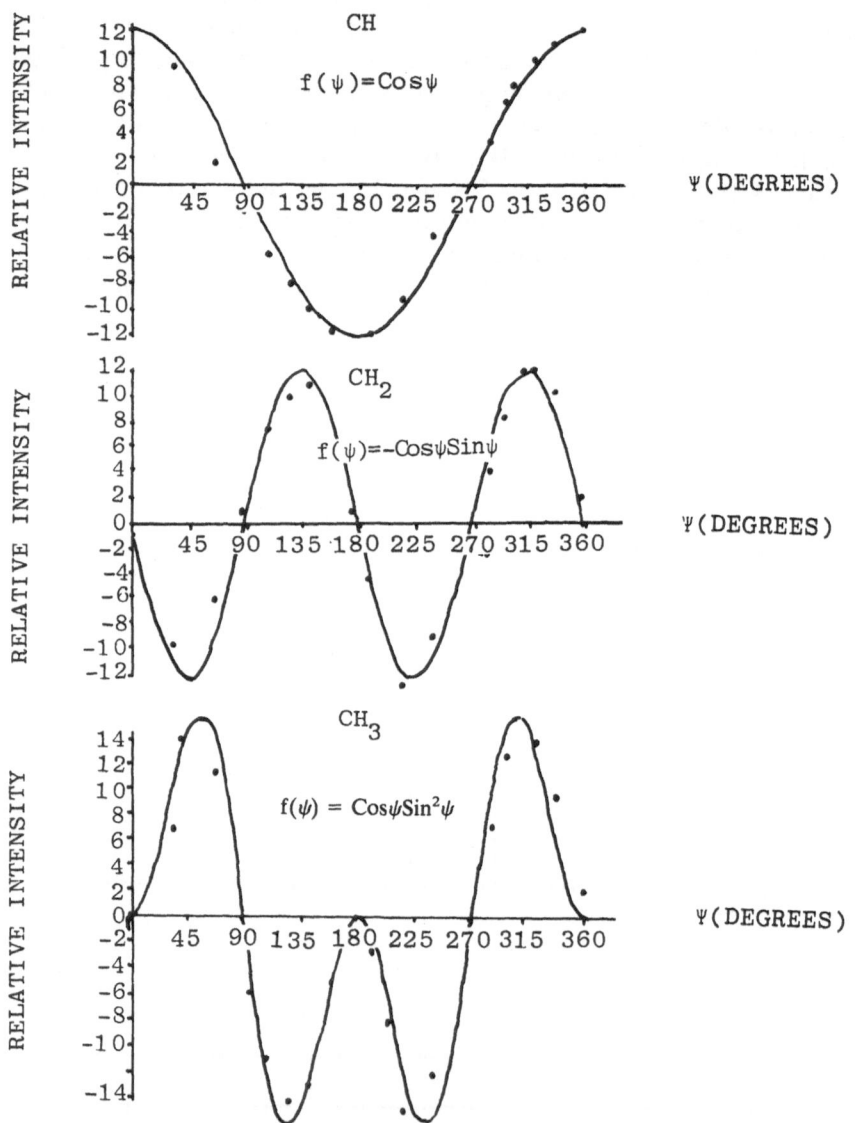

Figure 5.26. Variation of relative signal intensities for carbon resonances of CH, CH$_2$, and the low-field CH$_3$ group of 2-butanol with the variation of the phase angle of the multiple-quantum conversion pulse P$_4$. From [127].

5.1.3.2.1 Phase Oscillations to Maximize Editing (POMMIE). A recent modification* of DEPT employs a new pulse sequence shown in Figure 5.25. This pulse sequence eliminates most error signals arising from pulse in-homogeneity and removes ^{13}C signals arising from natural ^{13}C population differences. The variations of the relative signal intensities for CH, CH_2, and CH_3 carbon resonances as the phase angle of the multiple-quantum conversion pulse P_4 is varied is illustrated in Figure 5.26.

5.2 CARBON–CARBON CONNECTIVITY BY INADEQUATE SPECTRA

5.2.1 INADEQUATE Spectra

One of the most exciting developments in recent years is the ability to observe carbon–carbon couplings. A program developed recently, called the "*I*ncredible *N*atural *A*bundance *D*ouble *Qu*antum *T*ransfer *E*xperiment" (INADEQUATE) allows a study of carbon–carbon couplings by recording ^{13}C satellite spectra without perturbations from the stronger signals obtained from molecules containing a ^{13}C nucleus adjacent to a ^{12}C nucleus.

The observation of ^{13}C–^{13}C coupling constants can provide valuable information about the type of coupled carbons since the magnitude of the couplings is dependent on the state of hybridization of the coupled carbons. Furthermore, ^{13}C–^{13}C couplings can establish carbon–carbon connectivities and hence help in structure elucidation. The chances of finding one ^{13}C atom at a particular position in a molecule are low (1.1%) but the chance of finding two adjacent ^{13}C nuclei are far lower (1.1% × 1.1% = 0.012% or one molecule in ten thousand). The ^{13}C–^{13}C couplings therefore appear only as very weak satellite peaks which are masked by the much more intense peaks of molecules with single ^{13}C atoms. Clearly, if one could develop a pulse sequence which would suppress the undesired intense signals of single ^{13}C nuclei, it would allow one to observe the weaker satellite signals of two adjacent ^{13}C nuclei. This can be done by creating a 90° phase difference between the transverse magnetizations of the intense ^{13}C signal and the weaker ^{13}C–^{13}C satellite signals. This is achieved by generating a *double-quantum coherence* by incorporating $90_x^\circ – \tau – 90_x^\circ$ pulses in the pulse sequence used for the purpose (Figure 5.27). The protons are subjected to broad-band decoupling through-out the experiment.

Application of a 90_x° pulse to the ^{13}C nuclei aligns them along the y'-axis [Figure 5.27(a')]. During the subsequent $1/(4J)$ delay period, ^{13}C–^{13}C coupling causes the magnetization vectors of the coupled ^{13}C nuclei to fan out,

---

* J.M. Bulsing, W.M. Brooks, J. Field, and D.M. Doddrell, Polarisation transfer via an intermediate multiple quantum state of maximum order, *J. Magn. Resonance* **56**, 167 (1984).

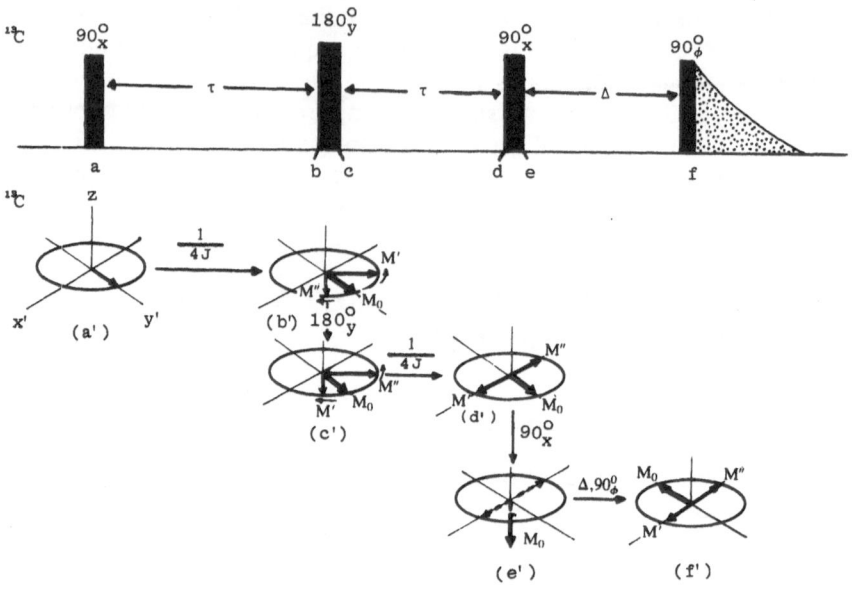

Figure 5.27. The INADEQUATE pulse sequence. Since at the end of the pulse sequence, the intense magnetization M_0 of the single ^{13}C nucleus is 90° out of phase with respect to the weaker magnetizations, M' and M'', of the adjacent ^{13}C nuclei, the latter can be detected without interference from the former. As M' and M'' are out of phase in (f), an additional sequence $1/(4J)-180_x^\circ-1/(4J)$ can be applied to refocus them.

while the magnetization M_0 of the more abundant uncoupled ^{13}C nucleus remains static during this period. At the end of this $1/(4J)$ delay period a 90° phase difference exists between the two moving vectors, M' and M'' [Figure 5.27(b')]. A nonselective 180_y° pulse is next applied which serves to refocus chemical shift and inhomogeneity contributions but does not change the direction of rotation of the moving magnetization vectors [Figure 5.27(c')] so that at the end of the next $1/(4J)$ delay period, they have acquired a phase difference of 180° [Figure 5.27(d')]. Throughout this period the magnetization vector M_0 for the intense uncoupled ^{13}C nucleus has remained static along the y'-axis. The second 90_x° pulse is now applied which produces coherence between $\alpha\alpha$ and $\beta\beta$ states, i.e., generates double quantum coherence, and changes the direction of M_0 so that it becomes aligned with the $-z$-axis [Figure 5.27(e')]. After a delay Δ ($\sim 10\ \mu s$), a 90° pulse of a variable phase (90_ϕ°) is applied which transforms the transverse magnetization of ^{13}C satellites to detectable magnetization [Figure 5.27(f')]. The doublet components have opposite phases which results in the $^{13}C-^{13}C$ doublet signals also having an anti-phase relationship. This can be corrected by incorporating an additional spin-echo sequence ($\tau - 180_x^\circ - \tau$, where $\tau = \frac{1}{4}J_{CC}$) before the 90_ϕ° pulse. It is important that the detector phase is properly adjusted in order to allow the ^{13}C–^{13}C magnetization to be recorded. The relative phase shifts of the main signal, satellites, and detector are given in Table 5.1.

The delay τ between pulses is dependent on the coupling constant J_{CC}, and its value may be optimized to correspond to the C–C couplings to be observed. Some typical J_{CC} values are presented in Table 5.2. Since there is only a 1×10^{-4} chance of finding two ^{13}C nuclei linked to each other (and it is only these nuclei with which we are concerned when determining C–C couplings) the procedure suffers from the fact that it requires large sample quantities (0.5–1 g) to record ^{13}C satellite spectra within a reasonable time, even on the more powerful instruments. Normally the broad-band decoupled ^{13}C spectra and the INADEQUATE spectra are recorded under identical conditions. Figure 5.28 (upper part) shows two peaks corresponding to carbons A and B obtained in a section of the broad-band decoupled spectrum of a compound. In the lower section of the same figure the two ^{13}C satellite doublets arising from the coupling of carbon A with carbon B are seen. Since the magnitude of the couplings of the two carbons are the same (i.e., $J_{AB} = J_{BA}$), it can easily be verified which carbon atoms are coupled to which others in a spectrum by looking for *pairs* of equally coupled doublets in the INADEQUATE spectrum. It may be noted that the calculated chemical shifts of carbons obtained from the INADEQUATE spectrum will differ slightly from the chemical shifts obtained from the broad-band decoupled spectrum since in the former we are looking at two ^{13}C nuclei adjacent to one another while in the latter case we are looking at one ^{13}C nucleus adjacent to a ^{12}C nucleus. The differences in chemical shifts $\Delta\delta(A_B)$ and $\Delta\delta(B_A)$ are due to this isotope effect. The use of INADEQUATE in structural assignments is shown for *n*-octanol in Figure 5.29. There are eight signals for the eight carbon atoms in the broad-band decoupled spectrum. In the corresponding INADEQUATE spectrum shown in the top trace, each signal is split into a

Table 5.1. Relative phase shifts for main signal, satellites, and detector

ϕ^a	S_0^b	S^c	ψ^d
$+x$	$-y$	$+x$	$+x$
$+y$	$+x$	$-y$	$-y$
$-x$	$+y$	$-x$	$-x$
$-y$	$-x$	$+y$	$+y$

^{a} Phase shifts of read pulse.
^{b} Main signal.
^{c} Satellite magnetization.
^{d} Selective detection of ^{13}C–^{13}C satellite by the INADEQUATE experiment.

Table 5.2. One-bond $^{13}C-^{13}C$ spin–spin coupling constants

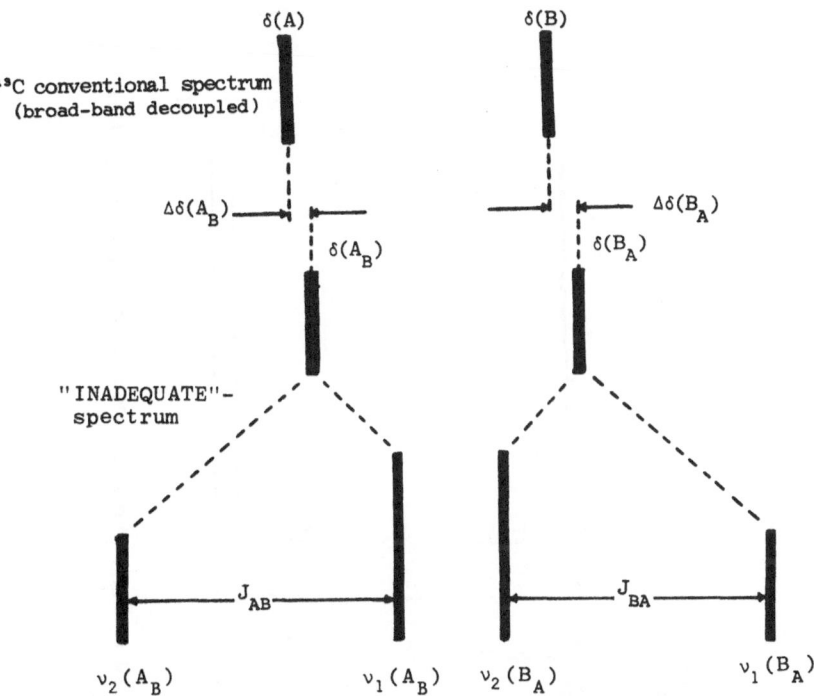

Figure 5.28. Spectral parameters in an INADEQUATE spectrum: $\delta(A_B)$ = chemical shift of carbon A in molecules with a neighboring ^{13}C nucleus B; $v_n(A_B)$ = resonance frequency of the nth satellite of carbon A in the doubly labeled isotopomer AB; $\Delta\delta(A_B)$ = isotope shift of carbon A induced by the neighboring ^{13}C nucleus B. From [96].

doublet but with one of the two peaks in each doublet appearing inverted. The expanded regions for carbons 1, 2, and 6 (Figure 5.30) show the coupling constant measurements. Thus the fact that the coupling constant obtained from the splitting of the C-1 signal (which is $J_{1,2}$) is equal to the coupling constant obtained from the splitting of the C-2 signal ($J_{2,1}$) establishes that it is these two carbons which are coupled with (and hence adjacent to) one another. When looking at carbon 2 in the INADEQUATE spectrum, it is seen that there is only a single inverted peak in the doublet but there are two upright peaks in the other half of the doublet. The explanation is that carbon 2 is coupled to two different carbon atoms (1 and 3) and, therefore, two different coupling constants will be found here ($J_{2,1}$ and $J_{2,3}$). By similar reasoning the entire carbon framework can be worked out. The analysis of INADEQUATE spectra is best carried out using a computer. A combination of INADEQUATE with INEPT has been developed to achieve increased sensitivity. A two-dimensional version of the INADEQUATE experiment has been developed, which is described in Section 5.6.1.

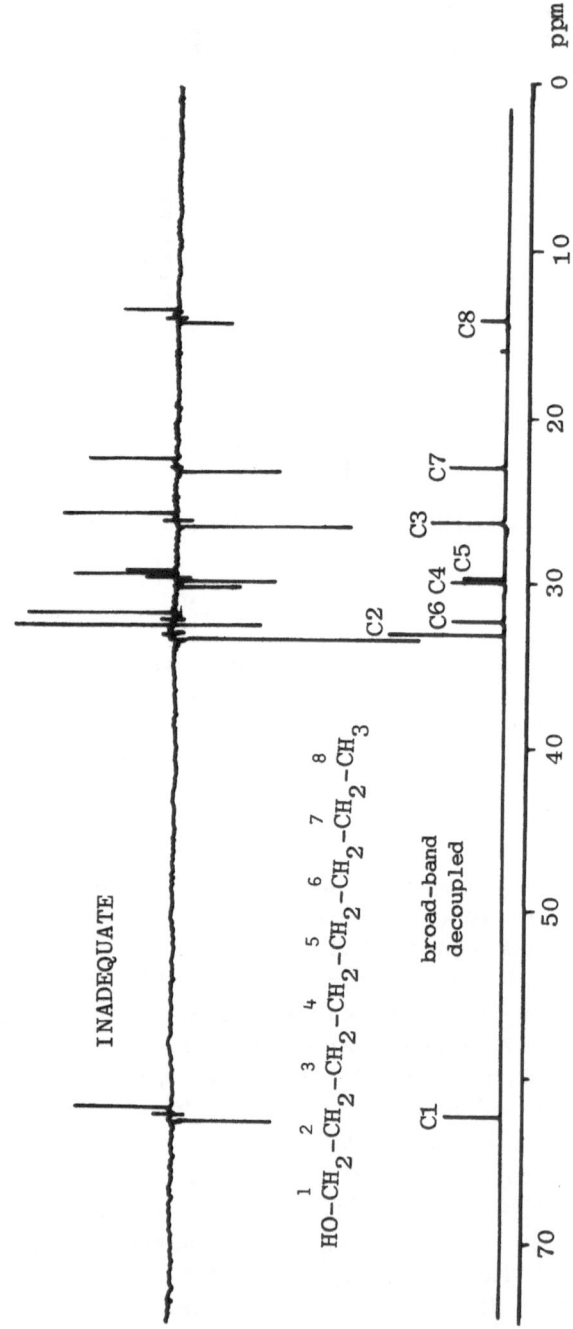

Figure 5.29. *Bottom:* ^{1}H-decoupled ^{13}C NMR spectrum (50.3 MHz) of *n*-octanol (90%). *Top:* INADEQUATE spectrum. From [96].

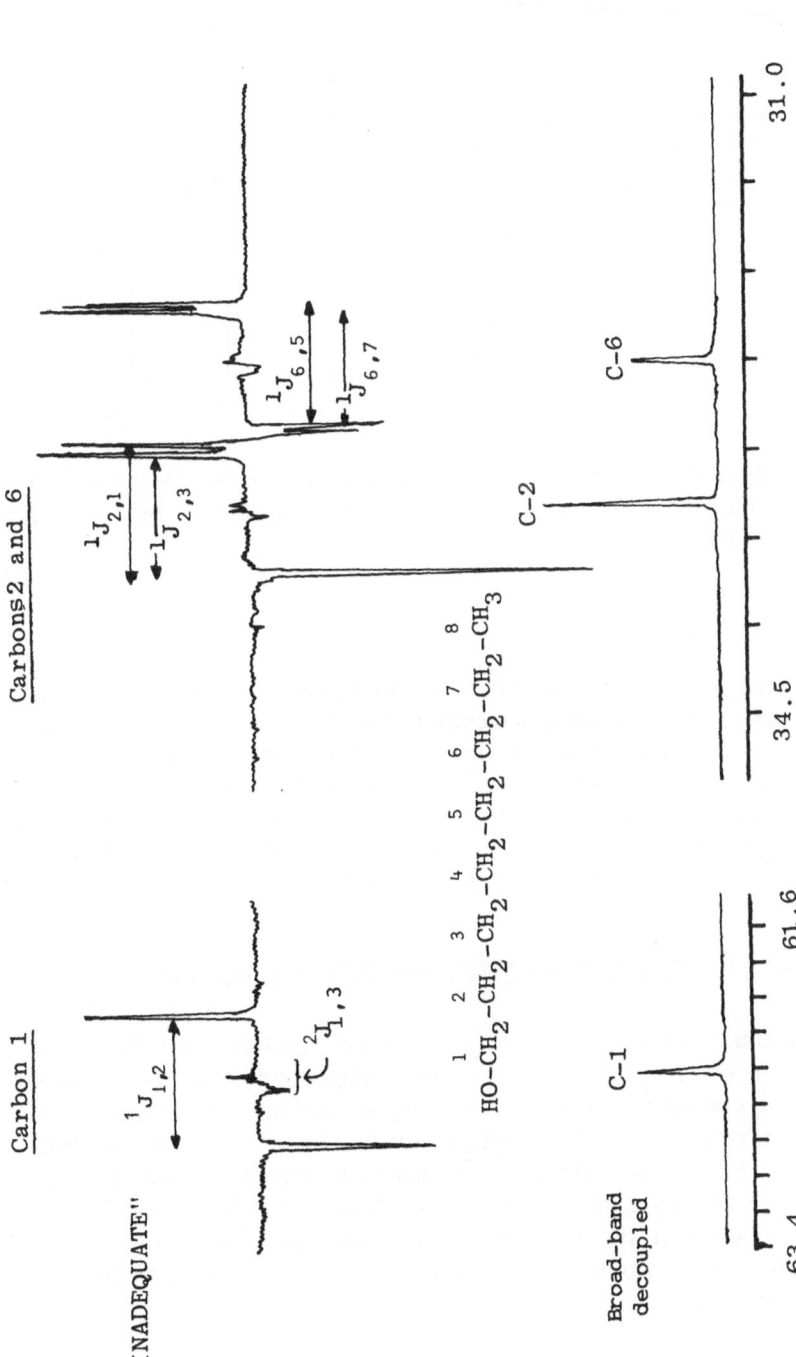

Figure 5.30. Expanded regions of the conventional and IN-ADEQUATE spectra of *n*-octanol shown in Figure 5.29. *Left*: Carbon 1 with its two satellites spaced by $^1J_{1,2}$; in the center, $^2J_{1,3}$ is just visible. *Right*: Carbons 2 and 6 with all satellites. The advantage of displaying the INADEQUATE spectrum with opposite phases for corresponding doublet peaks is evident. The satellites can be assigned to the corresponding carbons without ambiguity. From [96].

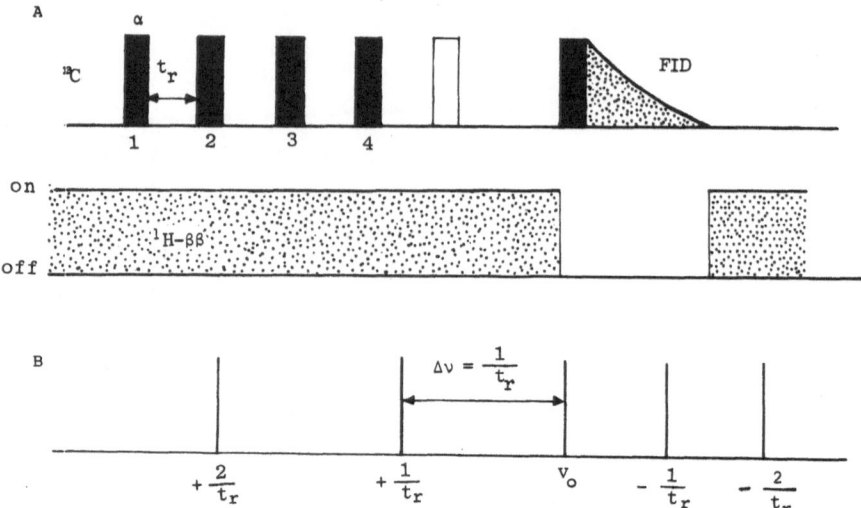

Figure 5.31. (A) Pulse sequence used in the DANTE experiment. The small pulse spacings t_r result in the frequency spectrum in (B); ^{13}C resonances can be selectively excited by this procedure.

5.2.2 DANTE Spectra

As the spectral window in FT-NMR is wide on account of the broad regions excited by pulses, it is difficult to achieve selective excitation of individual ^{13}C resonances. A tailored excitation procedure has been developed which involves application of a "pulse train," i.e., a series of pulses with very small pulse angles α and small pulse delays t_r (Figure 5.31). The series of small pulses can be regarded as selective pulse transmitters, and by correct choice of α and t_r, it is possible to selectively excite ^{13}C resonances.

5.3 TWO-DIMENSIONAL NMR SPECTROSCOPY

A serious problem often faced in "one-dimensional" spectroscopy is that there can be serious overlapping of peaks, making assignments of chemical shifts and coupling constants difficult. In one-dimensional spectroscopy, a short high-frequency pulse is used to excite all nuclei of one type. The resulting transverse magnetization (which is obtained with respect to only one time variable, say t_2) contains information relating both to the chemical shifts and coupling constants of the nuclei being irradiated and represents a time-domain signal, $S(t_2)$. Fourier transformation converts it into a frequency-domain spectrum $S(\omega)$:

$$S(t_2) \xrightarrow{\text{F.T.}} S(\omega)$$

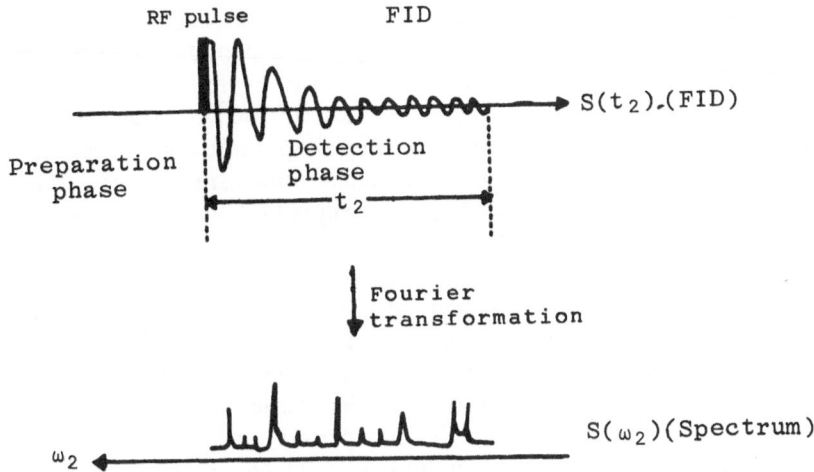

Figure 5.32. One-dimensional NMR spectrum.

where t_2 is the time for signal acquisition. The production of transverse magnetization may be called the *preparation phase* and the recording of the FID the *detection phase* (Figure 5.32). The idea of two-dimensional NMR spectroscopy was introduced by Jeener in 1971 but it was only several years later that it gained popularity. Jeener's idea was to introduce a second time period (t_1) between the preparation and detection periods, which is called the *evolution period*. During this evolution period the NMR receiver is not active but the nuclear motions are made to undergo some prescribed motions which eventually influence the signal $S(t_2)$.

The time axis in the two-dimensional experiment may therefore be divided into three (sometimes four) periods (Figure 5.33). The preparation period usually consists of a delay time which is long compared to the longest T_1 to allow the nuclei to reach thermal equilibrium and to establish reproducible starting conditions. The preparation period may also consist of a series of pulses to align the nuclei in some desired manner. It is followed by one or more RF pulses which disturb the equilibrium polarization state established in the

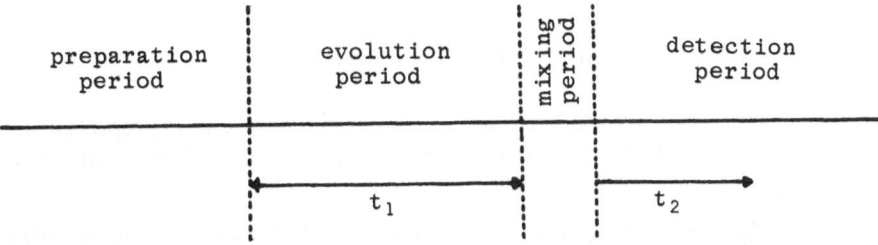

Figure 5.33. Subdivision of the time axis in a two-dimensional experiment.

preparation period. The NMR experiment gains a second dimension by varying the duration of the subsequent evolution period, t_1 which causes a periodic charge in signal amplitudes. During this evolution period, the x, y, and z components of the magnetization are allowed to evolve under the forces acting on the nuclei, which include spin–spin interactions between the coupled nuclei. In many experiments, e.g., heteronuclear shift correlation and J-spectroscopy, the evolution period is interrupted midway by the application of a nucleus-selective 180° pulse.*

At the end of the evolution period, t_1, the existing magnetization is recorded in the form of an FID during the detection period, t_2. The evolution period t_1 is increased ("incremented") gradually and for each t_1, a separate FID is recorded (Figure 5.34). One thus obtains a series of FIDs, each of which may be considered as a function of both time periods, i.e., $S(t_1, t_2)$. Fourier transformation of each FID after the time period t_2 affords a corresponding spectrum of lines (in the frequency domain, ω_2) which may be regarded as a function of both time t_1 and the frequency ω_2. This may be represented as:

$$S(t_1, t_2) \xrightarrow{\text{F.T. over } t_2} S(t_1, \omega_2)$$

The lines in the spectra obtained as a result of the Fourier transformations occur at the same frequencies as in the normal one-dimensional spectra but their intensities vary with changes in t_1. Thus if we arrange the same line in each spectrum in horizontal rows ("arrays") one behind the other (but slightly shifted to the right to allow for clarity of viewing), then it will be seen that each line decreases in intensity from a maximum value through zero to a minimum (negative) value, and then grows back again to a positive value. Thus if we look "up" each row, the array of lines resembles an FID. This is shown in Figure 5.35. The intensity variation of a single line obtained by lining up 800 spectra is shown in Figure 5.36. The large number of spectra obtained after Fourier transformations at different t_1 values can be similarly arranged in parallel arrays one behind the other so that when each peak in the spectrum is viewed in all the arrays, its intensity variations over different t_1 values has the appearance of an FID running at a right angle to the spectra in the frequency domain ω_2. The new data matrix obtained by aligning the spectra in parallel arrays $|S(t_1, \omega_2)|$ is subjected to a column-wise second Fourier transformation:

$$S(t_1, \omega_2) \xrightarrow{\text{F.T. over } t_1} S(\omega_1, \omega_2)$$

This results in a two-dimensional spectrum (with ω_1 and ω_2 representing the two axes), in which the peaks in the spectrum rise up as hills in the third

---

* The evolution period may be followed by a *mixing period* which consists of applying a single nonselective pulse (as in homonuclear shift correlated spectra or homonuclear multiple-quantum spectra) or by a number of nucleus-selective pulses (as in heteronuclear shift correlated spectra).

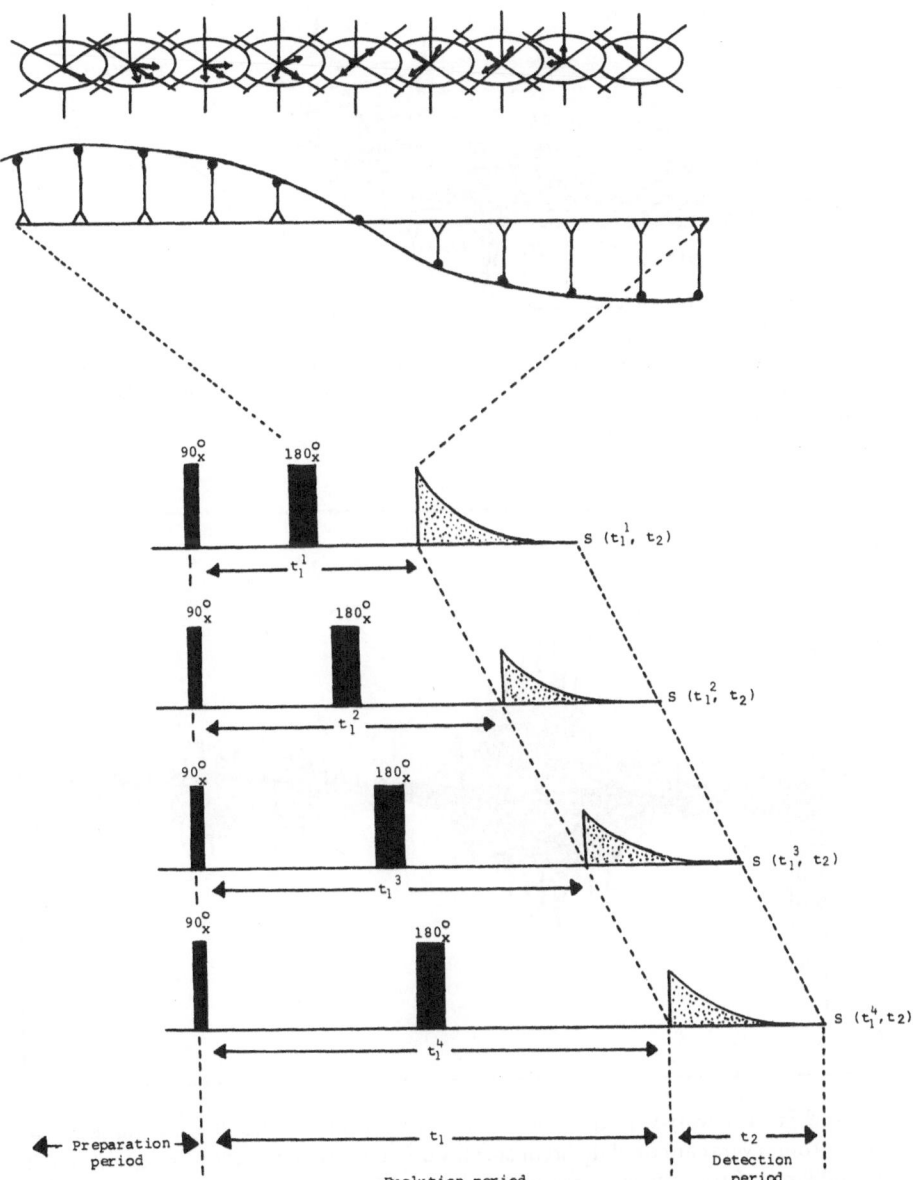

Figure 5.34. 2D J-Resolved NMR spectroscopy; the evolution time t_1 is gradually increased while the detection time t_2 is kept constant.

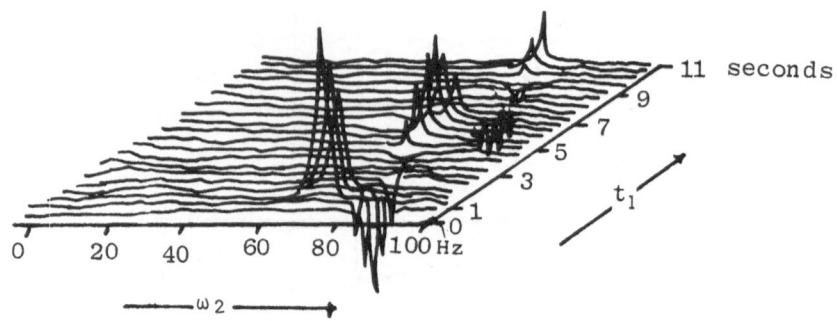

Figure 5.35. Evolution of peak intensities in the NMR spectrum as a function of t_1. Fourier transformation of the cross section parallel to the t_1 axis would afford the two-dimensional spectrum.

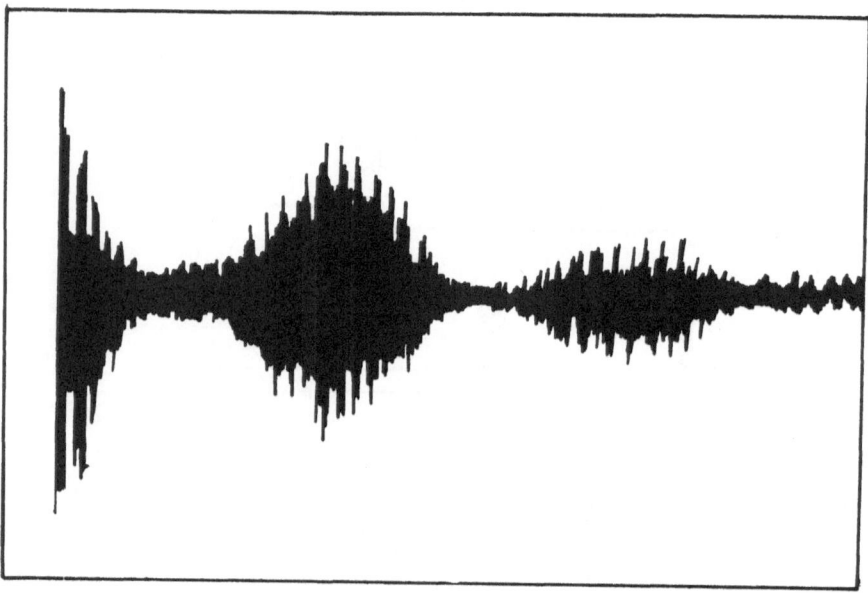

Figure 5.36. Intensity variations observed when a *single* spectral line is arranged in an array from several hundred different spectra obtained at increasing t_1 values. Note the resemblance of the pattern obtained to a normal FID. Fourier transformation of this pattern would afford a 2D spectrum.

dimension above the $\omega_1\omega_2$ plane. The position of these peaks in both coordinates ω_1 and ω_2 depends on the conditions to which the nuclear spins are subjected during the evolution and detection phases. The ω_1 axis in the 2D spectrum contains information about the coupling interactions with other nuclei during the evolution period, while the ω_2 axis contains information about the precession frequencies (and hence the chemical shifts) of the nuclei (Figure 5.37).

It is important to appreciate that the coupling information is retained in the detection period, in spite of the fact that broad-band decoupling is on during detection. This is so because coupling leads to a periodic change in the amplitude of transverse ^{13}C magnetization during the evolution period. This is exemplified by examining the spin evolution for CH, CH_2, and CH_3 fragments during time t_1 (Figure 5.38). Figure 5.38A shows the behavior of the magnetization vectors of a doublet with respect to time after the application of a 90_x° pulse. Below the magnetization vectors is shown the modulation of the signal intensity over time t_1. Since it is the net magnetization along the y axis which is detected as a signal, this magnetization will vary depending on the position

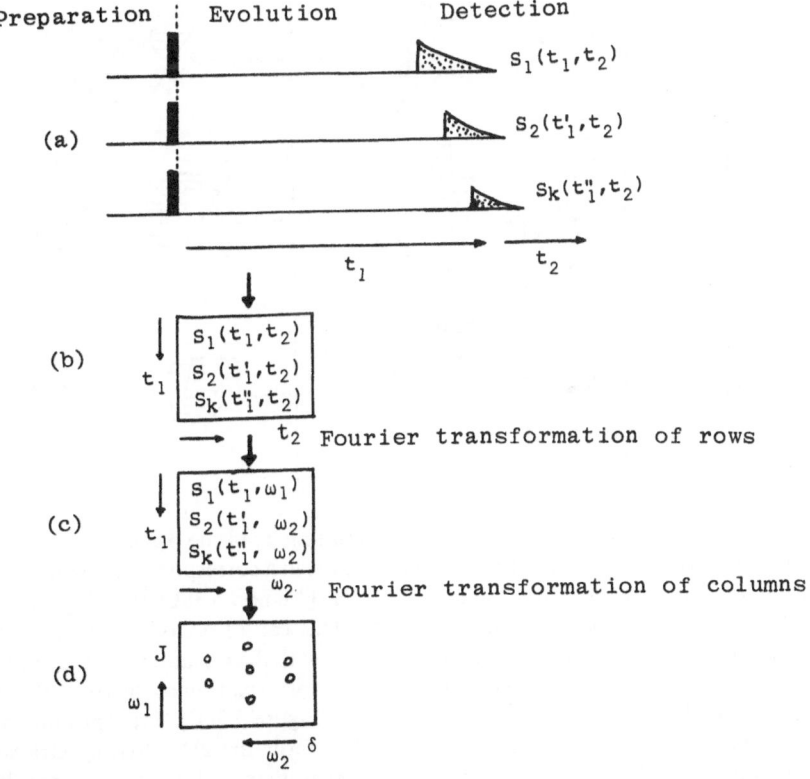

Figure 5.37. A schematic presentation of the mechanics of obtaining a 2D-NMR spectrum: (a) FIDs are recorded at differing values of t_1; (b) the FIDs are schematically shown as signals in t_1/t_2 dimensions; (c) Fourier transformation of the rows of FIDs afford a series of spectra in t_1 and ω_2 dimensions; (d) transposition and Fourier transformation of columns of FIDs afford the 2D-NMR spectrum with signals in the ω_1 and ω_2 dimensions. The chemical shifts lie along one axis while the coupling constants lie along the other axis.

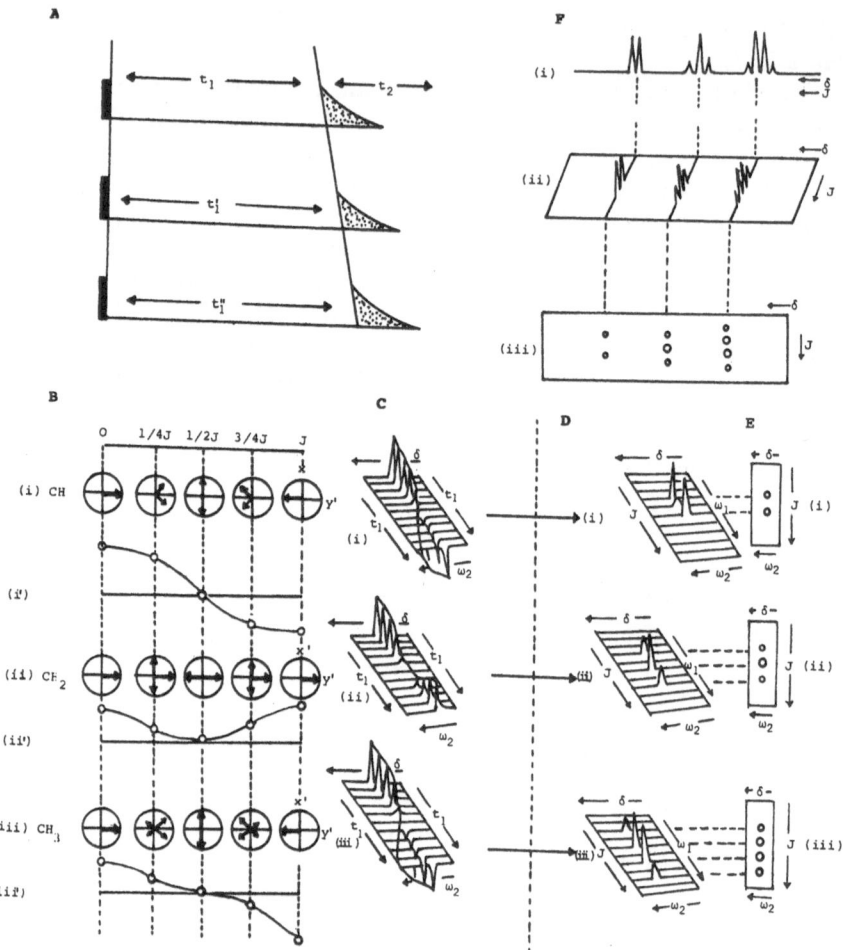

Figure 5.38. The gradual change of the evolution time t_1 in the basic pulse sequence used in 2D NMR spectroscopy (Figure 5.38A) results in a corresponding change in the position of the magnetization vectors for CH [Figure 5.38B(i)], CH_2 [Figure 5.38B(ii)], and CH_3 [Figure 5.38B(iii)] groups. This causes periodic changes in the amplitude of signals for CH, CH_2, and CH_3 groups with the variation of the evolution time, t_1. [Figure 5.38B(i'), (ii'), and (iii')]. Fourier transformation of the rows of FIDs gives a series of spectra in t_1 and ω_2 dimensions (Figure 5.38C). Transposition and Fourier transformation of the columns of FIDs affords the 2D NMR spectra with signals in the ω_1 and ω_2 dimensions. These are shown as stacked plots in Figure 5.38D (i), (ii), and (iii). They may also be presented as an overhead view of the peaks in the form of contour plots (Figure 5.38E (i), (ii), and (iii). Figure 5.38F(i) shows a 1D spectrum with the information of chemical shifts and coupling constants both contained on the same axis. Figure 5.38F (ii) and (iii) show the 2D J-resolved spectra with the multiplets having been effectively rotated about their chemical shifts so that the chemical shifts lie on one axis while the coupling constants lie on the other axis.

of the magnetization vectors at the end of each evolution period, t_1. As the values of t_1 are varied, one "catches" the magnetization vectors in their various positions, resulting in a modulation of the signal amplitudes with changing t_1 values. The first series of Fourier transformations of the corresponding FIDs would afford the spectral peaks in which the horizontal axis contains the chemical shift information while the vertical axis defines the evolution period t_1 [Figure 5.38D, diagram (i)]. A second Fourier transformation, perpendicular to the time axis t_1, is then carried out which results in a two-dimensional plot [Figure 5.38E, diagram (i)] in which the horizontal axis still contains the chemical shift information but the vertical axis now contains the coupling information. Thus in a CH group, Fourier transformation of the modulating amplitudes would afford a doublet with lines at $\pm J/2$.

In the case of a triplet (Figure 5.38B), an analogous situation prevails except that there are now three magnetization vectors, and the time-domain functions of the triplet are different since the vectors in the triplet oscillate at twice the frequency of the doublet. Fourier transformation of the phase-modulated signals now affords a triplet with lines on the ω_1 axis at $+J$, 0, and $-J$. Similarly the four oscillating magnetization vectors of the CH_3 group would give rise to the two-dimensional plots shown in Figure 5.38D, diagram (iii). It is clear from this example that the information about the scalar coupling constant is given by the modulation of signals which results from the periodic perturbation of the system during the evolution period t_1. The t_1 dimension thus contains coupling information while the t_2 dimension contains the chemical shift information.

Two-dimensional NMR spectra may be represented in a number of ways. One method is to arrange them in a series of "stacked plots" as shown in Figure 5.39 for the quartet of the methylene group of ethanol. Another method employs contour plots.

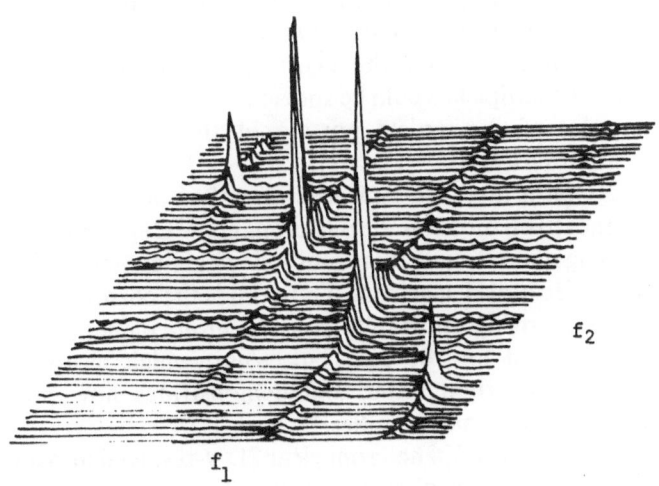

Figure 5.39. A quartet arranged in a 2D array of peaks.

There are two broad classes of 2D-NMR spectra. In the *J-resolved* spectra, the peaks in a spectrum are so spread that the chemical shifts are found in one dimension while the coupling constants occur in the second dimension. The second class covers *correlated* spectra which are basically correlation diagrams between two spectra. Thus the ^{1}H-NMR spectrum may lie along one axis while the ^{13}C-NMR spectrum lies on the other axis and the contours obtained at the cross-points allow one to correlate which protons in the ^{1}H-NMR spectrum are attached to which carbon atoms in the ^{13}C-NMR spectrum.

5.3.1   2D *J*-Resolved Spectroscopy

In the one-dimensional NMR experiment, the spectrum obtained contains information relating both to the chemical shifts of the nuclei being observed and to the couplings of these nuclei with other neighboring nuclei. In 2D *J*-resolved spectroscopy, the chemical shifts and coupling constants lie on two different axes which allows one to unambiguously assign both δ and J. Overlapping multiplets of nuclei with different chemical shifts can thus be easily resolved in the two dimensions, affording chemists a very powerful method for interpreting complex spectra.

5.3.1.1   Heteronuclear 2D *J*-Resolved Spectra

In heteronuclear 2D *J*-resolved spectra, the ^{13}C chemical shifts are recorded along the ω_2 axis while the CH coupling information is present along the ω_1 axis. This separation of chemical shifts from coupling constants provides a very useful method for determining the nature and number of nuclei coupled to one another. It may be noted that since the ω_2 axis contains *only* the chemical shift information, a multiplet in the normal 1D spectrum would afford a single peak in the *J*-resolved "projection" spectrum. This is illustrated in Figure 5.40. The upper trace shows the normal 1D spectrum of D-sucrose recorded on a 400-MHz instrument. The lower trace is the simplified *J*-resolved "projection" spectrum, which corresponds to the proton-decoupled ^{1}H-spectrum. The multiplets would be spread along the ω_2 axis in the 2D plot. The contour plot of the upfield portion of the *J*-resolved spectrum is presented in Figure 5.41.

Another example is the ^{13}C-NMR spectrum of 5α-androstane presented in Figure 5.42. Trace (i) on the left is the off-resonance ^{13}C spectrum which shows many overlapping multiplets, making multiplicity assignments difficult. Trace (ii) is the proton-decoupled spectrum. Trace (iii) is the corresponding stacked plot with the peaks rising up as hills from the $\omega_1\omega_2$ plane. Trace (iv) is the contour plot obtained for the same spectrum, which clearly shows the multiplicity of each carbon atom, and demonstrates how powerful this technique is for assigning multiplicities.

The pulse sequences used in heteronuclear 2D *J*-resolved measurement are similar to the APT sequence discussed earlier. The main difference from the

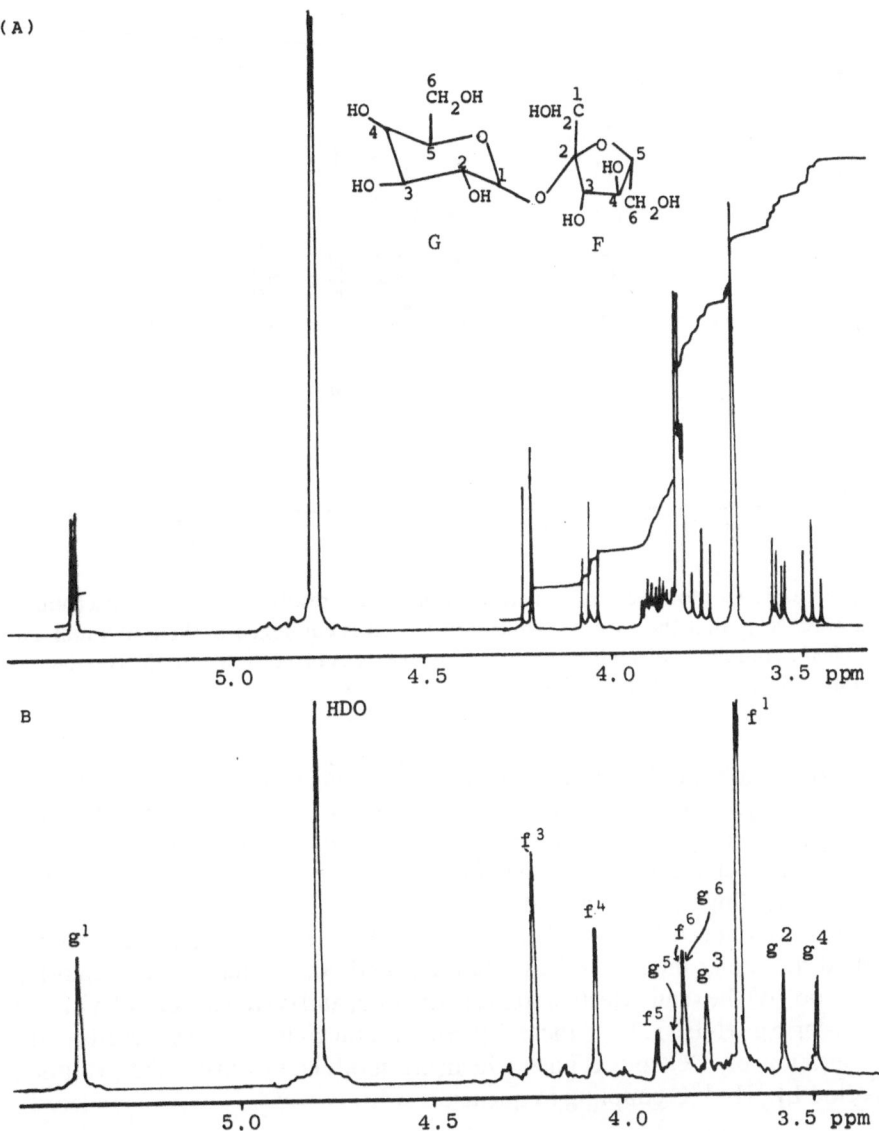

Figure 5.40. (A) Normal 1D spectrum of sucrose; (B) J-resolved projection spectrum, showing only singlets instead of multiplets.

APT experiment is that a large number of FIDs (e.g., 800), are recorded at different values of t_1 which is increased gradually (for instance, from zero to one second) over the 800 spectral recordings. Fourier transformation affords a corresponding number of spectra in which the intensity of each ^{13}C spectral line varies (or is "modulated") sinusoidally, and the arrangement of the spectra

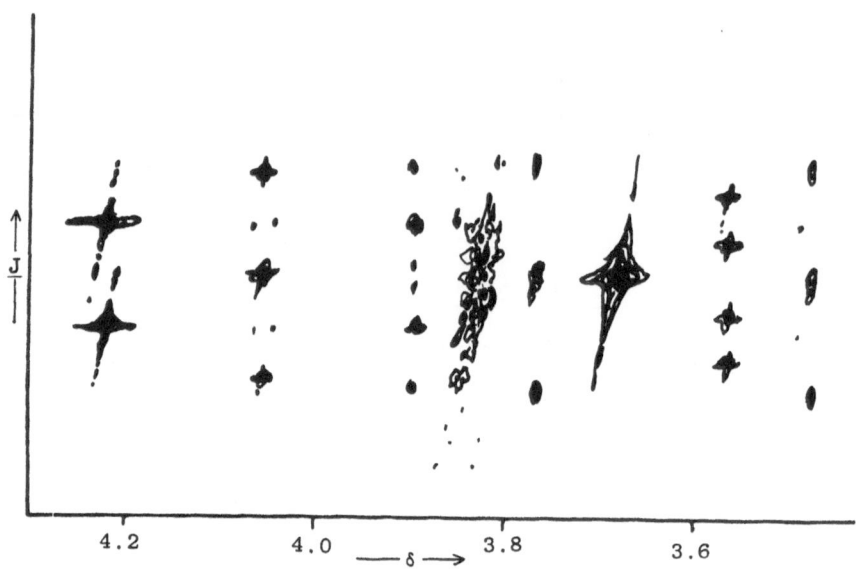

Figure 5.41. Contour plot of upfield portion of a J-resolved ^{1}H-NMR spectrum of sucrose. Note that the chemical shift information is contained in the horizontal axis while the coupling information is in the vertical axis. A projection onto the vertical axis would afford the single peaks found in Figure 5.40B.

in parallel arrays affords rows of lines of varying intensities (Figure 5.38). These arrays can be Fourier transformed to afford the multiplet structures along the ω_2 axis. The maximum value of the coupling constant which can be observed is determined by the spectral width in the ω_2 dimension, and is equal to the sampling rate per second.

There are several types of heteronuclear 2D J-resolved experiments: (i) the gated decoupler method, (ii) the spin-flip method, (iii) the selective spin-flip method, (iv) the semi-selective spin-flip method, and (v) the use of polarization transfer, e.g., INEPT. In all these experiments, the pulse sequence results in the generation of ^{13}C spin echoes which are modulated during the evolution period by ^{1}H$-^{13}$C coupling frequencies.

5.3.1.1.1 The Gated Decoupler Method.   This is the simplest and the most widely used of the presently available methods for heteronuclear 2D J-resolved spectroscopy. The pulse sequence employed and its effect on ^{13}C magnetization of a CH group is shown in Figure 5.43. Nuclear Overhauser enhancement of the signal is provided during the preparation period by ^{1}H decoupling. This is followed by a 90° ^{13}C pulse which serves to bring the magnetization of the ^{13}C nuclei from the z-axis to the y'-axis. Since proton decoupling is applied during the first half of the evolution period, t_1, the two components of the ^{13}C doublet do not separate during this period, but rotate at the same angular velocity $\Omega/(2\pi)$ [Figure 5.43(b)]. At the midpoint of the

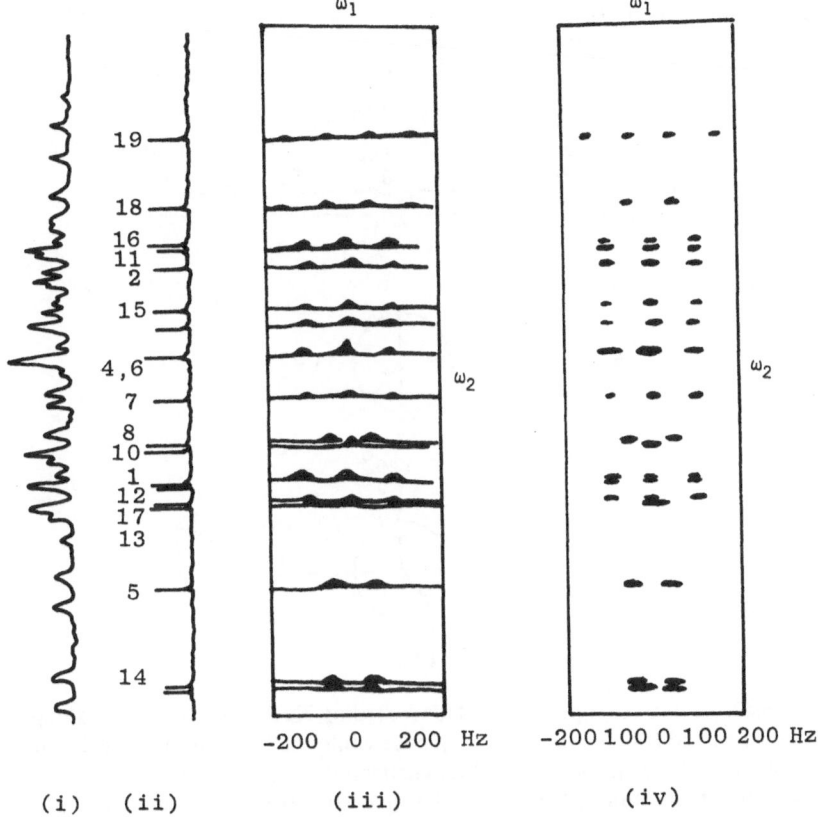

Figure 5.42. (i) Normal 1D proton-coupled ^{13}C-spectrum of 5α-androstane. (ii) 1D proton-decoupled spectrum of 5α-androstane. (iii) Stacked trace plot of 2D J-resolved spectrum of 5α-androstane obtained with gated decoupler sequence. (iv) Contour plot of 2D J-resolved spectrum of 5α-androstane obtained with gated decoupler sequence.

evolution period (i.e., at $t_1/2$) a 180°_x pulse is applied to the ^{13}C nuclei along the x'-axis, which causes the chemical shift vector to invert across the x'-axis and adopt a mirror image position [Figure 5.43(c)]. The proton decoupler is now turned off so that the coupling effects come into play, with the result that the two magnetization vectors associated with the doublet components separate and precess away with angular velocities of $\Omega \pm \pi J$ [Figure 5.43(d)]. The separation between the two vectors at any point of time is determined by the magnitude of the coupling constant, J_{CH}, and after time t_1 they will not refocus but will differ in phase by $\pm \pi J t_1/2$. Since proton decoupling is switched on at time t_1, the divergence does not continue and both vectors precess with a uniform angular velocity Ω. During acquisition with the broad band decoupling turned on, the vector sum of the two components is observed. This will depend on the extent of divergence between the components of the

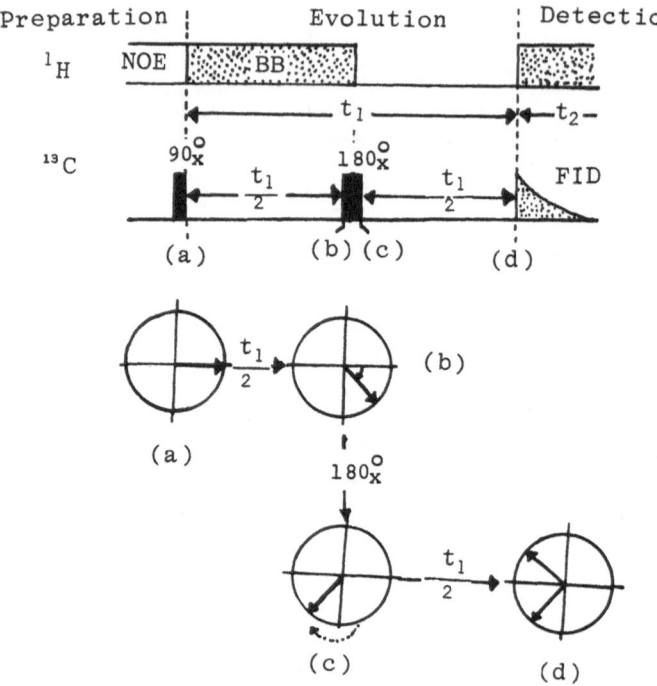

Figure 5.43. Gated decoupler method for recording 2D J-resolved ^{13}C-NMR spectra. The upper portion shows the pulse sequence while the lower portion represents the effect of the pulse sequence on ^{13}C magnetization of a CH group; ^{1}H decoupling during the preparation period provides nuclear Overhauser enhancement. The position of the two magnetization vectors in (d) will be dependent on t_1; the sum of the two vectors which is detected will depend both on t_1 and the magnitude of the coupling constant, J. The signals are then said to be "J-modulated."

doublet, which in turn depends on the magnitude of the corresponding coupling constant. A number of spectra obtained at gradually increasing values of t_1 therefore affords modulation of the signal amplitudes with the couplings ("J-modulation"). Fourier transformation gives the coupling constants along the ω_1 axis and the chemical shifts along the ω_2 axis. Since couplings are present only during half the evolution period, they are recorded with a 50% reduction in magnitude. This can make small couplings difficult to resolve.

5.3.1.1.2 The Spin-Flip Method. In an earlier section it was shown that if a 180°_x pulse is applied to both ^{1}H and ^{13}C nuclei at time τ after a 90° ^{13}C pulse, then during the subsequent period the two ^{13}C magnetization vectors of a CH system do not refocus but continue to diverge [see Figure 5.6(c)]. The spin-flip method utilizes the same principle. The pulse sequence and its effect on ^{13}C magnetization vectors of a ^{13}C–^{1}H system are shown in Figure 5.44. A 90°

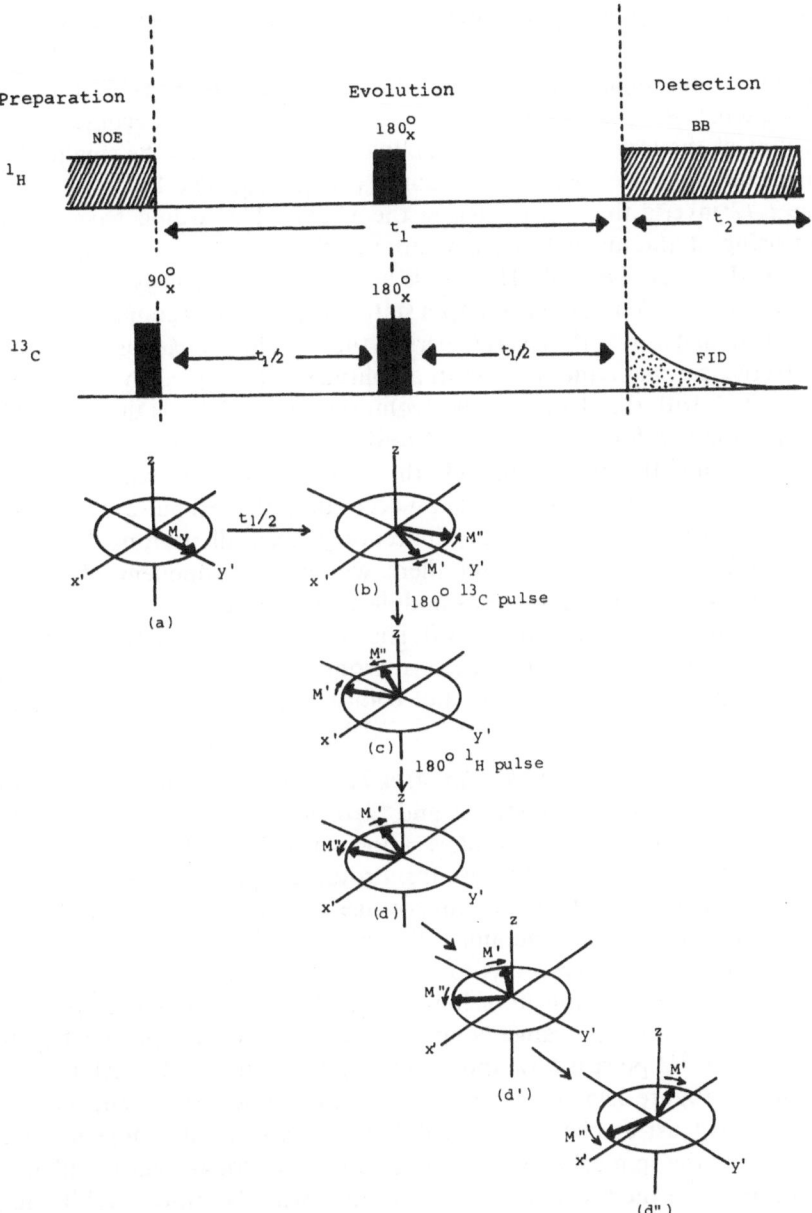

Figure 5.44. The spin-flip method for recording 2D J-resolved ^{13}C-NMR spectra. The application of the 180° ^{13}C pulse as well as 180° ^{1}H pulse in the center of the evolution period causes the two ^{13}C magnetization vectors of the CH group to adopt mirror image positions across the x'-axis, as in (c), as well as to change the direction of rotation, as indicated in (d). During the second half of the evolution period they diverge from each other (as shown in (d'), (d''), etc. The extent of this divergence and hence the signal intensity is dependent on the magnitude of the coupling constant, J_{CH}. Fourier transformation of the various rows of "J-modulated" FIDs at different t_1 values affords a series of spectra with varying signal intensities in the t_1 and ω_2 dimensions. Fourier transformation of the rows affords the 2D J-resolved spectra.

^{13}C pulse first aligns the ^{13}C magnetization along the y'-axis. During the subsequent $t_1/2$ period, the ^{13}C magnetization vectors become split on account of the larger one-bond coupling ($^1J_{CH}$) as well as the smaller long-range couplings over two or three bonds ($^2J_{CH}$, $^3J_{CH}$). The $180°_x$ ^{13}C pulse at time $t_1/2$ inverts the vectors across the x'-axis. This would have led to a refocusing of the magnetization vectors at the end of the subsequent $t_1/2$ period. However if a $180°$ ^1H pulse is applied simultaneously with the $180°$ ^{13}C pulse, it would exchange the spin states of the protons coupled to the ^{13}C nucleus so that during the last $t_1/2$ period, they would not refocus but continue to diverge, causing a modulation in amplitude of the signal by the coupling constant, J, with the change of the evolution time t_1. This is the basis of the spin-flip method for recording 2D J-resolved spectra. Since the modulation occurs during the full t_1 period, the J values appear with the actual magnitudes. A disadvantage is that an accurate $180°_x$ (^1H) pulse is required in this method since if there is any inaccuracy, then all protons will not be inverted by the $180°_x$ pulse and there will be a component of the ^{13}C magnetization which will not be J-modulated, resulting in the formation of an artifact peak on the ω_1 axis at $\omega_1 = 0$. The method works best with first-order (weakly coupled) spectra. For these reasons, the gated decoupler method described previously is the method of choice when recording 2D J-resolved spectra.

5.3.1.1.3 The Selective Spin-Flip Method.

Measurement of long-range ^{13}C–^1H coupling constants (^2J, ^3J, and ^4J) of organic molecules, can provide important information on the conformation in which a molecule exists and the relative orientation of attached groups. Proton-coupled ^{13}C spectra often do not allow measurement of such long-range couplings because of the large number of overlapping couplings. Selective 2D heteronuclear J-resolved NMR spectroscopy offers a method for determining conformations through measurement and assignment of long-range ^{13}C–^1H couplings. Suppose the ^{13}C nucleus has a long-range coupling interaction with a proton H_A. If a selective $180°_x$ ^1H pulse is now applied, it inverts the inner ^{13}C satellites of the proton H_A (Figure 5.45A). This results in an elimination of all other coupling interactions (just as chemical shift differences or field inhomogeneities are removed by the spin-echo experiment) and only the long-range couplings of proton H_A with the ^{13}C nucleus are detected. Modulation of the ^{13}C signal amplitudes occur only by the long-range couplings with changing values of t_1, so that Fourier transformation affords a 2D plot containing chemical shifts on one axis and ^{13}C–^1H long-range couplings on the other axis. Thus, in the case of the $1 \rightarrow 4$ linked disaccharide β-methylcellobioside [Figure 5.45B], the 2D experiment allowed measurement of a coupling of 4.2 Hz between H-1' and C-4 across the glycosidic linkage.* From the value of this coupling constant, the torsional angles ϕ and ψ can be calculated via a Karplus-type relationship.

---

* M.J. Gidley and S.M. Bociek, Long range ^{13}C–^1H coupling in carbohydrates by selective 2D heteronuclear J-resolved N.M.R. spectroscopy, J. Chem. Soc., Chem. Commun., 220 (1985).

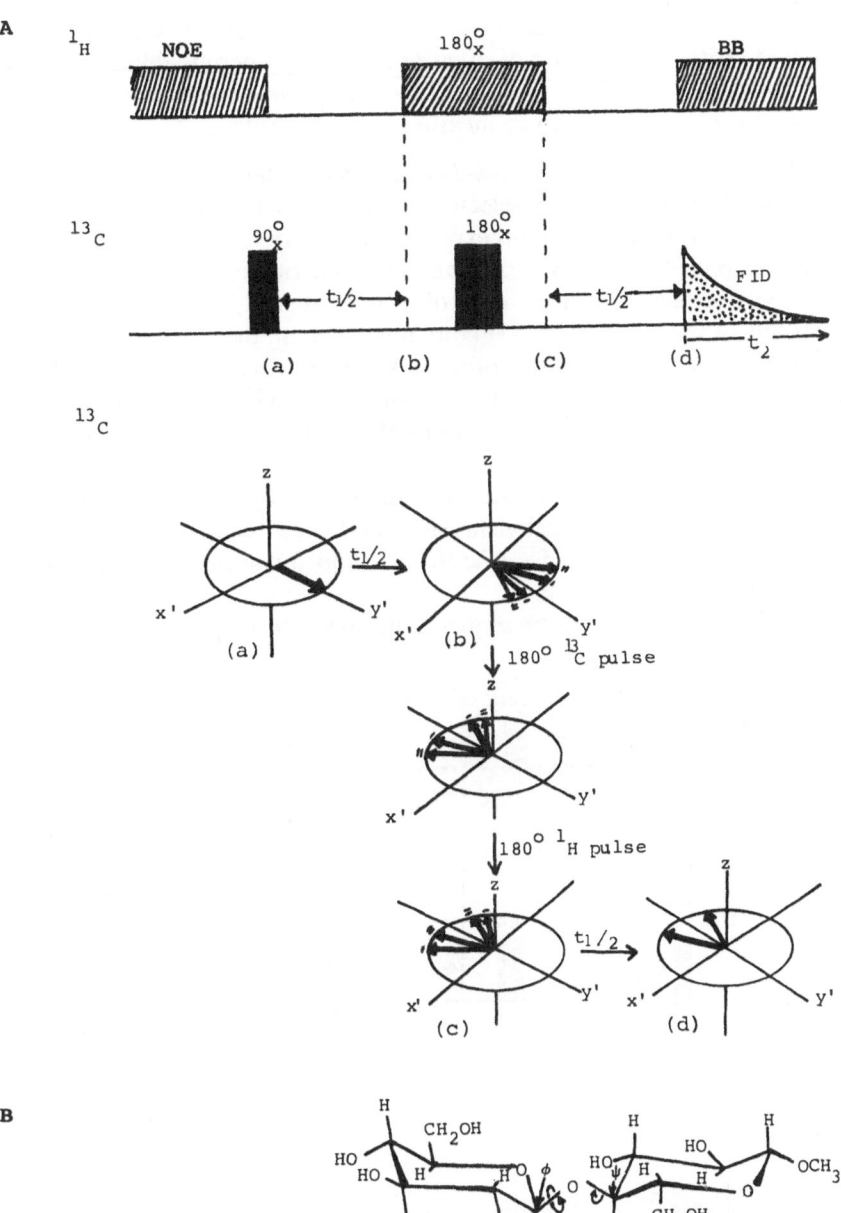

Figure 5.45. (A) The selective spin-flip method for recording heteronuclear 2D J-resolved spectra in which the ^{13}C signals are modulated by long-range coupling constants. The pulse sequence is shown above while its effect on the ^{13}C magnetization vectors is shown below it. The selective 180° ^{1}H pulse results in elimination of the large one-bond coupling constants, J_{CH}. (B) β-Methylcellobioside with the torsional angles ϕ and ψ indicated.

A disadvantage of the method is that overlapping of ^{13}C satellites of the proton H_A with ^{13}C satellites of other protons is not permissible, as it would cause unwanted signals to appear. This means that spectra of only well-separated proton signals can be measured satisfactorily.

5.3.1.1.4   The Semi-Selective Spin-Flip Method.   This method is of a greater general applicability than the selective spin-flip experiment described above. In the semi-selective proton-flip experiment (Figure 5.46), the selective 180° (^1H) pulse applied in the center of the evolution period is replaced by a $90^\circ_x-\tau-180^\circ_x-\tau-90^\circ_x$ sequence. If the value of τ is kept at $1/(2J)$, where J is the magnitude of the one-bond C–H coupling constant, then the pulse sequence results in inversion of only those protons which are directly coupled to the ^{13}C nucleus. The spin-echo is modulated with the frequency of the direct C–H coupling constant with the variation of the evolution time, and the resulting two-dimensional spectrum obtained on Fourier transformation shows singlets, doublets, triplets, and quartets unaffected by long-range couplings for quaternary, methine, methylene, and methyl carbons, respectively. This experiment therefore provides a useful method for accurately determining direct C–H coupling constants.

A modification of this pulse sequence has been developed which results in

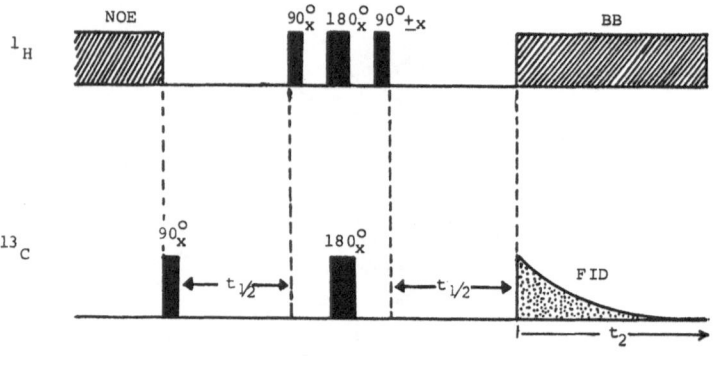

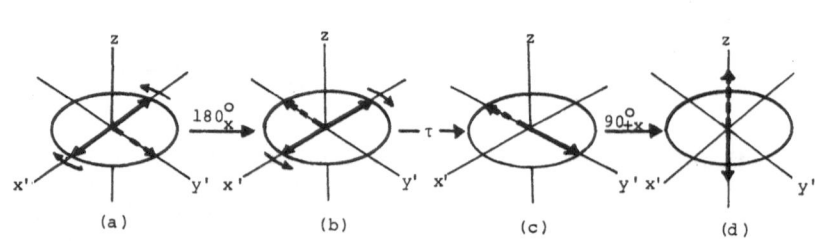

Figure 5.46. The semi-selective spin-flip method for recording heteronuclear 2D *J*-resolved spectra in which the ^{13}C signals are modulated only by the direct one-bond C–H coupling constants with changing values of t_1. The pulse sequence is shown above, while its effect on the ^{13}C magnetization vectors is shown below it.

only $^{13}C-^1H$ long-range couplings being recorded. In this alternative selective proton inversion sequence $90^\circ_x - 1/(2J) - 180^\circ_x - 1/(2J) - 90^\circ_{-x}$ is applied. This differs from the previous pulse sequence in that the second 90° pulse sequence in the ^1H range is applied along the $-x$-axis. This flips only those protons which are *not* directly bound to ^{13}C nuclei and therefore shows up only $^{13}C-^1H$ long-range couplings.

5.3.1.1.5   2D J-Resolved Spectra with INEPT.   Another method used to obtain heteronuclear 2D *J*-resolved spectra utilizes polarization transfer techniques (INEPT) and is comparable to the spin-flip method. The pulse sequence used is

$$^{13}C: \qquad 180^\circ_x \qquad 90^\circ_x - t_1/2 - 180^\circ_x - t_1/2 - FID(t_2)$$

$$^1H: 90^\circ_x - \tau - 180^\circ_x - \tau - 90^\circ_y \qquad 180^\circ_x \qquad BB$$

The first 2τ preparation period is used to transfer magnetization from the ^1H to ^{13}C nuclei. The evolution period t_1 is identical to that in the spin-flip method. The method suffers from the disadvantage that the multiplet patterns are distorted as in 1D INEPT (e.g., central lines of triplets missing) and artifact peaks appear due to the inaccuracies associated with the setting of the 180° ^1H pulse. The advantages include the appearance of *J* couplings with their full magnitudes and good sensitivity enhancement.

5.3.1.2   Homonuclear 2D *J*-Resolved Spectra

In homonuclear 2D *J*-resolved spectra, the coupled nuclei are of the same type. Thus the ^1H chemical shifts may be aligned along the ω_2 axis while the $^1H-^1H$ couplings are spread on the ω_1 axis. The basic pulse sequence used in homonuclear 2D *J*-resolved spectroscopy is that shown in Figure 5.47A. The sequence results in the generation of a spin-echo at time t_1 after the initial 90°_x pulse. The effect of this sequence on the two magnetization components H_A and H_B of a ^1H nucleus coupled to another hydrogen atom in two different spin states α and β is as follows: the 90°_x pulse moves both component vectors, H_A and H_B, from the z-axis to the y'-axis [Figure 5.47B, (a')]. During the subsequent diverging delay, the two vectors move away from one another [Figure 5.47B, (b')], H_B moving away faster than H_A from the y'-axis. The 180°_x pulse rotates the H_A and H_B vectors so that they adopt mirror image positions across the x'-axis [Figure 5.47B, (c')]. The 180°_x pulse, however, also "re-labels" the two vectors* [Figure 5.47B, (c'$_2$)] so that the faster vector H_B

---

* Thus if component A of the ^1H nucleus was coupled to the α-state of the neighboring ^1H nucleus before the 180°_x pulse, it switches over to a state in which it is coupled to the β-state of the neighboring ^1H nucleus after the 180°_x pulse. A similar inversion occurs with component B. This inversion is accompanied by an exchange of angular velocities. Thus if component A was rotating with a Larmor frequency $\Omega_A/2\pi - (J/2)$ before the 180°_x pulse, after the 180°_x pulse it precesses with the frequency $\Omega_A/2\pi + (J/2)$ (which was the Larmor frequency of component B). Similarly, component B adopts the frequency of A. This means that a "re-labeling" has taken place so that A can be written instead of B and vice versa. This re-labeling takes place because the 180° pulse is applied nonselectively to all the spins in the homonuclear coupled system.

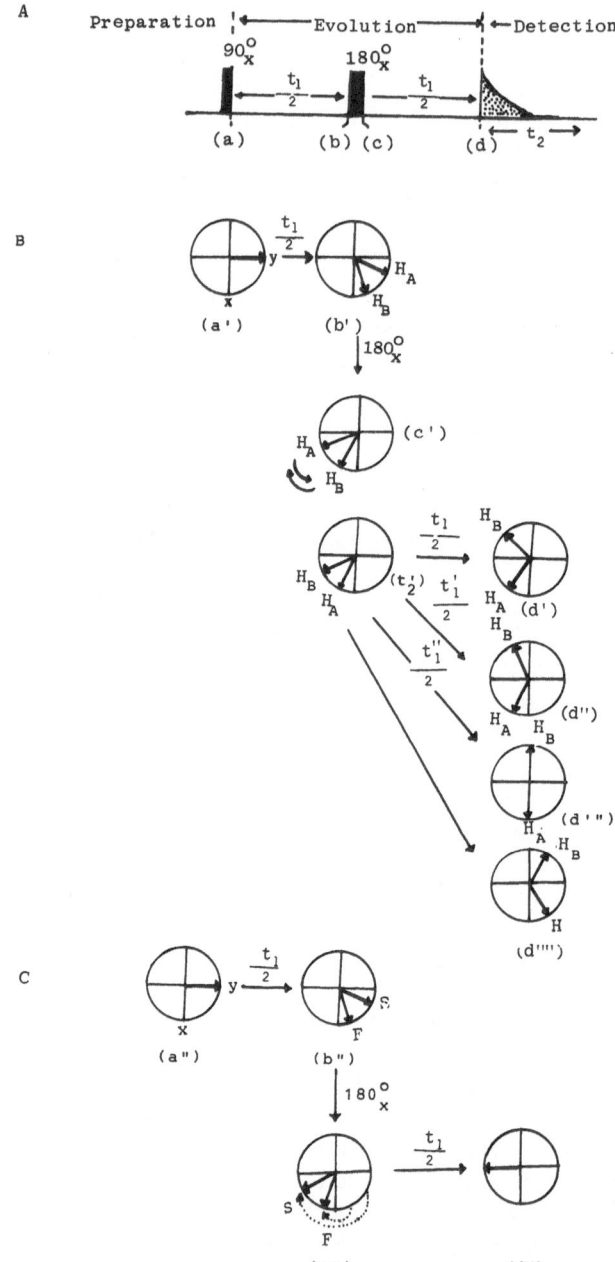

Figure 5.47. (A) Pulse sequence used in homonuclear 2D J-resolved spectroscopy. (B) Effect of the pulse sequence on a ^{1}H doublet; the component vectors of the doublet are represented as H_A and H_B. The 180°_x pulse rotates both vectors by 180° across the x'-axis (c'); simultaneously an exchange of their Larmor frequencies takes place so that the faster vector H_B lies ahead of the slower vector H_A and they continue to diverge during the subsequent $t_1/2$ period. At various values of t_1 (i.e., t_1, t'_1, t''_1) the vectors H_A and H_B are present in different positions (d', d'', d''', d''''). Their vector sum thus varies with t_1 and a "J-modulated" spectrum is obtained. (C) If no coupling is present the vectors refocus at time t_1 after a 90°_x pulse. This serves to eliminate any field inhomogeneities or chemical shift difference effects.

continues to move away from H_A and after time t_1 vector H_A will have a phase $+\pi J t_1$ while H_B will have a phase $-\pi J t_1$. This divergence during the second $t_1/2$ period applies only to the vectors of the multiplet component. The vectors associated with the chemical shift of each spin undergo a refocusing (i.e., gain phase coherency) during this $t_1/2$ period with the result that after time t_1, one obtains a spectrum in which normal phased signals appear at the chemical shift frequencies (singlets, center lines of triplets, etc.) but all other signals show phase distortions (J-modulations).

Figure 5.47C shows the fate of the magnetization of the two vectors if no coupling is present. The $90°_x$ pulse aligns the vectors along the y'-axis and during the subsequent $t_1/2$ period they diverge from each other, the faster F lying ahead of the slower one, S [Figure 5.47C, (b″). The $180°_x$ pulse rotates each of them by 180° across the x'-axis [Figure 5.47C, (c″)]. As the two vectors are not connected by spin–spin coupling, no exchange of their Larmor frequencies takes places as happened in Figure 5.47B, and the faster vector F which lies behind the slower vector S catches up with it during the subsequent $t_1/2$ period, producing a spin-echo [Figure 5.47C, (d″)]. This refocusing results in elimination of effects of chemical shift and field inhomogeneities at time t_1 after the $90°_x$ pulse.

This difference in behavior during the pulse sequence between the vectors associated with the chemical shift and those associated with the coupling constants is shown in Figure 5.47C and B, respectively. Since the multiplet components are modulated as a function of t_1, a number of spectra are obtained at varying values of t_1 and the coupling modulation of echoes at different t_1 values is recorded. It may be noted that echo modulation occurs only in homonuclear coupling since it is only in this case that a nonselective 180° pulse is applied to *both* nuclei. In the heteronuclear case the other nucleus coupled to the nucleus under observation does not flip its state, and if echo modulation is desired, than a 180° pulse must be applied to that nucleus also. Since the decoupler is off when the FID is being recorded, the ω_2 dimension contains information about both the proton chemical shifts and the coupling constants, while the ω_1 dimension contains information about the coupling constants only. This results in the multiplets lying along 45° diagonals and a projection along the ω_1 axis before rectification would result in a normal coupled spectrum. However, projection at 45° to the ω_1 axis would result in a fully proton-decoupled spectrum with each proton affording a single peak. A computer correction is usually automatically carried out so that the ω_1 axis contains only the coupling information while the ω_2 axis contains the chemical shifts. This is illustrated schematically for an AX_2 spin system in Figure 5.48 and for the vinyl protons of vinyl acetate in Figure 5.49. A further example is provided by the homonuclear 2D J-resolved spectrum of the methylene group of ethanol (Figure 5.50). Again the multiplet is slanted at an angle of 45° with respect to the ω_1 axis but through an appropriate computer correction this is rectified so that there is no spreading of the chemical shift along the ω_1 axis. On the right side of Figure 5.50 are shown corresponding contour drawings,

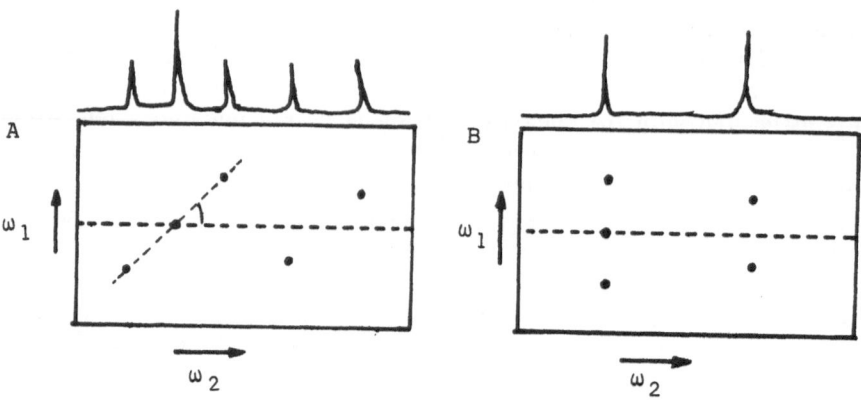

Figure 5.48. Schematic 2D homonuclear J-spectrum of an AX_2 spin system: (A) normal; (B) after tilting by 45° so that contours appear in vertical lines.

which are often more convenient and informative as the overlappings of "hills" on one another are avoided. These contours may be considered to be obtained by looking down at the hills from the top at various cross sections. If the cross section is made too high, then only the taller peaks will be observed. If the cross section is too low, then the contour plot will be complicated by the appearance of many of the very small peaks as well as of noise. The bottom right contour plot in Figure 5.50 shows the corrected contour obtained after a 45° rotation of the upper contour plot.

5.3.2 2D-Shift Correlated Spectroscopy

So far we have been concerned with separating the chemical shifts from the coupling constants so that the multiplicity of each carbon (or proton) can be determined. In complex organic molecules it is frequently necessary to find out which protons are coupled to which other protons or carbons in a molecule. This was normally carried out in the past by spin–spin decoupling but this has to be done for one proton at a time, making the whole process rather tedious. Moreover, it is difficult to selectively decouple just one proton since if there are other protons very close to the one which is being decoupled, they too will come under the influence of the decoupling frequency, which is never very sharply focused. The advent of two-dimensional shift correlated spectroscopy has allowed chemical shift correlations to be made (a) between similar nuclei, e.g., protons coupled to one another (*2D-homonuclear shift correlated spectroscopy*) or (b) between dissimilar nuclei, e.g., protons coupled to carbons (*2D-heteronuclear shift correlated spectroscopy*). The spectra appear on the two axes of the 2D contour plot and the cross peaks appear in the plot at points corresponding to the chemical shifts of the coupled nuclei. The pulse

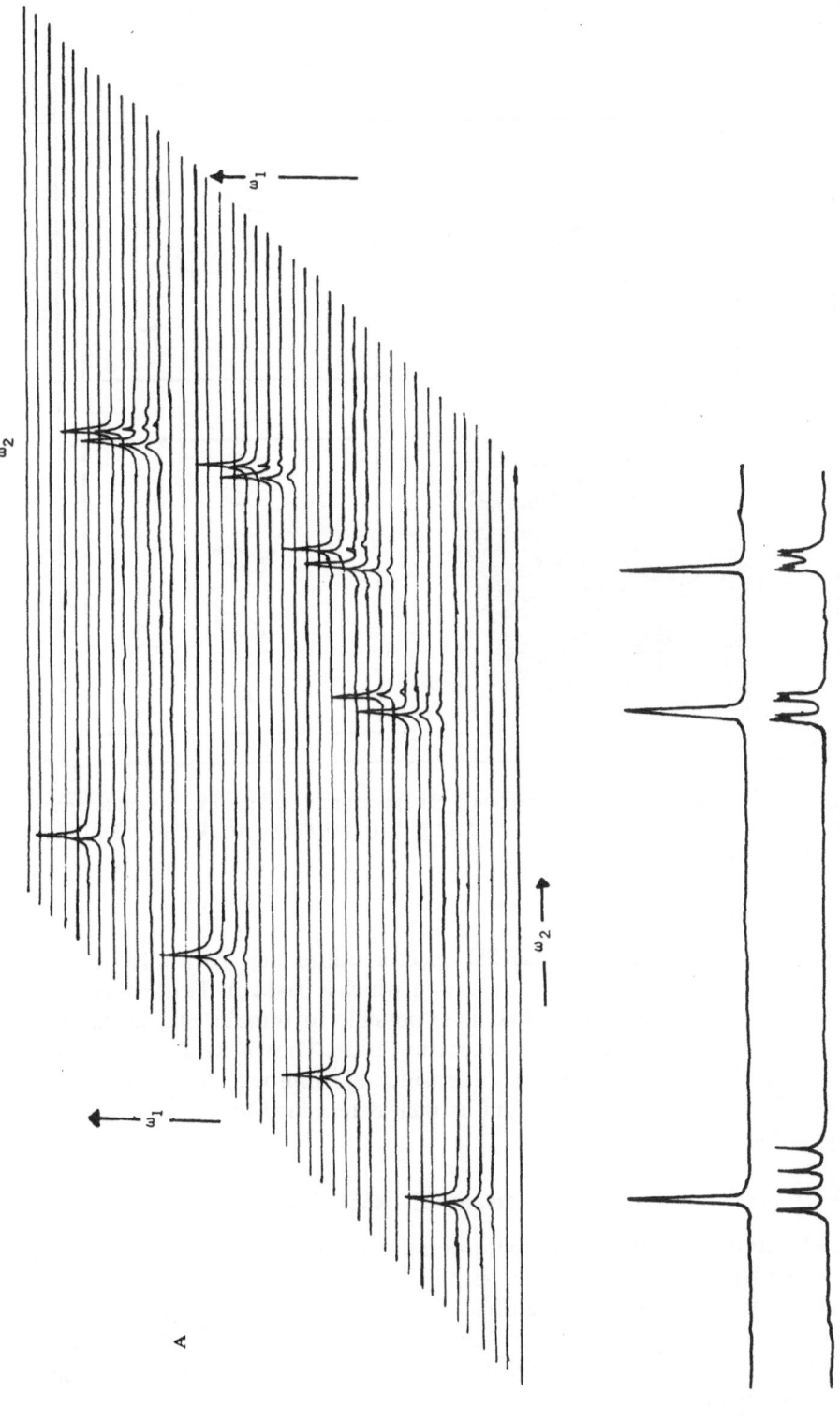

Figure 5.49. (A) Homonuclear 2D *J*-resolved spectrum of the vinyl protons of vinyl acetate. (B) Projection at 45° to the ω_1 axis yields a proton-decoupled spectrum. (C) Projection in the ω_1 direction (i.e., vertically) results in a proton-coupled spectrum.

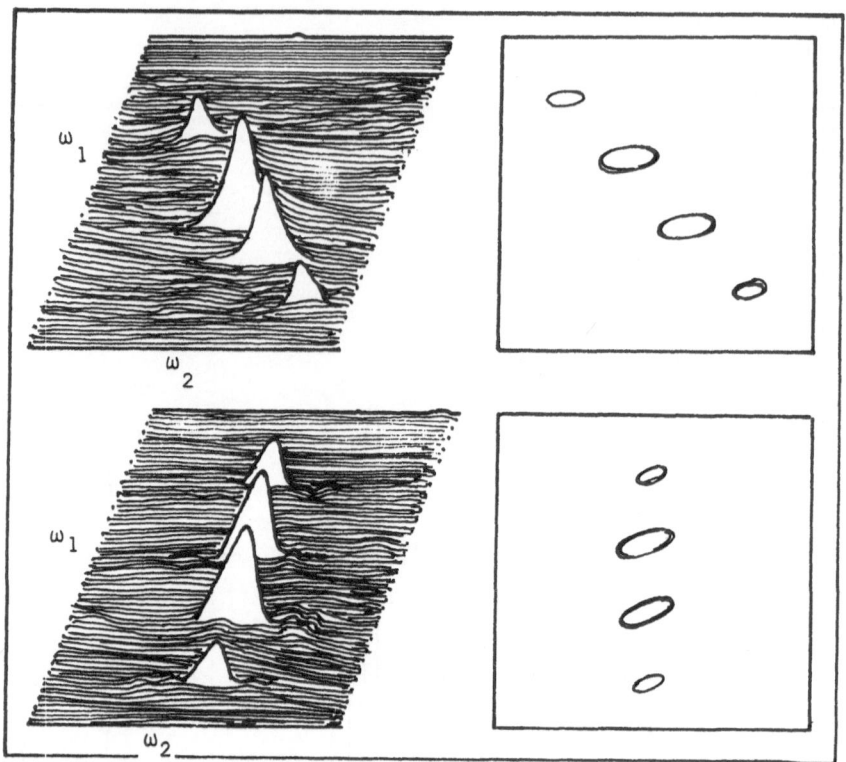

Figure 5.50. Homonuclear 2D J-resolved ^{1}H spectrum of ethanol in the methylene region. *Above*: Stacked plot (left) and contour plot (right). *Below*: Stacked plot (left) and contour plot (right) after 45° rotation.

sequences used to obtain these 2D spectra utilize the cross-polarization phenomenon discussed in an earlier section, which is based on the discovery by Maudsley and Ernst in 1977 that it is possible to transfer magnetization between nuclei of different gyromagnetic ratios. Thus if magnetization is transferred from ^{1}H to ^{13}C, there is a considerable improvement in sensitivity since the strength of the ^{13}C signal is then determined by the proton spin population.

5.3.2.1   Heteronuclear Shift Correlated 2D-NMR Spectra

The heteronuclear 2D J-resolved spectra described earlier arose through the modulation of the ^{13}C signal amplitudes by the frequency of the C–H coupling constant, $^1J_{CH}$. In heteronuclear 2D-shift correlated spectra, on the other hand, the ^{13}C signals in a compound are modulated by the Larmor frequencies (i.e., chemical shifts) of the protons to which the respective carbon atoms are bonded.

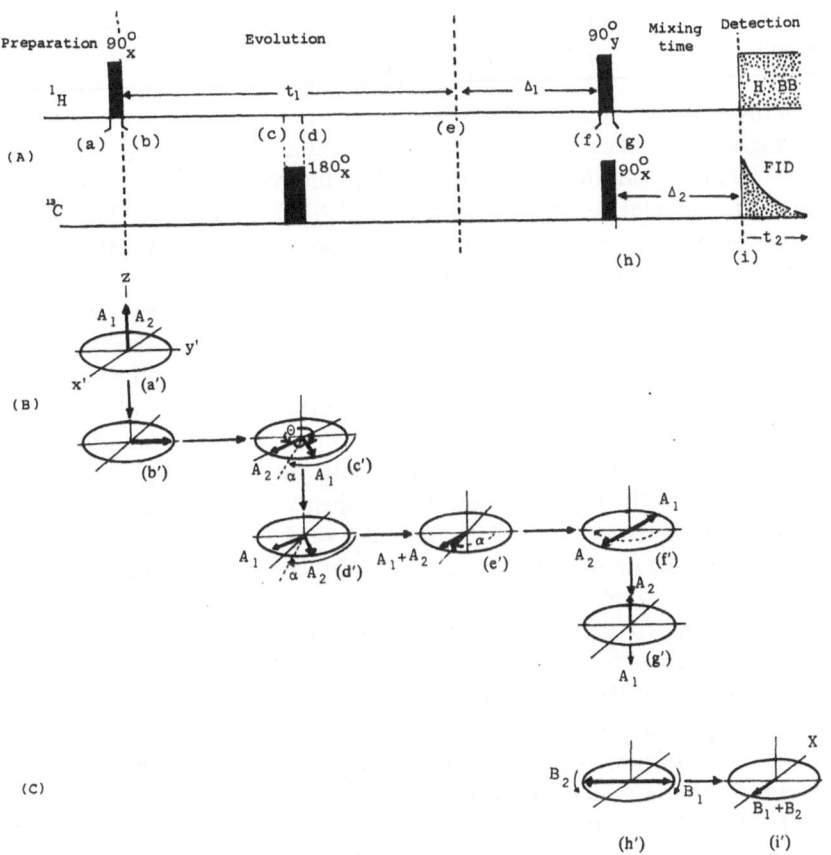

Figure 5.51. Pulse sequence (A) and vector diagrams (B, C) of a heteronuclear 2D-shift correlated experiment. The magnetization vectors A_1 and A_2 represent the doublet for the ^1H nucleus in a CH group. The phase modulation of the ^{13}C signals is dependent on the angle α.

The basic pulse sequence employed in heteronuclear 2D-shift correlated spectroscopy and its effect on ^1H and ^{13}C magnetization vectors are shown in Figure 5.51. Experimentally shift correlated spectra require an additional time interval, the "mixing time," between the evolution and detection periods. Let us consider a two-spin system, $^{13}C-^1H$. The application of a 90° proton pulse tips the proton magnetization so that it comes to lie along the y'-axis [Figure 5.51(b')]. This proton magnetization may be regarded as being composed of two different vectors, one in which the protons are bound to ^{13}C nuclei in the α-spin state (designated A_1) and the other in which the protons are bound to ^{13}C nuclei in the β-spin state (designated A_2). During the subsequent interval, $\frac{1}{2}t_1$, the two magnetization vectors diverge from each other and precess away from the y-axis at different angular velocities* [Figure 5.51(c′)]. The fanning-out angle, θ, between the vectors A_1 and A_2 is determined by the magnitude of the coupling constant, while the angle α (representing the angle between the average position of the vectors and the y-axis) is determined by the chemical shift of the protons. The 180° ^{13}C pulse reverses the populations of the ^{13}C states, resulting in a corresponding interchange of the rotational frequencies (or spin labels) of the vectors A_1 and A_2 [Figure 5.51(d′)] so that in the subsequent $\frac{1}{2}t_1$ period, instead of diverging from one another, they converge and are refocused at the end of the evolution period, t_1 [Figure 5.51(e′)]. The angle α now contains the chemical shift information. The proton–carbon splittings are suppressed by this refocusing in the t_1 dimension, while proton noise decoupling during signal acquisition suppresses the splittings in the t_2 dimension. The variable evolution time t_1 is followed by a constant mixing time Δ which serves to transfer the information about the proton chemical shifts contained in the phase angle α to the ^{13}C magnetization. The length of the mixing time is adjusted so that it is equal to the mean CH coupling constant, J. After time $\Delta_1 = 1/(2J)$, the two vectors are found to be oriented on the $y′$-axis in opposite directions (i.e., angle $\theta = 180°$). The second $90°_y$ "mixing" or proton polarization pulse transfers magnetization from the protons to their ^{13}C partners to which they are coupled and significantly enhances the sensitivity of the ^{13}C signals. This $90°_y$ pulse rotates those components of the magnetization which lie along the $+y′$-axis to the $+z$-axis. This polarization of protons leads simultaneously to polarization of the ^{13}C magnetization, and a $90°_x$ pulse on the ^{13}C nuclei aligns this magnetization along the $y′$-axis [Figure 5.51(h′)]. During the subsequent delay, $\Delta_2 = 1/(2J)$, the ^{13}C vectors refocus and may be detected as singlets if proton decoupling is applied during detection. The values of the two delays introduced, Δ_1 and Δ_2, are calculated to allow 180° relative phase rotations between the components in order to prevent them from canceling each other. The measured signal intensity is dependent on the polarization which in turn depends on the Larmor frequency (and hence the chemical shift) of the protons. Thus as the evolution period t_1 is varied, there is a corresponding change in signal intensity (amplitude modulation) which is not related to the coupling (as was the case in 2D J-resolved spectra) but to the chemical shift of the other nucleus. The amplitude of the ^{13}C signal is thus seen to be modulated at the proton chemical shift frequency with variation of t_1. The first Fourier transformation results in a spectral series in which the ^{13}C signals are modulated by the resonance frequencies (δ_H) of the protons to which they are coupled.

$$S(t_1, t_2) \xrightarrow{\text{FT}(t_2)} S(t_1, \delta_c)$$

---

* At any time t during the evolution period before the application of the 180° ^{13}C pulse, the angles by which the two vectors have precessed may be represented as $(\Omega_0 + \pi J)t$ and $(\Omega_0 - \pi J)t$ where Ω_0 is the chemical shift frequency of the protons.

The first delay, Δ_1, is optimized to a value of $1/(2J)$ for the CH_n system. The value of Δ_2 is set at $1/(2J)$ for CH alone, and at $1/(4J)$ for CH, CH_2, and CH_3 groups. A large number of spectra with varying values of t_1 are recorded. These can be arranged in a data matrix and a second Fourier transformation affords a 2D plot showing the correlation of ^1H and ^{13}C chemical shifts.

$$S(t_1, \delta_c) \xrightarrow{FT(t_1)} S(\delta_H, \delta_c)$$

Since the value of the coupling constant J_{CH} determines the final outcome of the experiment, there are two main types of 2D-heteronuclear shift correlated spectra which can be recorded. If the value is set between 125 and 140 Hz, then nuclei of sp^3- or sp^2-hybridized atoms with direct or one-bond couplings to protons are observed. This would afford cross peaks in the 2D contour plot, which allow one to determine which protons (in the ^1H-NMR spectrum drawn on one axis) are coupled to which ^{13}C nuclei (in the ^{13}C-NMR spectrum drawn on the other axis). If the value of J_{CH} is set at 5–10 Hz (corresponding to J_{CH} for nuclei separated by two or three bonds), then the longer-range couplings between the protons in the ^1H-NMR spectrum and ^{13}C nuclei in the ^{13}C-NMR spectrum can be identified by the presence of corresponding cross peaks in the 2D-NMR spectrum. This is illustrated by the two 2D-NMR plots for ethyl alcohol presented in Figure 5.52). In Figure 5.52A the J_{CH} value has been adjusted to 140 Hz. The two peaks on the horizontal axis are the ^{13}C chemical shifts of CH_3 and CH_2 groups while the vertical axis contains the ^1H-NMR spectrum. The cross peak A below the ^{13}C signal for the methyl group corresponds horizontally with the methyl protons in the ^1H-NMR spectrum of ethanol. This thus immediately tells us which carbon atom is connected to which protons. The second cross peak, B, similarly establishes a one-bond coupling correlation between the methylene carbon atom and the methylene protons.

In the second heteronuclear shift correlated 2D plot of ethanol (Figure 5.52B) the value of J is adjusted to 7 Hz so that only long-range couplings show up. The cross peak C now establishes an interconnection between the methyl carbon and the methylene protons, showing that a two-bond long-range coupling interaction in the range of 7 Hz exists between these nuclei. Similarly, cross peak D establishes a long-range coupling between the methylene carbon and the methyl protons.

The heteronuclear shift correlated 2D-NMR spectrum of a cyclic peptide, cyclo(Pro–Phe–Gly–Phe–Gly), is presented in Figure 5.53. The proton chemical shifts (0–5 ppm) are on the horizontal axis while the ^{13}C chemical shifts (20–70 ppm) are on the vertical axis. The cross peaks establish the interconnection between the chemical shifts of protons and the chemical shifts of carbon atoms to which those protons are attached.

The alternative older 1D procedure for determining CH connectivities was to record a series of selectively ^1H decoupled ^{13}C-NMR spectra. It suffered from the disadvantage that since protons may have close chemical shifts, it is

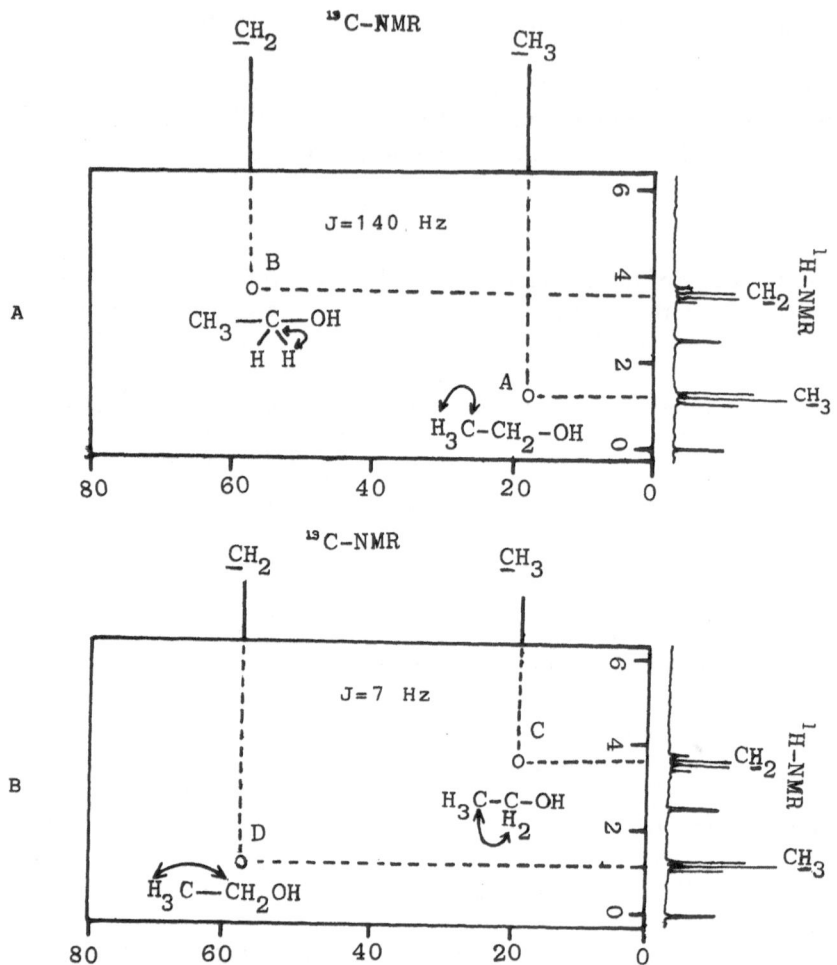

Figure 5.52. (A) Heteronuclear one-bond correlation plot for ethyl alcohol (J adjusted to 140 Hz). (B) Long-range heteronuclear correlation plot for ethyl alcohol (J adjusted to 7 Hz).

often difficult to irradiate them selectively without affecting other close-lying protons, so that ambiguities may result in the "selectively" decoupled ^{13}C-NMR spectra. Moreover, a large number of spectra with different proton irradiations need to be recorded if all the C–H connectivities are to be unraveled, a formidable task particularly in larger molecules. The heteronuclear 2D-shift correlation experiment described above is less ambiguous since any overlapping in the proton region is removed due to the spread of the cross peaks over the large ^{13}C chemical shifts. Moreover, in one experiment all C–H linkages can be identified, with the added advantage of a gain of sensitivity due to transfer of polarization from ^{1}H to ^{13}C nuclei.

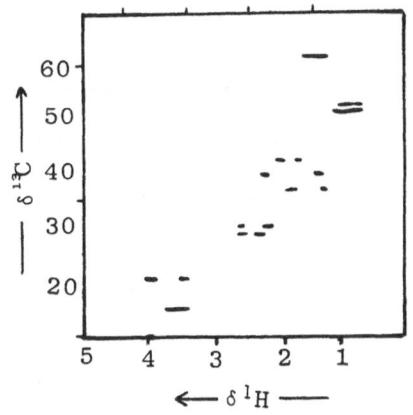

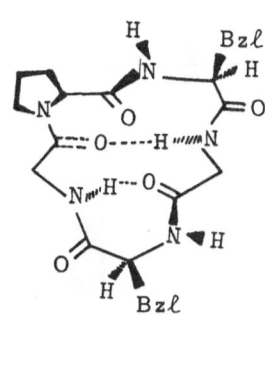

Figure 5.53. Heteronuclear shift correlated 2D-NMR spectrum of a cyclic peptide, cyclo(Pro–Phe–Gly–Phe–Gly).

5.3.2.1.1 Semiselective Refocused Heteronuclear Shift Correlation Spectroscopy. Semiselective refocusing has recently been employed in heteronuclear spectroscopy for detection of *proton-decoupled* proton spectra.* This allows detection of heteronuclear couplings without complications from proton–proton couplings, and affords spectra with increased sensitivity and resolution. The pulse sequence used is $90°(^1H)-\tau-180°(^1H)-180°(X)-\tau-90°(^1H)$ where X is the heteronucleus. This allows selective refocusing of only those protons which are directly coupled to the heteronucleus.

5.3.2.1.2 2D DEPT Heteronuclear Shift Correlation Spectroscopy. The DEPT pulse sequence relies on the fact that the intensities of $^{13}C-\{^1H\}$ signals for CH, CH_2, and CH_3 groups depend on the flip angle θ of the last ^1H pulse in the sequence. Recently, a 2D DEPT sequence has been described which allows presentation of DEPT spectra in a two-dimensional matrix.† The pulse sequences used are shown in Figure 5.54A and 5.54B. The 2D DEPT $^1H-^{13}C$ correlation maps for (a) all protonated carbons, (b) CH groups only, (c) CH_2 groups only, and (d) CH_3 groups only are presented in Figure 5.55(A–D).

5.3.2.1.3 XCOR Spectra. A recent improvement made in 2D-heteronuclear correlated spectroscopy is the use of a selective proton inversion pulse sequence which increases the sensitivity of the experiment by removing proton homonuclear coupling and allows easier correlation between ^1H and ^{13}C chemical shifts. The "XCOR" pulse sequence used for the purpose is shown in

---

* J.A. Wilde and P.H. Bolton, Suppression of homonuclear couplings in heteronuclear two-dimensional spectroscopy, *J. Magn. Resonance* **59**, 343 (1984) and references therein.
† T.T. Nakashima, B.K. John and R.E.D. McClung, Selective 2D DEPT heteronuclear shift correlation spectroscopy, *J. Magn. Resonance* **59**, 124 (1984) and references therein.

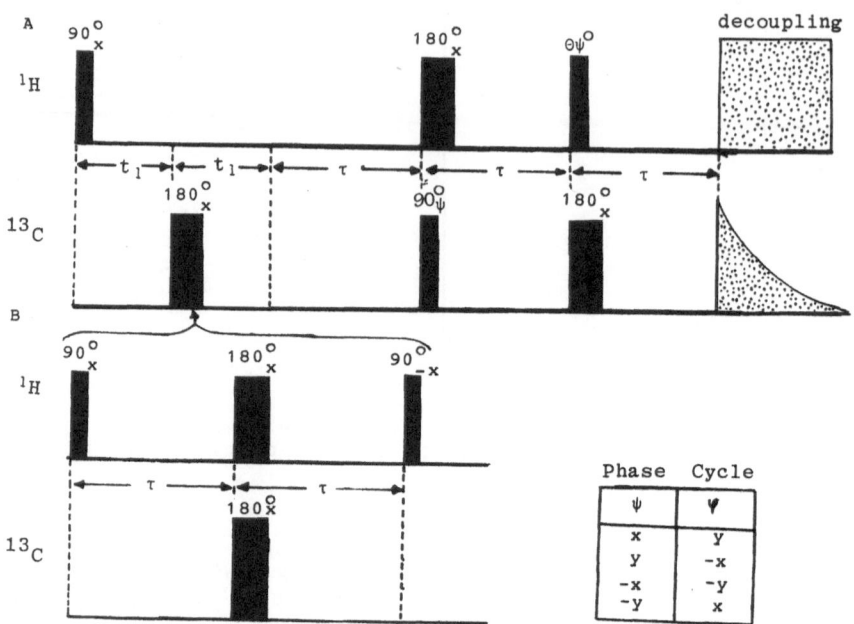

Figure 5.54. (A) Pulse sequence for 2D DEPT heteronuclear COSY experiment. (B) ^{1}H--^{1}H couplings between weakly coupled protons can be removed by inserting the pulse sequence shown in place of the 180° ^{13}C pulse in the center of the evolution period.

Figure 5.56. After a delay D_1, which allows the protons to reach equilibrium, a 90°_x pulse is applied. During the subsequent delay $D_2/2$, the proton magnetization is allowed to evolve. The protons are decoupled at this point by applying the selective proton inversion sequence: $90^\circ_x - 1/(2J) - 180^\circ_x - 1/(2J) - 90^\circ_{-x}$. The sequence is effectively equivalent to applying a 180° pulse to those protons which are not bound to ^{13}C atoms. A 180° pulse is applied to the ^{13}C nuclei at the same time as the 180° proton pulse. This causes the ^{13}C nuclei to invert their spin states. During the subsequent $D_2/2$ delay, the protons are again allowed to evolve. The values of delays D_3 and D_4 are adjusted to allow the observation of either the large C–H one-bond couplings or the smaller long-range couplings. The FID is then acquired with the proton decoupler switched on.

5.3.2.1.4　Heteronuclear Shift Correlation Spectroscopy via Long-Range Couplings (COLOC).　A new pulse sequence has recently been introduced which allows the use of small proton–carbon long-range couplings for the assignment of carbonyl groups.* Known as "Correlation Spectroscopy via

---

* H. Kessler, C. Griesinger, J. Zarbock, and H.R. Looslie, Heteronuclear shift correlation via small coupling constants, *J. Magn. Resonance* **57**, 331 (1984).

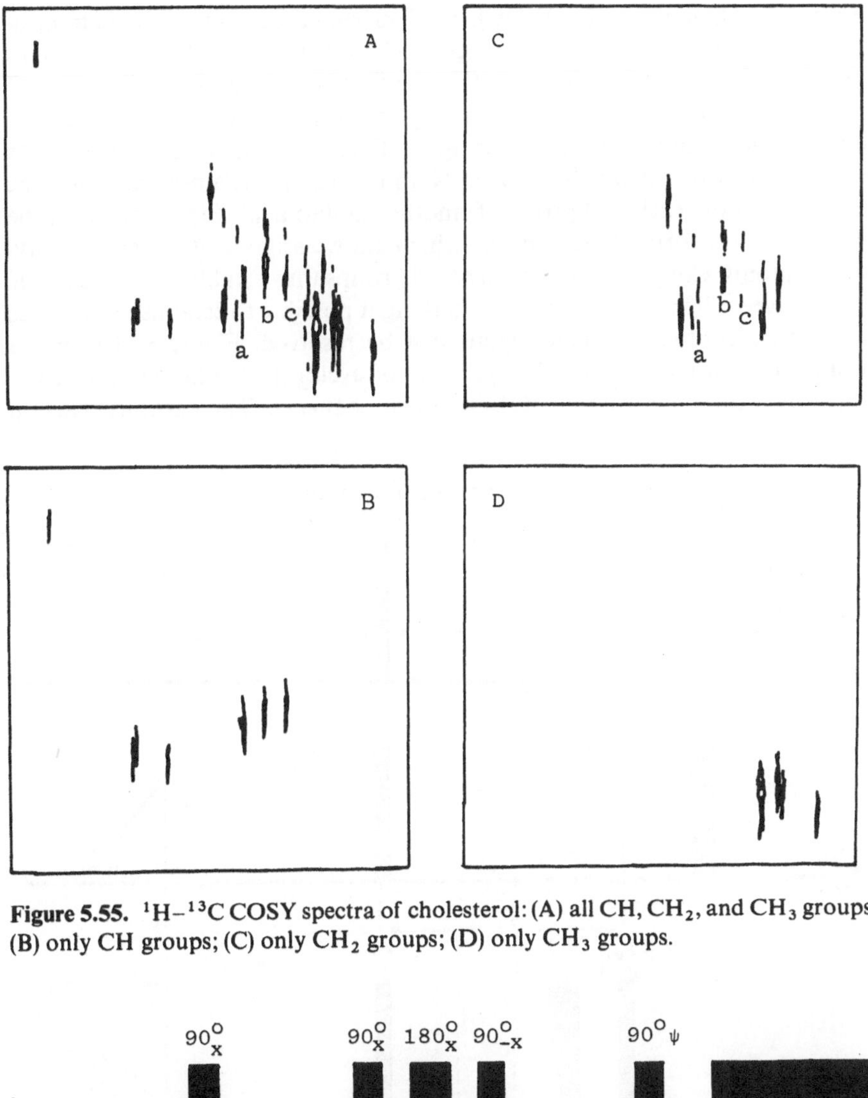

Figure 5.55. ^1H–^{13}C COSY spectra of cholesterol: (A) all CH, CH_2, and CH_3 groups; (B) only CH groups; (C) only CH_2 groups; (D) only CH_3 groups.

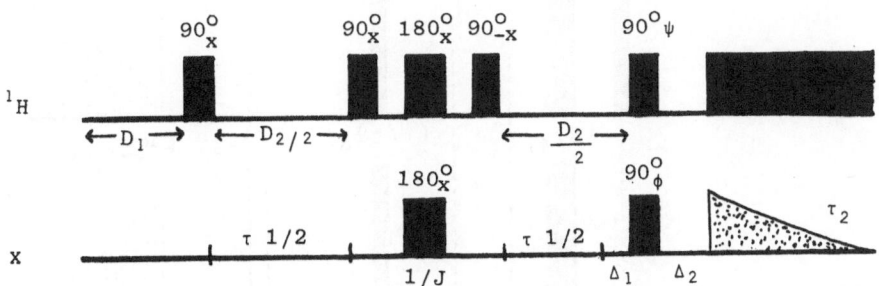

Figure 5.56. Pulse sequence diagram for the XCOR experiment.

Long-Range Couplings" (COLOC), it differs from the conventional heteronuclear shift correlation experiment (Figure 5.57A) in that evolution is allowed to
take place during the first delay, Δ_1 (Figure 5.57B). This results in reduction of
the proton transverse relaxation time, T_2, from $t_1 + \Delta_1$ to only Δ_1, and also
refocuses field inhomogeneities during t_1. Thus long-range couplings across
carbonyl groups in peptides between NH protons of one amino acid and the
protons α to the carbonyl group of another amino acid (Figure 5.58) can be
observed in the form of cross peaks which allow assignment of the carbonyl
carbon atoms. Couplings across carbonyl groups thus establish spectroscopic
links between different amino acids, making it possible to establish sequences
when the α-proton and NH signals can be resolved. Figure 5.59 shows a
contour plot of the H, C-COSY spectrum of cyclo(–Pro–Phe–D–Trp–Lys–
Thr–Gly–) obtained by the COLOC pulse sequence. The carbonyl carbon

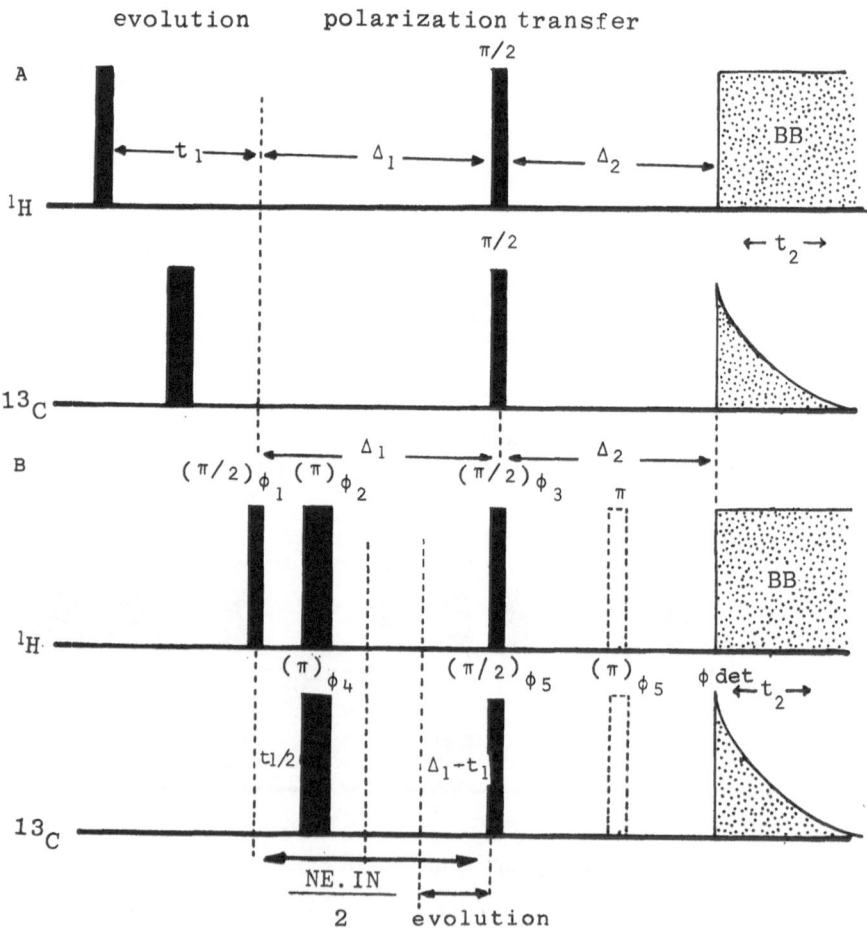

Figure 5.57. Pulse sequences for heteronuclear shift correlation experiment:
(A) normal mode; (B) employing homonuclear broad-band decoupling (COLOC).

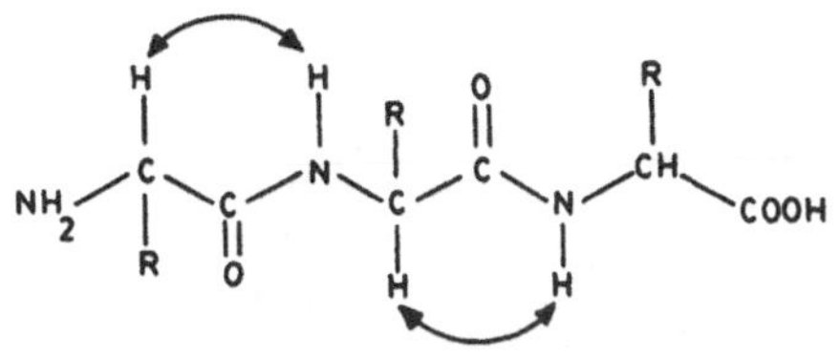

Figure 5.58. The COLOC pulse sequence allows long-range couplings (e.g., between NH and CH proton across the carbonyl group in the peptide) to be observed.

signals are on the vertical axis while the ^{1}H-NMR spectrum of the peptide is on the horizontal axis. If one looks down an imaginary horizontal line coinciding with the lysine carbonyl carbon, one finds two cross peaks, marked T (in the NH region) and K (in the region where H atoms attached to the carbon α to C=O occur). The cross peak K corresponds to the signal for the lysine α CH (see the proton assignments above the COSY diagram) while the cross peak marked T corresponds to the threonine NH signal, establishing that lysine is bound to threonine. Connectivities between other amino acids can be established in a similar manner.

5.3.2.2   Homonuclear Shift Correlated 2D-NMR Spectra

5.3.2.2.1   COSY Spectra.   In homonuclear shift correlated 2D-NMR spectra the coupling interactions between the same type of nuclei in a compound are recorded in the form of a two-dimensional contour plot. The contours situated on a 45° diagonal line of this plot correspond to an overhead view of the peaks in the 1D-NMR spectrum whereas the contours which appear away from this diagonal line show the coupling interactions between the protons. The top end of the diagonal line represents the upfield region whereas the lower end represents the downfield region. In order to find out which protons are coupled to which other protons, one simply has to draw two lines at 90° from each other starting at the contour cross peak and ending at the diagonal line. The two points on the diagonal line where the two lines cross it then correspond to the chemical shifts of the two coupled nuclei. An example is given in Figure 5.60. The NMR spectrum drawn on the horizontal axis shows four main groups of peaks labeled 1,2,3, and 4. The diagonal line passes through four corresponding groups of contours numbered (i), (ii), (iii), and (iv). There are three cross peaks designated as A, B, and C. Thus if one draws a line straight down from cross peak A, it can be seen that it meets the diagonal at point (i). If one draws another line from the same cross peak A but this time horizontally, it is seen to meet the diagonal at point (iv). One can therefore conclude that protons at point (i) (corresponding to the multiplet labeled 1) are coupled to the protons at point (iv) (corresponding to the multiplet labeled 4). Similarly, cross peak B demonstrates that that the doublet labeled 2 is coupled to the multiplet labeled 4. Thus all the coupling interactions in a compound can be discerned in one two-dimensional plot. The cross peaks A', B', and C' are mirror images of A, B, and C. The chemical shifts of the coupled protons

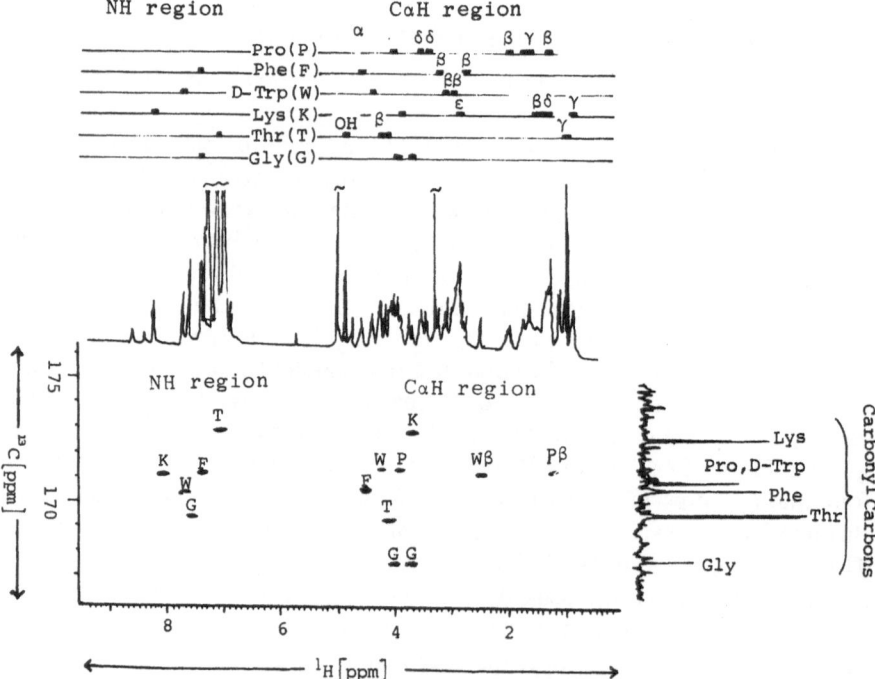

Figure 5.59. H, C-COSY spectrum of the carbonyl carbon atoms of cyclo[–Pro–Phe–D–Trp–Lys–Thr–Gly–] obtained by COLOC pulse sequence. The carbonyl carbons are shown on the vertical axis while the ¹H-NMR assignments of the *trans* isomer for the various amino acids in the peptide are shown above.

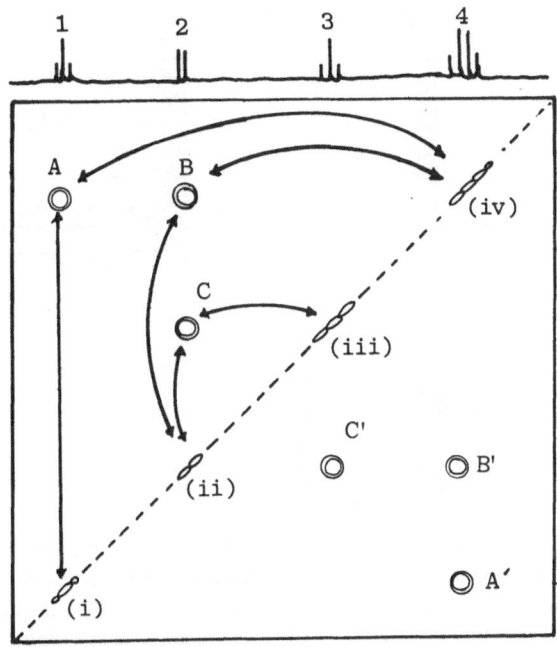

Figure 5.60. Example of a COSY spectrum. The contours on the diagonal correspond to the peaks in the 1D spectrum drawn above the 2D plot. Cross peaks A, B, and C establish the proton–proton connectivities.

can be read from the horizontal or vertical axes. Since multiplets due to the same type of nuclei at different chemical shifts are being correlated in terms of the scalar J-couplings, the above experiment is called a homonuclear shift correlated 2D spectroscopy experiment (COSY). It was first proposed by Jeener in 1971 and is essentially a magnetization transfer experiment.

The basic pulse sequence used in the COSY experiment is shown in Figure 5.61. The first 90° pulse along the x'-axis creates transverse magnetization (in the x',y'-plane). The magnetization components in the x',y'-plane precess around the z axis during the evolution period, t_1. The second 90° mixing pulse transfers magnetization amongst all the possible transitions in the coupled spin system. Magnetization transfer between transitions of noncoupled nuclei results in the contours appearing on or near the diagonal line, while magnetization transfer between coupled nuclei results in the contours appearing at positions away from the diagonal lines but at the intersections of their chemical shifts. In quadrature detection mode, the transmitter is centered on the ω_2 axis and phase cycling is carried out to distinguish between positive (P-type) and negative (N-type) signals. Longitudinal relaxation during t_1 may result in their being some z-magnetization when the second 90° pulse is applied, which can cause the appearance of undesired axial peaks at $\omega_1 = 0$. Phase alternation of the mixing pulse is therefore carried out to eliminate these. Thus at least four experiments are required for each value of t_1 to

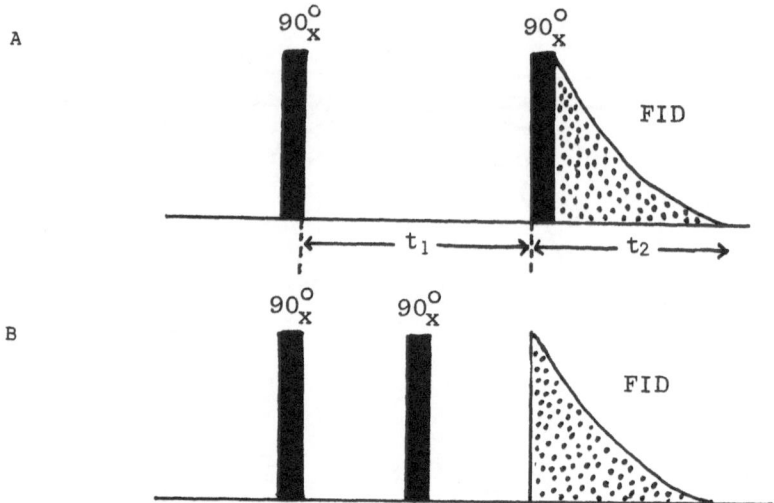

Figure 5.61. Pulse sequences for homonuclear 2D-shift correlated NMR spectroscopy; (A) *COSY*-90 pulse sequence; (B) *SECSY* pulse sequence. The second 90° "mixing" pulse causes an exchange of magnetization between the coupled nuclei. If the second pulse in (A) is a 45° pulse, then a simplified "*COSY*-45" spectrum is obtained with reduced intensities of cross peaks near the diagonal.

distinguish between positive and negative peaks and to eliminate axial peaks. A way around this is to use two identical pulsed field gradients just before and after the second 90° pulse, which ensure that only the coherence transfer echo component is being detected (Figure 5.62). The second field gradient also serves to defocus any magnetization giving rise to the axial peaks.

A variation of the COSY-90 experiment described above is to apply a 45°_x mixing pulse instead of the 90°_x second pulse. The COSY-45 sequence reduces the intensity of the components near the diagonal, and simplifies the 2D plot.

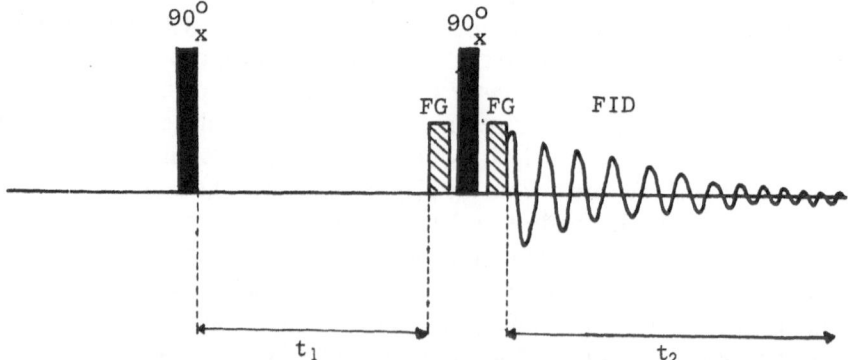

Figure 5.62. Alternative pulse sequence for homonuclear 2D-shift correlated spectroscopy. The two pulsed field gradients, FG, are applied to suppress axial peaks and to distinguish between positive (P) and negative (N) peaks.

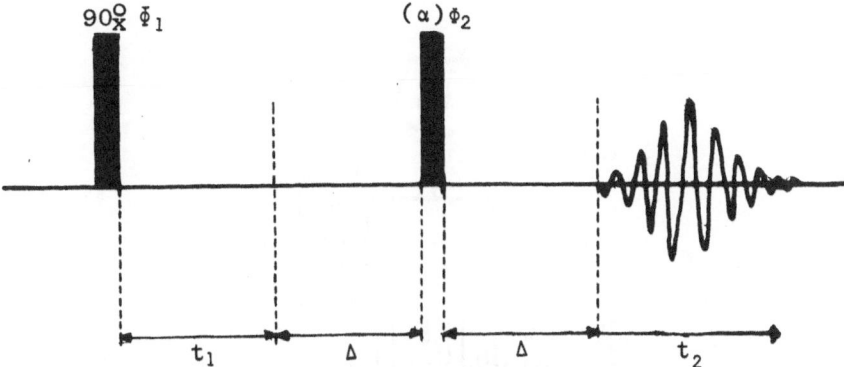

Figure 5.63. A variation of the COSY-90 experiment. Fixed delays Δ are inserted at the end of the evolution and before the detection periods to allow detection of long-range couplings.

Another variation of the COSY-90 experiment is to introduce a fixed delay Δ just before the mixing pulse. The sequence may then be represented as shown in Figure 5.63. The value of Δ is adjusted to about 0.5 s. This sequence shows up the long-range couplings.

A useful technique is to use broad-band decoupling in the ω_1 dimension. The pulse sequence shown in Figure 5.64 is used to achieve this. The fully coupled spectrum appears on one axis while the fully decoupled spectrum appears on the other axis. This method has the advantage that the chemical shifts of the individual protons can be read from the singlets on one axis and the proton–proton couplings can be easily identified. The Jeener spectrum of a tricyclodecane derivative is shown in Figure 5.65. The spectrum above the

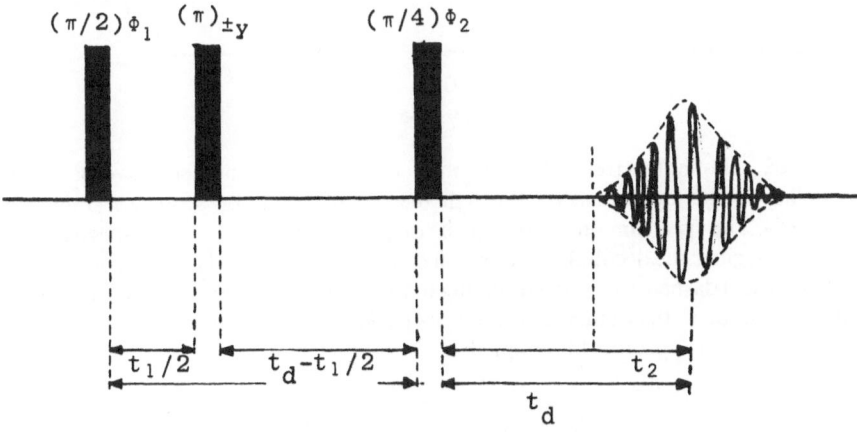

Figure 5.64. The pulse sequence shown results in decoupling in the ω_1 dimension. The delay t_d is fixed and only t_1 and t_2 are varied.

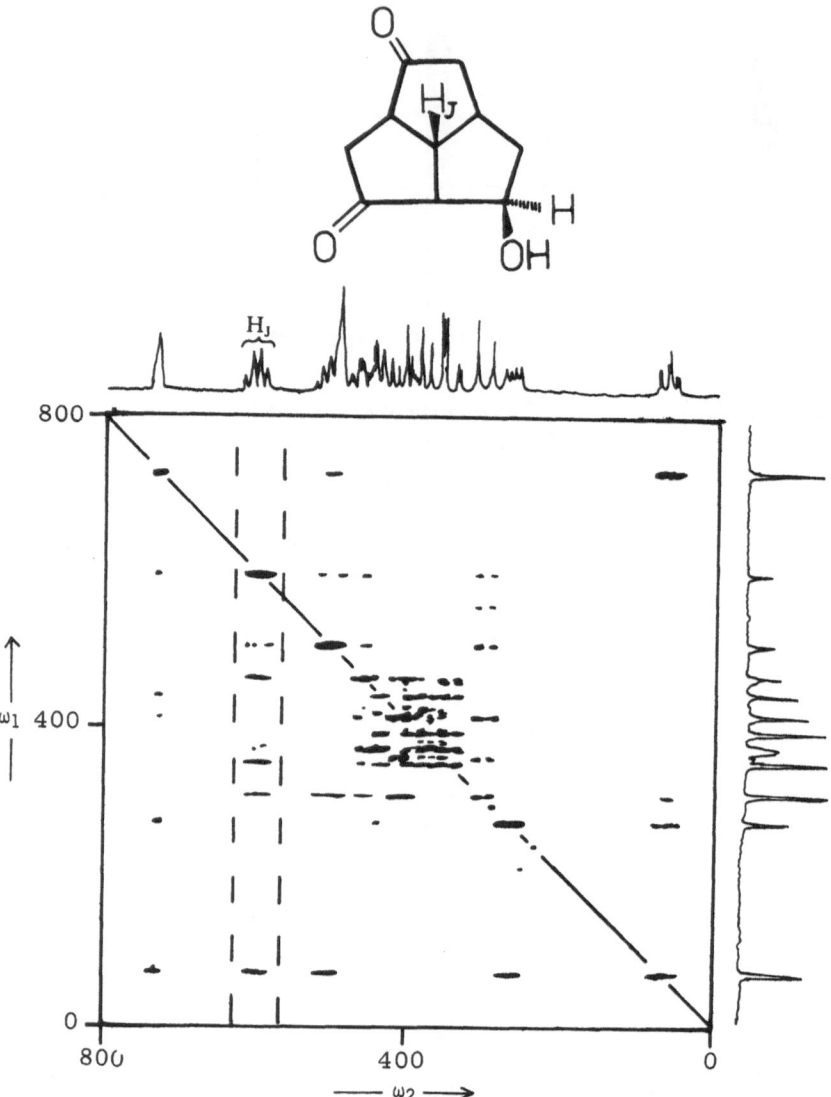

Figure 5.65. 2D-Shift correlated spectrum of the tricyclodecane derivative, showing decoupling in the ω_1 dimension through the use of the pulse sequence shown in Figure 5.64. The spectrum on the top of the diagram is the fully coupled spectrum; in the right margin is the decoupled projection of the 2D spectrum. The area enclosed by the two vertical dashed lines corresponding to the multiplet for proton J contains cross peaks. Projection to the right from these cross peaks affords the various chemical shifts of protons to which proton J is coupled.

2D contour plot is the fully coupled spectrum while the spectrum on the right of the 2D plot is the fully decoupled spectrum. If one wishes to see which protons are coupled to, say, the J protons, a vertical cross section spanning the multiplet is taken. Lines are projected to the right from the contours falling within this cross section so that they meet the singlets appearing in the fully decoupled spectrum on the right. It can be seen that such unambiguous assignments would be difficult if coupled spectra were present on both axes, particularly when there are several overlapping multiplets coupled to each other.

5.3.2.2.1.1 COSY via Zero-Quantum Coherence. It has recently been shown that by employing a modified pulse sequence (Figure 5.66), it is possible to excite zero-quantum coherence which can be used for mapping spin–spin couplings. The two-dimensional proton–proton zero-quantum spectrum of *n*-butanol with detection pulse P_4 set at 90° and 45° are shown in Figure 5.67A and 5.67B, respectively. The horizontal lines in Figure 5.67A or the diagonal lines in Figure 5.67B show the connectivities between coupled protons. A 45° observation pulse (as in Figure 5.64B) is preferable to a 90° pulse since there are only strong zero-quantum peaks between directly connected spins, and cross peaks due to long-range couplings are suppressed.

Zero-quantum coherence has also been recently used as an alternative to double-quantum coherence for establishing carbon–carbon connectivities. This is shown for the ^{13}C-NMR spectrum of *n*-butanol in Figure 5.67C. The diagonal lines establish the connectivities between coupled carbons. An advantage of the zero-quantum coherence spectra, which resemble the SECSY spectra in appearance, is that in the latter, one usually encounters a large ridge of unresolved peaks near $F_1 = 0$, but this is largely suppressed in zero-quantum spectra.

5.3.2.2.1.2 Two-Dimensional Chemical Shift Correlation Using Water Suppression Pulses. One of the problems in recording 2D-NMR spectra of dilute solutions is the strength of the solvent peaks, which often mask a region

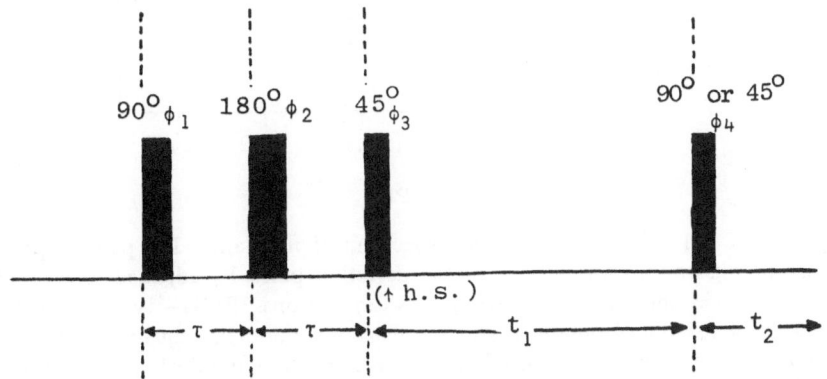

Figure 5.66. Pulse sequence used in zero-quantum coherence experiments.

of the NMR spectrum. An obvious method to overcome this is to use gated low-power irradiation of the solvent signals at all times during the spectral measurements except during the time periods t_1 and t_2. This approach, however, suffers from the drawback that sensitivity of the experiment is significantly reduced. A number of pulse sequences have been developed to achieve solvent suppression. A recent development has been to use the pulse sequence:

$$\alpha(X) - \tau - 3\alpha(-X) - \tau - 3\alpha(X) - \tau - \alpha(-X)$$

written in a shortened form as $1\bar{3}3\bar{1}$ (where α is the basic flip angle, bars indicate 180° phase shifts, and τ is the delay). Both the 90° pulses of the Jeener

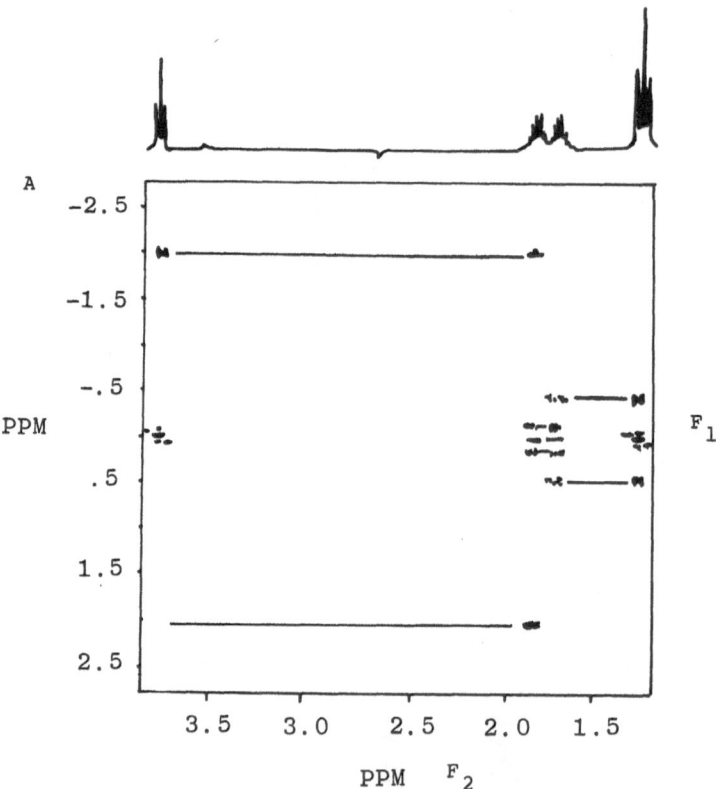

Figure 5.67. (A) ^{1}H–^{1}H zero-quantum spectrum of *n*-butanol. The pulse sequence shown in Figure 5.66 was used with the detection pulse P_4 adjusted to 90°. The horizontal lines show the connectivity between protons. (B) ^{1}H–^{1}H zero-quantum spectrum of *n*-butanol obtained with the detection pulse P_4 adjusted to 45°. The diagonal lines establish the connectivities between protons. (C) Carbon–carbon zero-quantum spectrum of *n*-butanol. The dotted vertical arrows indicate that the two ends of the diagonal lines establish the connectivity between carbon atoms "a" and "b". Similarly, carbon "b" is linked to "c" and carbon "c" to "d".

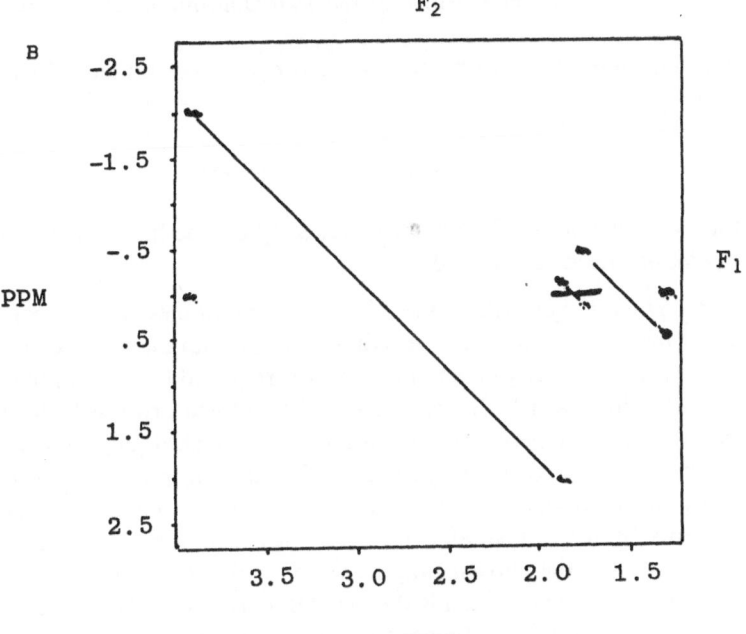

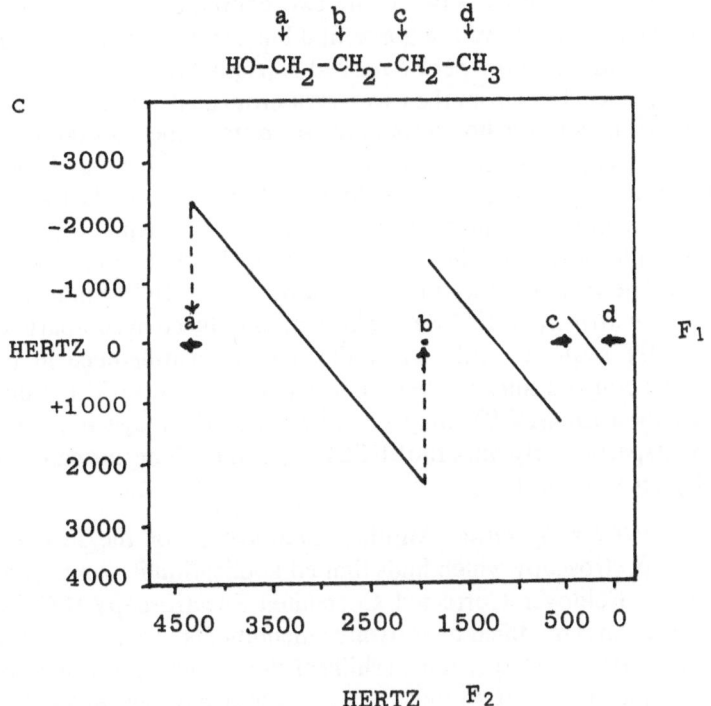

Figure 5.67. (*Continued*)

shift correlation sequence $[90°-t_1-90°-acq(t_2)]$ can be replaced by the pulse sequence:

$$1\bar{3}3\bar{1} - t_1 - 1\bar{3}3\bar{1} - acq(t_2)$$

to afford, for instance, a COSY spectrum of lysozyme in water in which the water peak has been suppressed.

5.3.2.2.2 SECSY Spectra. One of the disadvantages in the Jeener shift correlated spectra described above is that the computer memory requirements are very large, and if a high resolving power is required then the data may run into several million words. In cases where the coupled protons lie fairly close together, it is advantageous to use an alternative technique, known as *Spin Echo Correlated Spectroscopy* (SECSY). The mixing pulse is placed in the middle of the t_1 period and one can cut down the matrix storage space requirements by using a smaller ω_1 frequency (provided that the coupling protons fall within the chosen range). An example of a SECSY spectrum is presented in Figure 5.68, which is the ^1H shift correlated 2D-NMR spectrum of an annulene derivative. The contours corresponding to the normal spectrum appear on the central horizontal line at $\omega_1 = 0$ while the cross peaks appear above and below this line at distances which are equal to half the chemical shift differences between the two coupling nuclei. Thus the cross peaks marked H_3H_4 above the horizontal line are connected by means of an oblique line with the cross peaks marked H_3H_4 below the horizontal line. If one drops vertical lines from the two sets of cross peaks located at the ends of the oblique line onto the horizontal line (see dotted lines in Figure 5.68), they are seen to intersect the horizontal line at the chemical shifts of the coupled protons, H_3 and H_4. It may be noted that all oblique lines make the same angle with the horizontal line and that the cross peaks lie at equal distances above and below the horizontal line, which greatly simplifies the recognition of couplings. The various coupling interactions are shown in Figure 5.68.

The data matrix in the SECSY experiment is reduced in comparison to that in the COSY sequence only when the maximum difference in frequency between the coupled nuclei is less than half the spectral width. Since this is not often the case, SECSY may not have any advantage over the COSY experiment, particularly since the SECSY experiment is accompanied by a loss of resolution and sensitivity.

5.3.2.2.3 FOCSY Spectra. Another modification of the Jeener 2D-shift correlated spectroscopy which finds limited application in some special cases is known as Foldover Corrected Correlated Spectroscopy (FOCSY). The COSY spectrum is modified by software amendments to correct folding errors along the ω_1 axis, and the chemical shifts of the normal spectrum appear on a horizontal line at $\omega_1 = 0$, so that the 2D contour plot resembles the SECSY spectrum. Again no reduction in data matrix size is achieved unless the largest chemical shift difference between coupled nuclei is smaller than half of the spectral width.

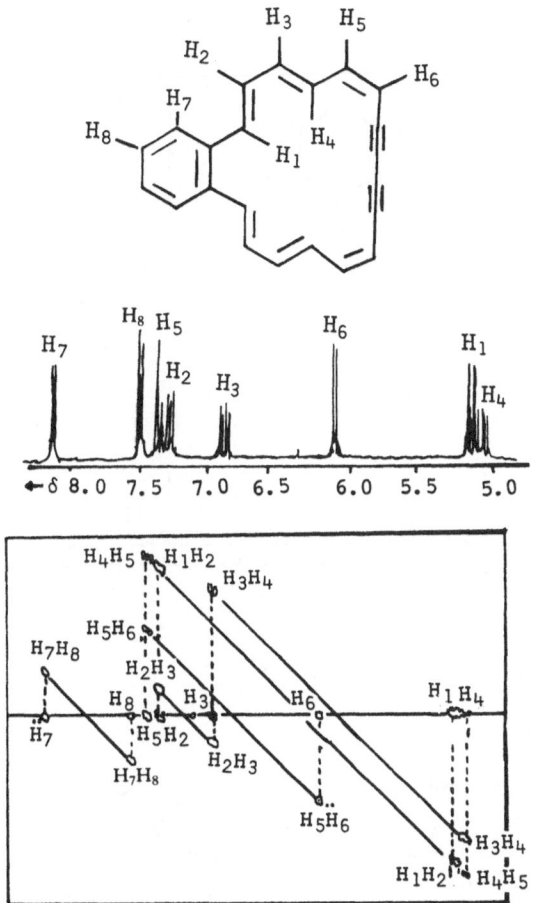

Figure 5.68. Homonuclear spin-echo correlated 2D spectrum (SECSY) of 9, 11-didehydro[18]annulene derivative.

5.3.2.2.4 DISCO spectra. Accurate values of coupling constants are conventionally obtained by double resonance methods or by 2D *J*-resolved spectroscopy. However, in complicated spin systems with a large number of couplings, these methods fail. It is well known that the cross peaks in ^{1}H, ^{1}H-COSY spectra contain information about the size of a coupling constant. A procedure has been developed in which phase-sensitive COSY spectra are recorded, and by using *differences* and *sums* of traces of cross peaks or diagonal peaks in the *COSY* spectra (DISCO), the values of the coupling constants can be obtained.*

The power of correlated spectroscopy in structure elucidation is further

---

* N. Kessler, A. Muller, and H. Oschkinat, Differences and sums of traces within COSY spectra (DISCO) for the extraction of coupling constants: "decoupling" after the measurement, *Magn. Resonance in Chemistry*, **23** (10), 844 (1985) and references therein.

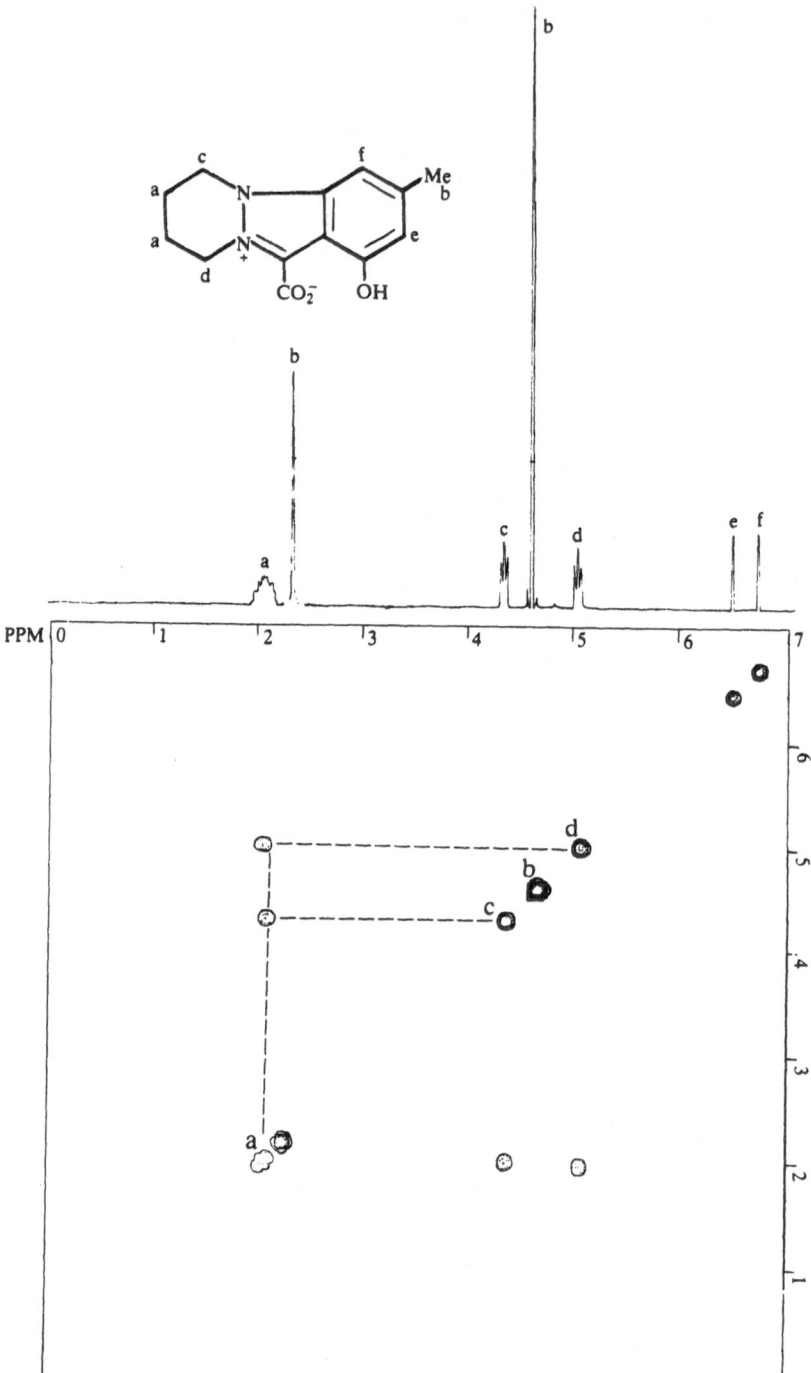

Figure 5.69. 2D-Homonuclear shift correlated ¹H-NMR spectrum of nigellicine, a new alkaloid isolated from *Nigella sativa* ("kalonji"). The circles on the diagonal line correspond to the peaks of the ¹H-NMR shown on the horizontal axis but give a view of them from above. The cross peaks above and below the diagonal establish the connectivities between "a" and "c"/"d".

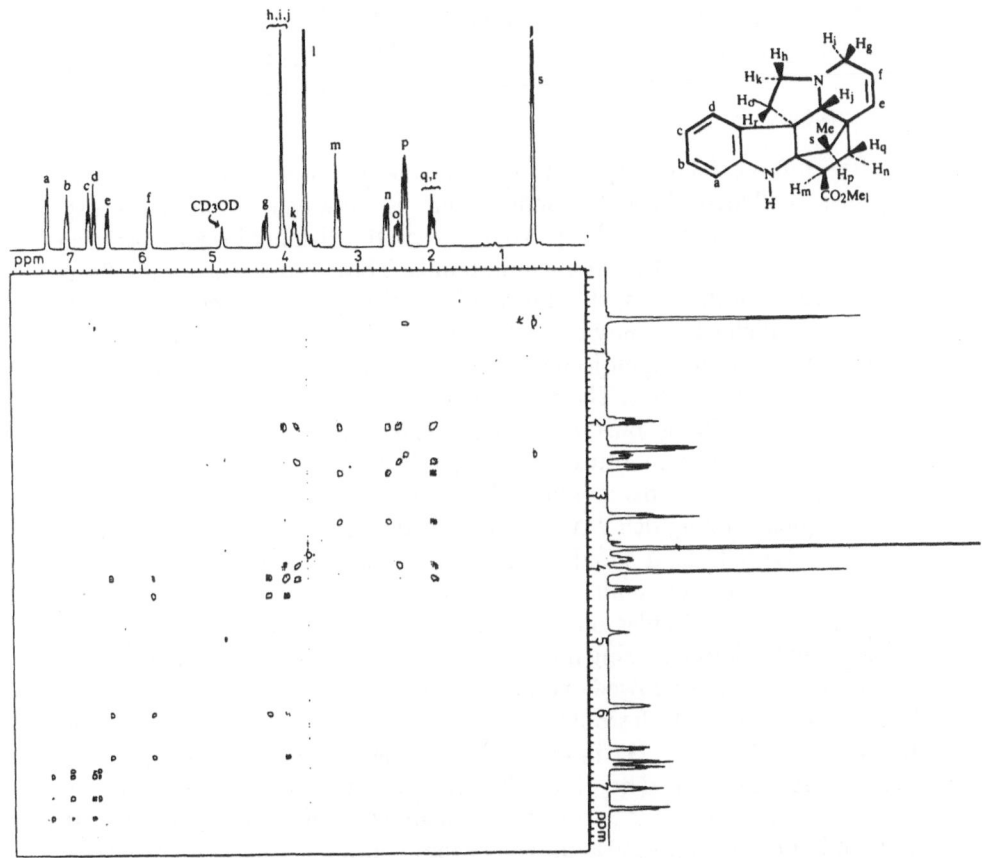

Figure 5.70. 2D-Homonuclear shift correlated ¹H-NMR spectrum of *19S-vindolinine*. The cross peaks establish the connectivities between the protons.

shown by the 2D plots for nigellicine and 19S-vindolinine, alkaloids isolated by Atta-ur-Rahman and coworkers from *Nigella sativa* and *Catharanthus roseus*, respectively. The two downfield triplets "c" and "d" in nigellicine appear at first sight to be coupled to one another but the 2D-homonuclear correlated spectrum (Figure 5.69) establishes that this is not so, and that both triplets are actually coupled to the upfield four-proton multiplet, "a". Examination of the structure of nigellicine shows why this should be the case. The farthest downfield triplet, "d", belongs to the two methylene protons α to the quaternary nitrogen while the other triplet, "c", is due to the methylene protons adjacent to the neutral nitrogen atom. The upfield multiplet "a" is due to the four protons of the central methylene groups.

The 2D-homonuclear correlated NMR spectrum of 19S-vindolinine is presented in Figure 5.70. The two ¹H-NMR spectra appear on the two axes, and each proton in this fairly complex molecule has been assigned on the basis of the coupling cross peaks.

5.4 CHEMICAL SHIFT CORRELATIONS THROUGH CROSS-RELAXATION AND EXCHANGE

5.4.1 NOESY Spectra

The 2D-shift correlated spectra discussed above were based on scalar couplings mediated through bonds. There is, however, a dipolar cross-relaxation mechanism (NOE) which depends on the distance in space between the nuclei, and not on the intervening bonds. These cross-relaxation effects are very useful in the study of conformations of large molecules, and signals arising from such effects may be correlated in 2D plots. Nuclei may also exchange their positions in molecules, and such exchanges can also be studied by 2D-NMR spectroscopy. The pulse sequence which allows the NOE effects to become active after a constant mixing time is presented in Figure 5.71. The first 90_x° pulse results in transverse magnetization, which precesses in the x',y'-plane during t_1. This is followed by a 90_x° mixing pulse. The value of the subsequent mixing delay Δ is critical to the success of this experiment. To study NOE effects, Δ should be approximately equal to the spin–lattice relaxation time (T_1) for small and medium-sized molecules, but it should be shorter for large molecules on account of negative NOE and rapid cross-relaxation in the latter. During this mixing time, magnetization transfer due to dipolar interactions is allowed to take place. The value of the mixing time may be varied between 0 (which would give the normal COSY spectrum) and, at the other extreme, $1/J(^{13}C, ^1H)$, in which case the heteronuclear shift correlated spectrum is obtained. The third 90_x° pulse serves to transform the z-magnetization produced into transverse magnetization, which is detected. If the 90_x° mixing pulse is placed at the end of the evolution period (Figure 5.71A), then the NOESY spectrum obtained is very much like a COSY spectrum in appearance (except, of course, that the cross peaks correspond to dipolar couplings through space). If, however, the 90_x° mixing pulse is applied at $t_1/2$ (Figure 5.71B), then the NOESY spectrum resembles the SECSY matrix form. Figure 5.71C shows the pulse sequence used for obtaining 2D-exchange spectra in ^{13}C-NMR spectroscopy. Chemical exchange within a molecule will also give rise to cross peaks when the NOESY sequence is applied. This is shown in the 2D-exchange spectrum recorded for the keto–enol tautomerism of acetylacetone (Figure 5.72). The intensity of the cross peaks depends on the ratio of the mixing time t_m (which is constant during 2D measurements) to the reciprocal of the rate of exchange (or dipolar cross-relaxation time, if NOE effects are being studied). Thus in suitable cases it is possible to separate the effects operating through the two mechanisms (i.e., cross-relaxation and exchange) by an appropriate choice of t_m.

5.4.2 Two-Dimensional Heteronuclear NOE (HOESY)

The homonuclear 2D experiment (NOESY) developed by Ernst and coworkers has been extended to heteronuclear 2D NOE (HOESY) experi-

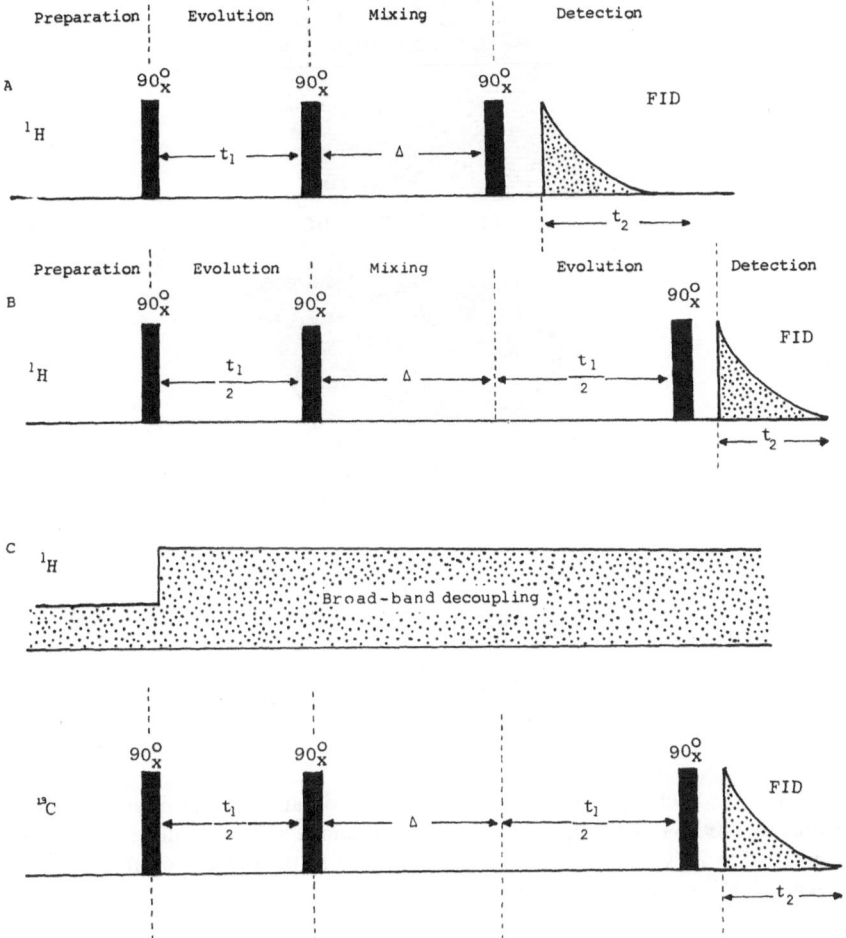

Figure 5.71. Pulse sequence for 2D-exchange NMR spectroscopy. (A) NOESY sequence for recording 2D ^{1}H-NOE spectra. The spectra are recorded in the COSY matrix form. (B) NOESY sequence as above but the second 90°_x pulse is applied at $t_1/2$ [instead of the end of the evolution time t, as in (A)]. The mixing delay Δ is followed by the second half of the evolution time $t_1/2$. (C) Pulse sequence for 2D-exchange spectra in ^{13}C-NMR spectroscopy.

ments for the investigation of heteronuclear dipolar interactions. The pulse sequence for the heteronuclear 2D NOE experiment and its effect on ^{1}H magnetization vectors are shown in Figure 5.73. During the initial equilibrium period, the equilibrium magnetization of ^{1}H and ^{13}C spins is established. The 90° (^{1}H) pulse flips the ^{1}H magnetization into the x',y'-plane. During the

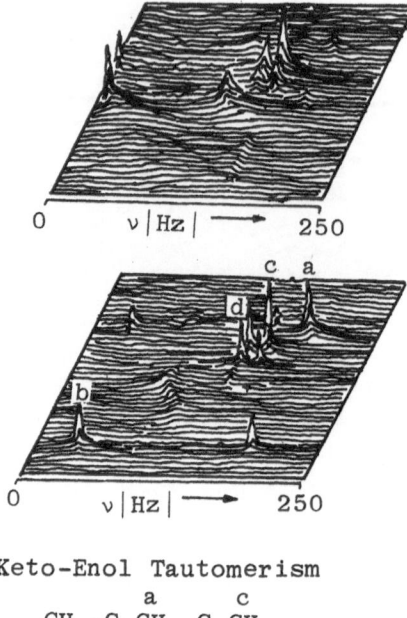

0   ν│Hz│ ⟶   250

c   a

d

b

0   ν│Hz│ ⟶   250

Keto-Enol Tautomerism

a   c
$CH_3-\underset{O}{\overset{}{C}}-CH_2-\underset{O}{\overset{}{C}}-CH_3$

$CH_3-\underset{O}{\overset{}{C}}-\underset{b}{\overset{}{C}}H=\underset{\underset{d}{OH}}{\overset{}{C}}-CH_3$

Figure 5.72. 2D-Exchange spectrum for the keto–enol tautomerism of acetylacetone measured by NOESY. The upper spectrum was obtained at 40°C while the lower one was recorded at 28°C. The broad signal corresponds to the water protons. The cross lines represent directly connected protons.

subsequent $t_1/2$ time period, the two ^1H vectors α and β precess and acquire a relative phase angle of $\pi J t_1$. A 180° ^{13}C pulse then serves to interchange the spin labels, and the two vectors therefore refocus after the subsequent $t_1/2$ time period, so that any coupling between ^1H and ^{13}C is removed. The second 90° ^1H pulse creates longitudinal ^1H magnetization and during the subsequent "mixing period," Δ, exchange of magnetization occurs between the ^1H and ^{13}C spins through dipolar interactions. The interval Δ is kept constant and the signal is recorded immediately after the 90° ^{13}C (observe) pulse. The experiment is repeated for a set of equally incremented values of t_1 as in other 2D experiments, and the data matrix $S(t_1 t_2)$ affords the desired 2D frequency-domain spectrum.

The HOESY spectrum is beneficial in establishing the spatial proximity between carbons and protons. This is particularly useful in assigning quaternary carbons which, due to their inefficient relaxation, normally afford weak signals. However, their NOEs are quite sizable due to dipolar relaxation

A

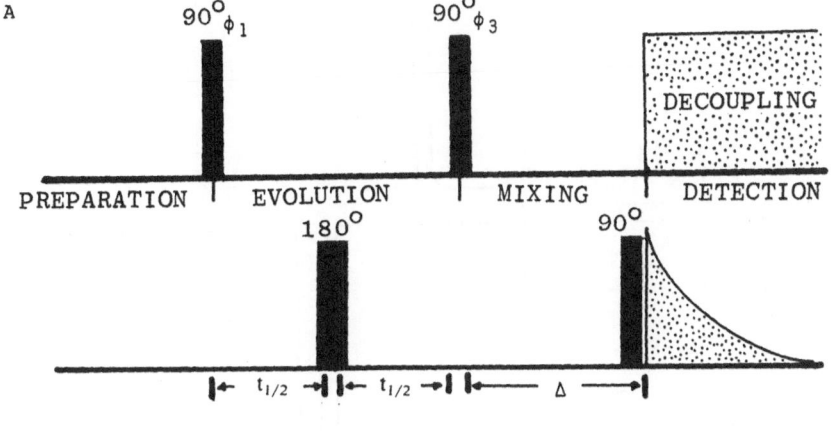

B

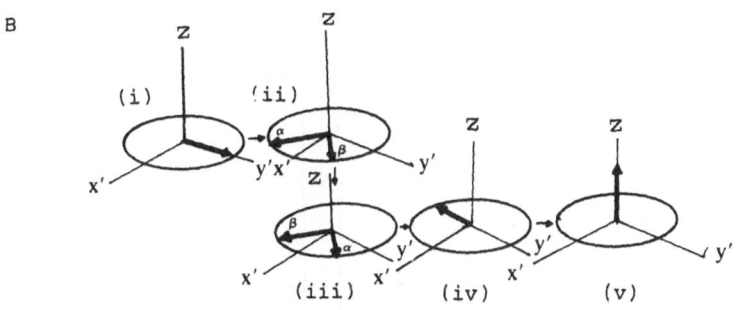

Figure 5.73. (A) Pulse sequence for the heteronuclear 2D NOE (HOESY) experiment. (B) Effect of the pulse sequence on ^1H magnetization vectors.

through protons on neighboring carbon atoms. A HOESY spectrum of camphor is shown in Figure 5.74.* The ^{13}C-NMR spectrum and the ^1H-NMR spectrum are drawn on the two axes of the 2D plot. The contours represent NOE interactions between ^1H and ^{13}C nuclei. The three cross peaks which have been marked show the NOE interactions between the quaternary carbons and the protons on the neighboring methyl groups.

5.4.3 Relayed NOESY

NOE cross peaks often occur in crowded regions, which make the assignments difficult. A new pulse sequence has been recently described which combines relayed magnetization transfer through the coupled spins with magnetization transfer through space by the nuclear Overhauser effect to afford a two-

---

* C. Yu and G.C. Levy, Two dimensional heteronuclear NOE (HOESY) experiments, *J. Am. Chem. Soc.* **106**, 6533 (1984) and references therein.

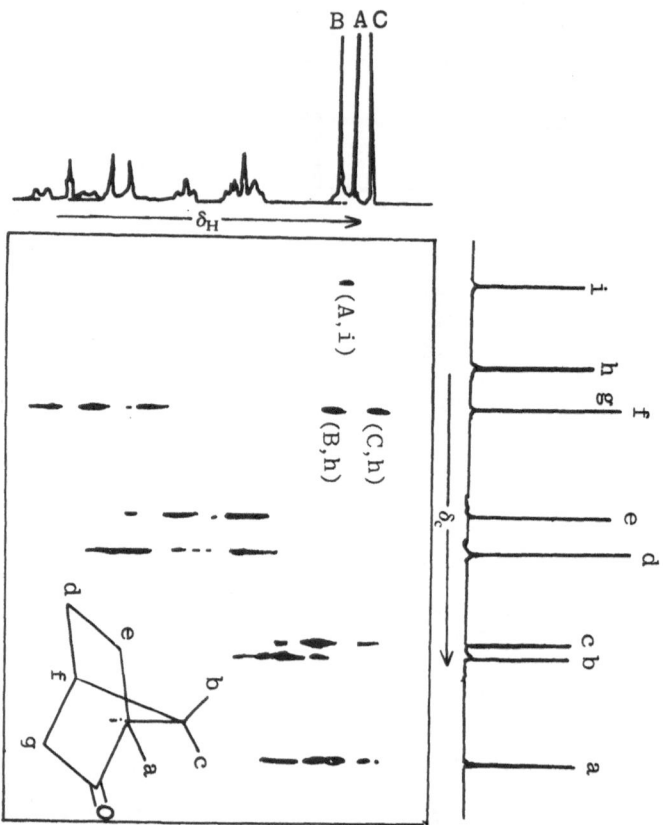

Figure 5.74. HOESY spectrum for camphor. The cross peaks arising from interactions between quaternary carbons and nearby protons are indicated. Peaks marked A, B, and C are those of methyl protons. The corresponding carbon atoms are labeled a, b, and c in the ^{13}C-NMR spectrum.

dimensional relayed NOESY spectrum.* Relayed coherence transfer spectroscopy involves transfer of magnetization from nucleus A to nucleus B, and then in a second transfer step to a third "spy" nucleus C which is detected. If nucleus A is several bonds away but in spatial proximity to nucleus B (or if nucleus B is similarly spatially close to nucleus C), then one of the two magnetization transfer steps can involve an incoherent transfer of magnetization via NOE.

The pulse sequence used for relayed NOESY (Figure 5.75A) is identical to the normal NOESY pulse sequence until the third 90° pulse. During the preceding period, τ_m, magnetization is transferred from the spatially remote spin (H_r) to the neighbor spin (H_n) via NOE. The third 90° pulse converts the

* G. Wagner, Two dimensional relayed coherence transfer-NOE spectroscopy, *J. Magn. Resonance* **57**, 497 (1984).

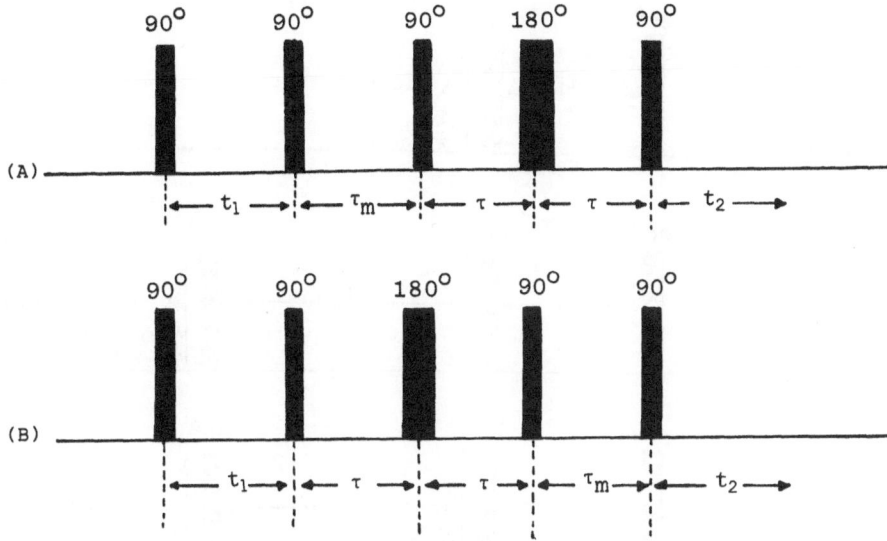

Figure 5.75. Pulse sequences used for relayed NOE: (A) an incoherent transfer of magnetization occurs, followed by a coherence transfer of magnetization; (B) a coherence transfer takes place first, and is then followed by an incoherent transfer of magnetization.

longitudinal magnetization of H_n into transverse magnetization, which is allowed to dephase during the subsequent 2τ period due to J coupling. The last 90° pulse transfers magnetization from H_n to the "spy" nucleus being observed, H_s, and this is detected during t_2.

$$H_r(\omega_1) \xrightarrow{\text{NOE}} H_n \xrightarrow{J} H_s(\omega_2)$$

The 2D spectrum contains three types of cross peaks: relayed NOE connectivities between remote nuclei, NOE connectivities between neighbor nuclei, and J connectivities between neighbor nuclei.

An alternative pulse sequence (Figure 5.75B) can also be used in which the magnetization transfer through J coupling between H_r and H_n occurs prior to the magnetization transfer through space (NOE) between H_n and H_s.

5.4.4 Simultaneous 2D Correlated (COSY) and 2D Nuclear Overhauser Enhancement (NOESY) Spectroscopy (COCONOSY)

In spite of the powerful applications of the COSY and NOESY sequences in structure elucidation of complex molecules, one of the drawbacks, specially with biological samples, is the long acquisition times which are sometimes necessary, and the tendency of samples to deteriorate over long acquisition

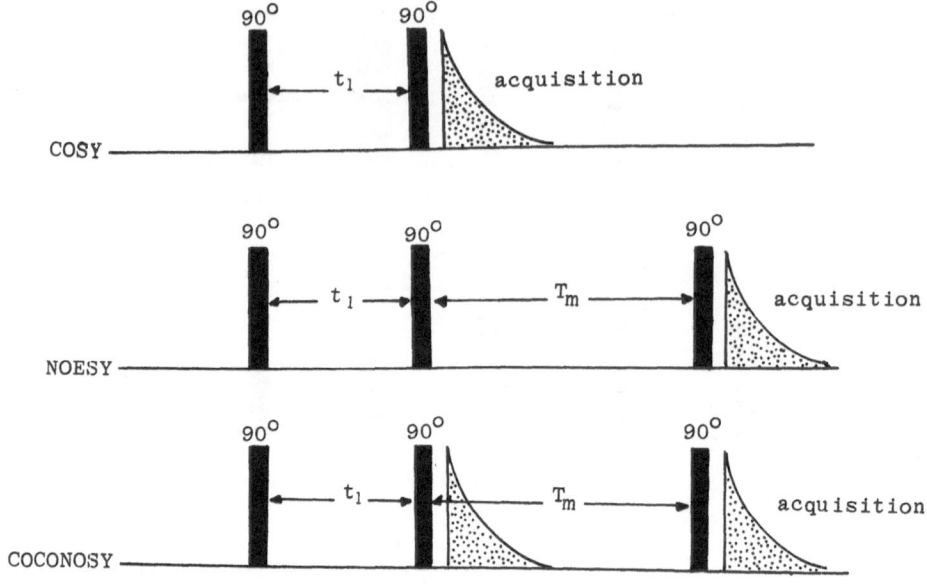

Figure 5.76. COSY, NOESY, and COCONOSY pulse sequences.

periods. The desirability of recording COSY and NOESY spectra under *identical* conditions has led to the development of a new pulse sequence, "COCONOSY," which is shown in Figure 5.76.* This involves an extension of the normal COSY sequence, after acquisition, by adding an extra 90° pulse and a separately stored data acquisition, thus allowing simultaneous recording of COSY and NOESY data in a single COCONOSY (*Com*bined *Co*rrelated and *N*uclear *O*verhauser Enhancement *S*pectroscop*y*) experiment.

5.5   TWO-DIMENSIONAL HETERONUCLEAR RELAYED COHERENCE TRANSFER (RCT) SPECTROSCOPY

The most informative source for deducing connectivities in organic molecules are proton and carbon coupling constants. There are three main methods which can be used for determining proton and carbon connectivities: (A) Homonuclear proton 2D-shift correlation spectroscopy combined with

---

* C.A.G. Haasnoat, F.J.M. Van de Ven, and C.W. Hilbers, COCONOSY combination of 2D correlated and 2D nuclear Overhauser enhancement spectroscopy in a single experiment, *J. Magn. Resonance* **56**, 343 (1984).

heteronuclear 2D carbon–proton shift correlation experiment — the first of these experiments shows which protons are coupled to which other protons while the second establishes the coupling of protons to carbon. An analysis of the results of these two experiments allows one to indirectly deduce carbon–carbon connectivities to protonated carbon atoms. (B) An alternative procedure for directly determining carbon–carbon connectivities is by the INADEQUATE method. (C) A new technique which has been recently developed for establishing connectivity between two remote nuclei within a given spin system is known as Relayed Coherence Transfer (RCT) spectroscopy. The magnetization (or coherence) is transferred in two steps. In the first step, magnetization is transferred from one nucleus (A) to a second nucleus (M) by scalar coupling interaction. This transferred magnetization is then again transferred from the second nucleus to a third nucleus (X). Thus the first two nuclei (A and M) may be protons while the third nucleus X may be ^{13}C. Since it is the ^{13}C spectrum which is finally observed, the experiment is less dependent on a well-resolved proton spectrum. The original relay experiment required a complex sixteen-step pulse sequence but this has recently been simplified to a four-step sequence shown in Figure 5.77. The effect of the pulse sequence on the magnetization vectors of protons A and M is also shown in Figure 5.77.

Let us consider a fragment C^AH^A–C^BH^B. Relayed magnetization transfer can occur by two distinct pathways: $H^A \rightarrow H^B \rightarrow C^B$ or $H^B \rightarrow H^A \rightarrow C^A$. In the relayed 2D spectra, these pathways would afford two cross peaks at $(\omega_1 = \delta_{HA}, \omega_2 = \delta_{CB})$ and $(\omega_1 = \delta_{HB}, \omega_2 = \delta_{CA})$, respectively. These cross peaks would be in addition to the two cross peaks arising from the conventional heteronuclear shift correlation experiment which gives peaks by single step magnetization transfer at $(\omega_1 = \delta_{HA}, \omega_2 = \delta_{CA})$ and $(\omega_1 = \delta_{HB}, \omega_2 = \delta_{AB})$. The four cross peaks would appear on the four corners of a rectangle, and the pattern would indicate that the two carbons corresponding to the cross peaks are bonded to one another.

In the conventional heteronuclear shift correlation experiment, the value of the first half of the mixing time Δ_1 was kept at $1/(2J_{CH})$ during which the ^{13}C satellites in the ^{1}H spectrum were allowed to acquire opposite phases (see Figure 5.51). Magnetization could then be transferred from the ^{1}H to the ^{13}C nuclei by simultaneous application of 90° pulses to both the ^{1}H and ^{13}C nuclei.* During the subsequent half of the mixing time Δ_2 the antiphase ^{13}C magnetization was converted to in-phase magnetization before detection.

---

* For a more detailed discussion on the mechanism of magnetization transfer see: (i) A.A. Maudsley, L. Müller, and R.R. Ernst, Cross-correlation of spin-decoupled NMR spectra by heteronuclear two-dimensional spectroscopy, *J. Magn. Resonance*, **28**, 463 (1977); (ii) G.A. Morris and L.D. Hall, Experimental chemical shift correlation maps from heteronuclear two-dimensional NMR spectroscopy 1. Carbon-13 and proton chemical shifts of raffinose and its subunits, *J. Am. Chem. Soc.*, **103**, 4703 (1981).

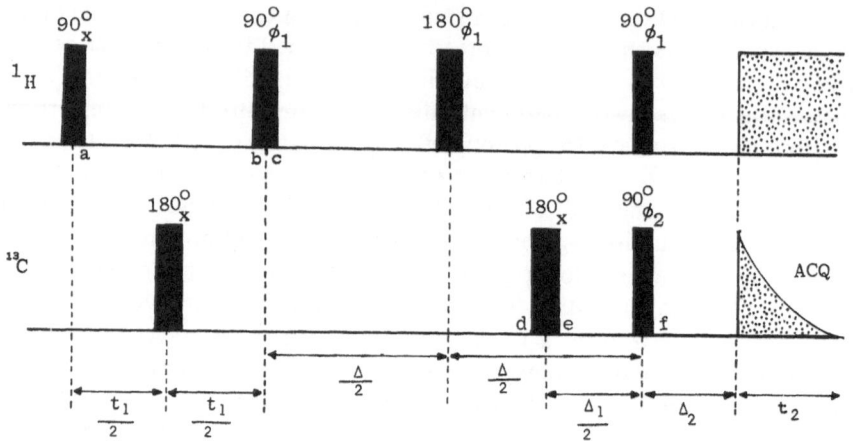

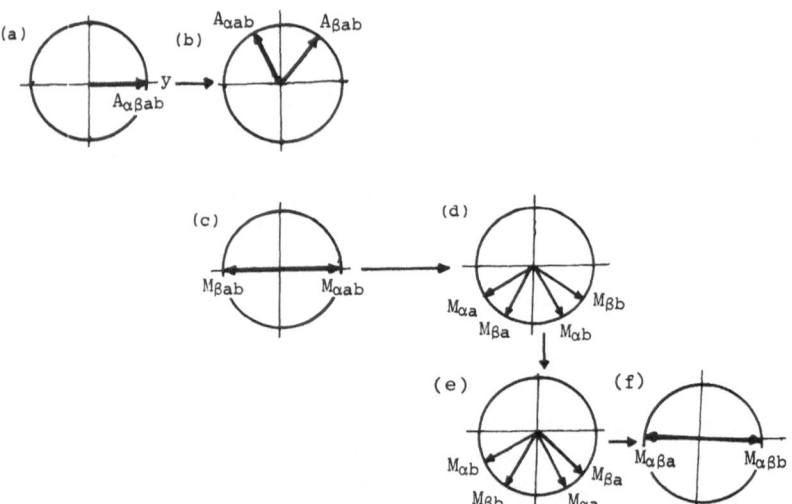

Figure 5.77. Pulse sequence for the heteronuclear Relayed Coherence Transfer (RCT) experiment, and the behavior of the magnetization vectors A and M. The indices on the nuclei A and M represent the spin states of the coupled proton (α or β) and the ^{13}C nucleus (a or b).

The first 90° proton pulse rotates the longitudinal magnetization to the y' axis [Figure 5.77(b)]. The 180° ^{13}C pulse at the center of the evolution period removes the overall effects of heteronuclear coupling which would have otherwise existed at the end of the evolution period. The second nonselective

90° proton pulse (which is the pulse which induces magnetization transfer from H_A to H_M) can be considered to be made up of a cascade of two hypothetical semiselective 90° proton pulses, applied to A and M protons one after the other. The hypothetical 90° H_A pulse changes the longitudinal magnetization of H_A, and since H_A is coupled to H_M, it results in contributions to the longitudinal magnetization of H_M. The hypothetical M pulse rotates this contribution of the longitudinal M magnetization so that it comes to lie along the y' axis. The transferred magnetization components are in antiphase just after the 90° magnetization transfer pulse and lie along the $+y$- and $-y$-axes [Figure 5.77(c)]. During the subsequent delay time Δ, two pulses are applied. A 180° proton pulse is applied at the center of this delay time which serves to suppress precession due to proton chemical shifts and heteronuclear coupling but does not affect precession due to J coupling. At time $\Delta_1/2$ [which is kept equal to $1/(4J_{MX})$], before the end of the Δ interval, a 180° ^{13}C pulse is applied which flips the spin state of the ^{13}C nucleus, and results in the heteronuclear M multiplet vectors becoming aligned along the $\pm y$-axis at the end of the Δ time interval [Figure 5.77(f)]. As in the INEPT experiment a 90° proton pulse is then applied simultaneously with the 90° ^{13}C pulse. This causes the second step in the transfer of magnetization to occur, i.e., transfer of magnetization from proton M to the ^{13}C nucleus. In order to achieve quadrature detection in both frequency domains (i.e., to position the ^1H and ^{13}C irradiating frequencies in the middle of their respective spectral ranges) the phases of the first proton pulse and the last carbon pulse are cycled through $x, y, -x$, and $-y$, and the signals are alternatively added and subtracted. By using this variation of the original relayed technique, an improvement in sensitivity of $\sqrt{2}$ may be obtained.

This brings us to the question: How much material is required in order to record conventional heteronuclear 2D-shift correlated spectra and relayed coherence transfer spectra? As mentioned in an earlier chapter, the sensitivity increases with increasing proton resonance frequency of the NMR spectrometer. On a 300-MHz instrument, one requires a minimum solution concentration of $0.02M$ (i.e., 0.05 mmol material or 20 mg of a compound of molecular weight 400) in order to record a 2D $^1H-^{13}C$ shift correlation spectrum over a 12 h period. Since the coherence transfer functions are below unity, about 4 times more substance is required in order to record a relayed spectrum (i.e., ~ 80 mg of a substance of molecular weight 400).

An example of the 2D-heteronuclear relayed coherence transfer experiment is provided by the 2D contour plot of 2-acetonaphthalene shown in Figure 5.78. The 1D proton and ^{13}C NMR spectra are drawn on the axes. The resonances due to direct (nonrelayed) coupling interactions are indicated with an asterisk, and they were determined by a conventional COSY experiment. The cross peak marked A shows a direct coupling between C-3 and the protons corresponding in position on the vertical axis to the cross peak A (see dashed line). The cross peak at B shows that the carbon marked 4 is *indirectly* coupled to the *same* protons. Therefore carbon 3 must be adjacent to carbon 4.

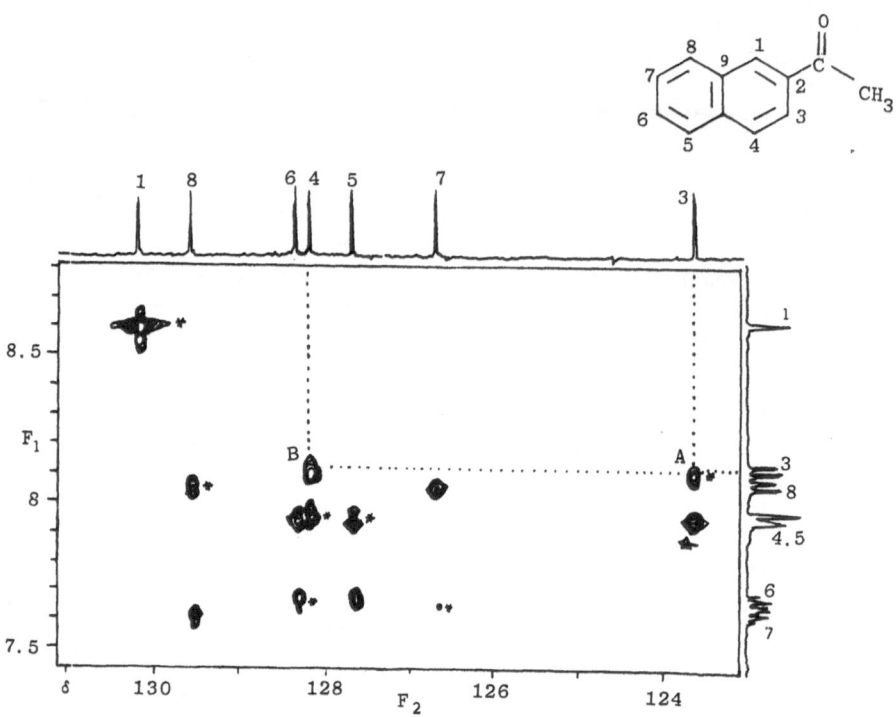

Figure 5.78. A 2D-heteronuclear relayed coherence transfer (RCT) spectrum of 2-acetonaphthalene. The cross peaks marked with an asterisk are due to nonrelayed magnetization, as determined by a normal heteronuclear shift correlated experiment.

Similarly, the connectivities between C-5 and C-6 or between C-7 and C-8 can be established from Figure 5.78.

5.5.1 Heteronuclear Relayed Proton Correlated Spectroscopy (HERPECS)

In the relayed coherence transfer experiment the first step is an initial homonuclear magnetization transfer (i.e., from proton A to proton B). The coherence is then relayed to a third nucleus, which may be a proton (homonuclear relay) or a different nucleus (e.g., ^{13}C, heteronuclear relay). Recently, another variation of the relay experiment has been reported which involves transfer of coherence to a heteronuclear spin as the first step (^1H–X heteronuclear shift correlation). The coherence is then relayed to a third nucleus, a proton (X–^1H coherence transfer), which is the nucleus observed.*

---

* M.A. Delsuc, E. Guittet, N. Trotin, and J.Y. Lallemand, Two-dimensional correlation spectroscopy with heteronuclear relay, *J. Magn. Resonance* **56**, 163 (1984).

This allows one to correlate protons which are not coupled to one another but share a common coupled partner through which coherence can be transferred (Figure 5.79A). The pulse sequence used is shown in Figure 5.80. Some of the methods for determining H–H, H–C, or C–C connectivities are summarized in Figure 5.79B.

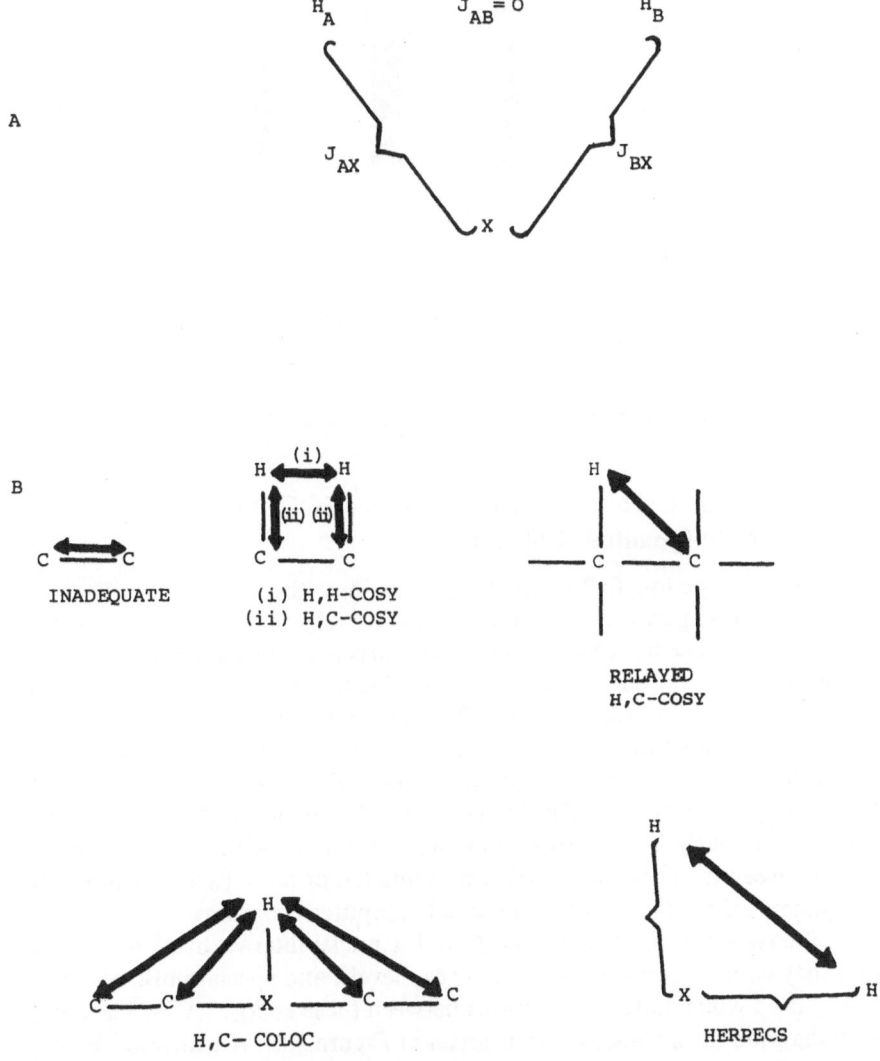

Figure 5.79. (A) In heteronuclear relayed proton correlated spectroscopy. (HERPECS), the coherence is transferred first from H_A to a heteronuclear spin, X, and is then relayed and detected at the third nucleus, H_B. (B) A schematic representation of INADEQUATE, H,H-COSY, and H,C-COSY, relayed H,C-COSY, H,C-COLOC, and HERPECS experiments.

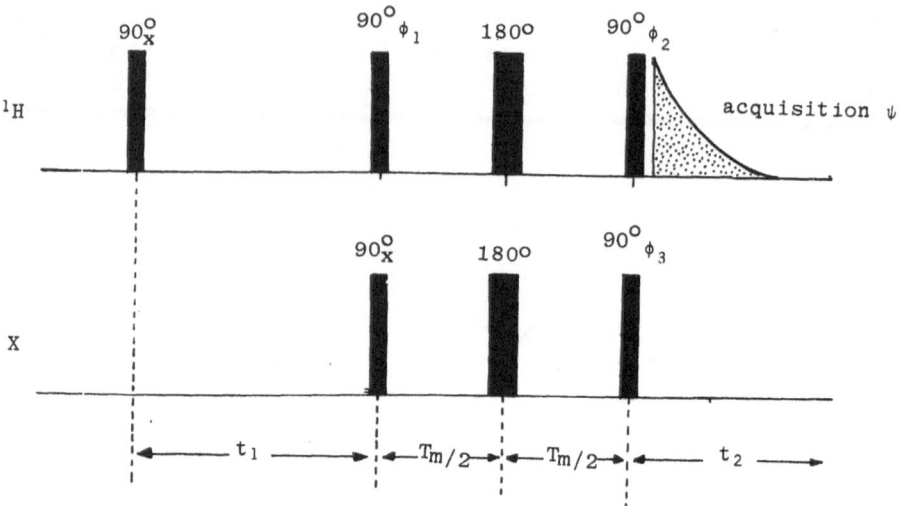

Figure 5.80. Pulse sequence for the HERPECS experiments.

5.6 DOUBLE-QUANTUM COHERENCE

5.6.1 Carbon–Carbon Connectivity Plot (CCCP) by Double-Quantum Coherence

In an earlier section (5.2.1) we have described the one dimensional IN-ADEQUATE experiment which allowed the study of ^{13}C–^{13}C couplings in natural abundance by recording the unperturbed satellite spectra. Since there is only 1% natural abundance of ^{13}C, the chances of finding two ^{13}C nuclei connected to one another is 0.01×0.01, i.e., one in ten thousand. To detect such weak interactions, it is necessary to suppress the 100 times stronger signals from molecules which contain only an isolated ^{13}C nucleus rather than two adjacent ^{13}C nuclei. The INADEQUATE assignments were made by allocating to each C—C bond two pairs of satellites with the same spacing, $^{1}J_{CC}$, between them. In molecules containing ten or more carbon atoms, such assignments are difficult without special computer programs.

In the two-spin AX system, the A and X nuclei interact in such a manner that they contain certain common energy levels, and special pulse sequences can be used which induce transitions between these energy levels. The energy level diagram for an AX system is given in Figure 5.81. A transition induced between levels 1 and 4 is a "double-quantum" transition. Application of a single nonselective pulse to a system in thermal equilibrium results in the creation of single-quantum coherence but if the system is in a nonequilibrium state, the pulse will result in the creation of multiple-quantum coherence.

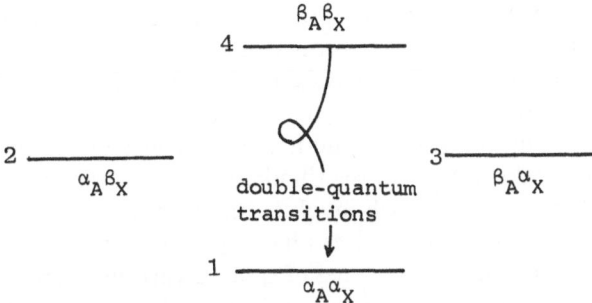

Figure 5.81. Energy level diagram of an AX spin system. A transition between levels 1 and 4 is a "double-quantum" transition.

Double-or multiple-quantum coherence cannot be detected as such but must be converted into single-quantum coherence (or observable magnetization in the x',y'-plane) by the application of a selective 180° pulse to the X spin. This results in the introduction of a phase shift of the signals which allows one to distinguish the pairs of doublets due to $^{13}C-^{13}C$ couplings from the undesired signals due to C nuclei not bound to other ^{13}C nuclei.

There are several methods for creating multiple-quantum coherence. A pulse sequence used for recording 2D-INADEQUATE spectra is given in Figure 5.82. The pulse sequence consists of two similar halves. During

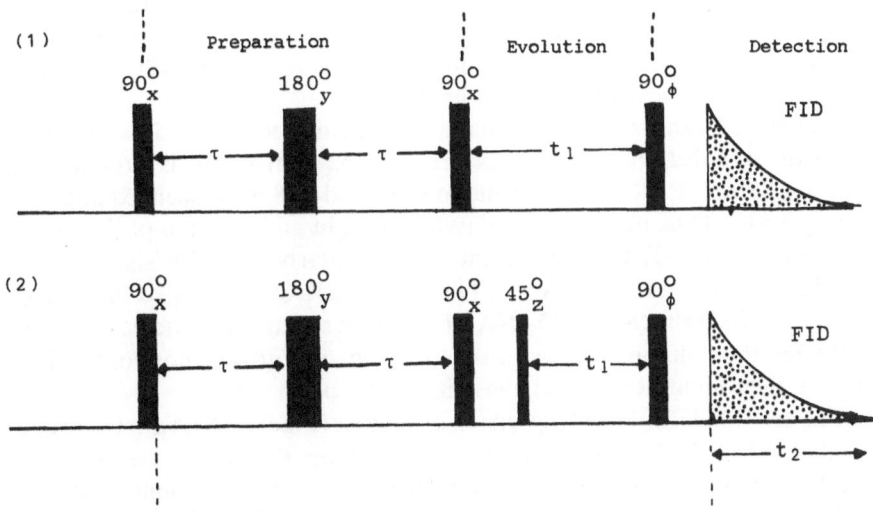

Figure 5.82. Pulse sequence for 2D-INADEQUATE experiment. The pulse sequences (1) and (2) are applied alternately to distinguish positive and negative peaks.

sequence (1), a normal 1D-INADEQUATE experiment is conducted with t_1 having a small fixed value. By the proper choice of $\tau = (2n + 1)/(4J_{CC})$, where $n = 0, 1, 2$, etc., a double-quantum coherence is created at the beginning of t_1. This double-quantum coherence evolves during t_1 according to the sum of the two coupled ^{13}C chemical shifts and is then converted back into transverse magnetization by the last 90° pulse. Elimination of N-type or P-type peaks is effected by phase cycling, which involves the introduction of a 45° pulse in sequence (2) and shifting the receiver phase by −90°. By recording a series of spectra at gradually increasing values of t_1, a 2D carbon–carbon connectivity plot can be created in which each pair of doublets from bonded carbon atoms present in the 1D-INADEQUATE spectrum are moved into the second dimension by an amount corresponding to the double-quantum frequency of the coupled spin system. This double quantum frequency depends on the shifts v_A and v_X of two bonded ^{13}C nuclei, as well as on the frequency v_0 of the excitation pulse which is kept in the center of the ^{13}C spectrum in the quadrature detection mode.

The first Fourier transformation in the t_2 domain affords a series of $^{13}C–^{13}C$ satellite spectra in which the signals are modulated by the corresponding double quantum frequencies. A second Fourier transformation in the t_1 domain results in a two-dimensional plot which contains chemical shift and $^{13}C–^{13}C$ coupling information on the t_2 axis and C—C scalar coupling information on the t_1 axis. Since bonded $^{13}C–^{13}C$ nuclei share a common double-quantum frequency, there will be two pairs of doublets for each $^{13}C–^{13}C$ system which will be found lying on a separate horizontal line, corresponding to the double-quantum frequency of the coupled nuclei. The experimental parameters are adjusted so that the two doublets are situated at equal distances on either side of a single diagonal line. This greatly facilitates identification and avoids errors due to artifacts (which will not occur symmetrically around the diagonal line).

Figure 5.83A shows the broad-band decoupled spectrum of 5α-androstane while the 1D-INADEQUATE spectrum of 5α-androstane is presented in Figure 5.83B. The ^{13}C doublets due to C-10 and C-8 have been expanded in Figure 5.83C. Thus the C-10 signal (set of upright and inverted peaks) shows couplings with C-1, C-19, C-9, and C-5. Similarly, the C-8 signal shows couplings with C-7, C-9, and C-14. The 2D-INADEQUATE spectrum shown in Figure 5.83D shows the 1D-INADEQUATE spectrum at the base and the peaks for the coupling carbon atoms located on *different* horizontal lines, making their identification very easy in the 2D plot.

The 2D-INADEQUATE spectrum of sucrose (Figure 5.84) with the broad-band decoupled ^{13}C-NMR spectrum drawn on the lower axis provides another example. One can easily assign the carbon–carbon connectivities by looking for the pairs of doublets lying on the same horizontal line and correlating them with the corresponding chemical shifts of the two connected carbon atoms.

Another example of 2D double-quantum coherence (2D-DQC) NMR is

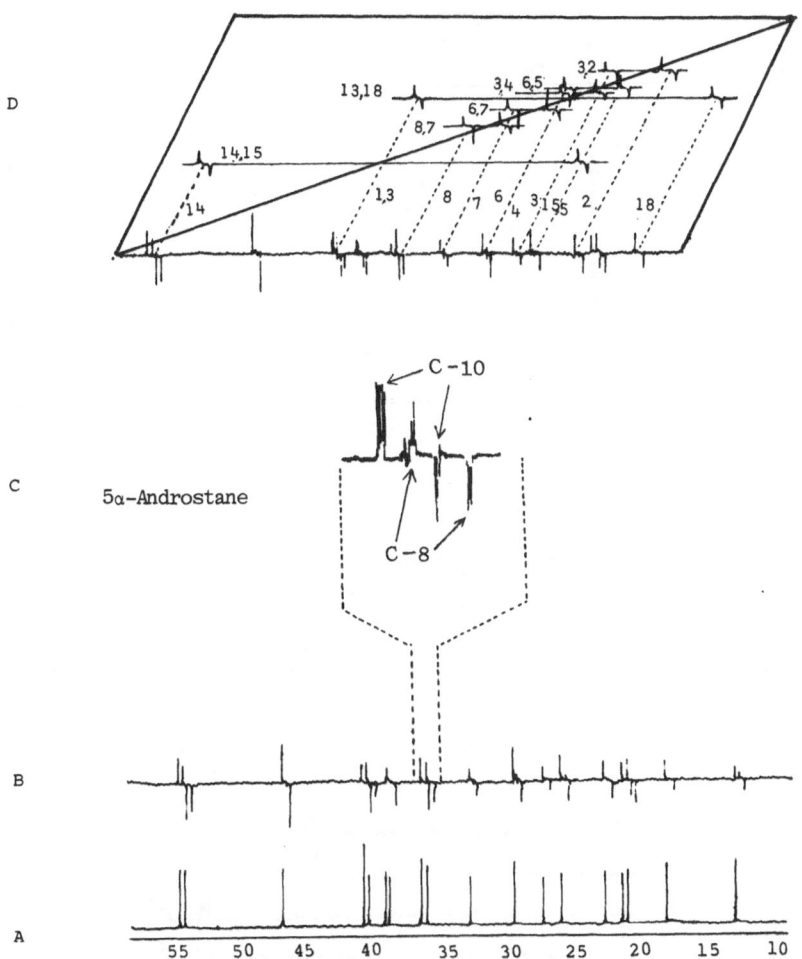

Figure 5.83. (A) Broad-band decoupled spectrum of 5α-androstane. (B) 1D-INADE-QUATE spectrum. (C) region of C-8 and C-10 expanded to show up $^{13}C-^{13}C$ couplings. (D) 2D-INADEQUATE spectrum of 5α-androstane showing the $^{13}C-^{13}C$ couplings lying on separate horizontal lines with the peaks located symmetrically on either side of a diagonal, thus greatly facilitating identification.

provided by ethyl butyl ketone. Figure 5.85A represents the one-dimensional broad-band decoupled ^{13}C spectrum with the six methylene and methyl carbon atoms identified. The 2D-DQC spectrum shown in Figure 5.85B contains two pairs of satellites on the same horizontal line for each carbon (except carbon atoms A and G, located at the end of the chain, which show only one pair of satellite peaks). Carbon atoms B and D, α to the carbonyl group, show paired satellites corresponding to the interaction with A and E, respectively, and unpaired satellites corresponding to the 1,3-interaction

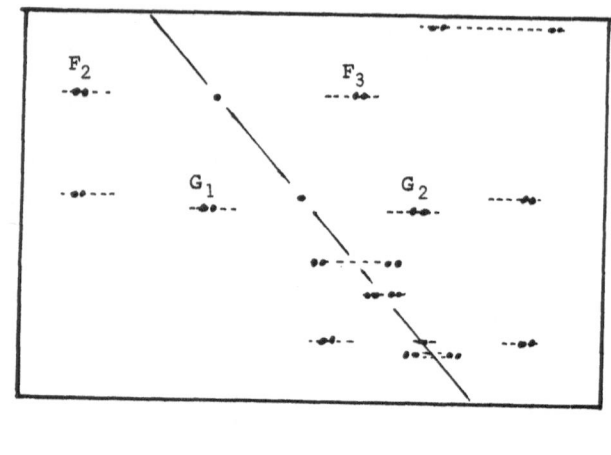

A

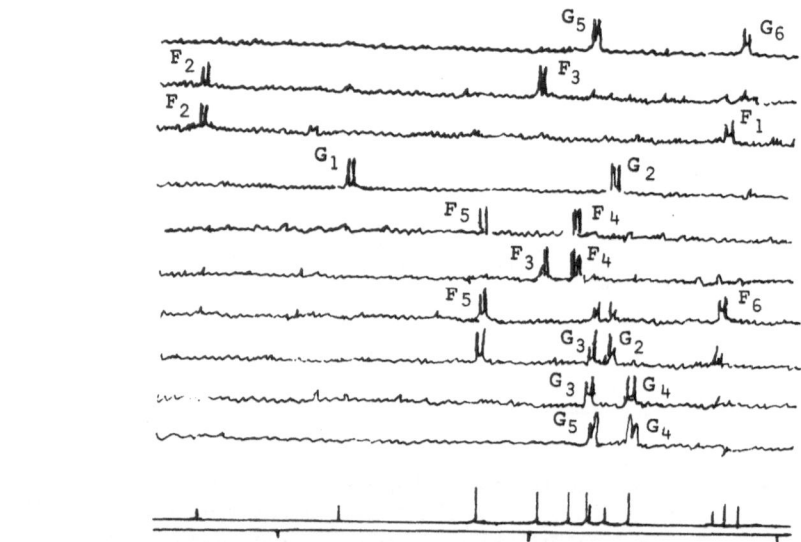

B

C

Figure 5.84. (A) ^{13}C 2D-INADEQUATE spectrum of sucrose. The coupled carbon atoms lie *symmetrically* on either side of a diagonal line. Pairs of coupled carbons appear as cross peaks on the *same* horizontal line, thus allowing them to be readily identified. (B) ^{13}C 2D-INADEQUATE spectrum but drawn as rows of matrix. (C) Broad-band decoupled ^{13}C-NMR spectrum.

between carbons B and D across the carbonyl group. By systematically proceeding from carbon D to E, then to F, and finally to G one can identify which carbon interacts with which other carbon atom in the butyl chain. Similarly, by proceeding from D to B, and to A it is possible to assign the ethyl carbon atoms and identify the interaction between D and B across the

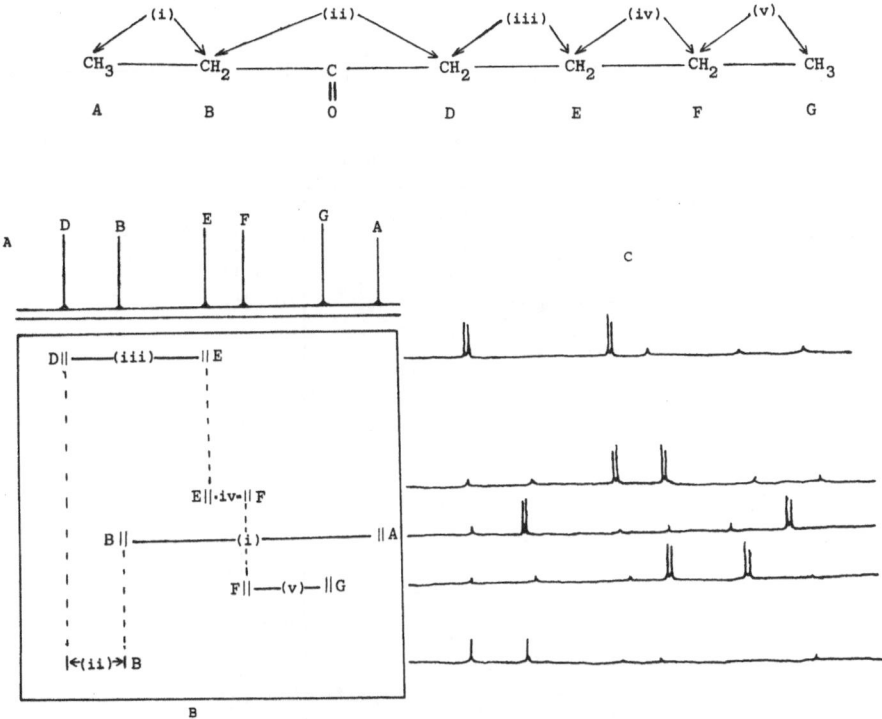

Figure 5.85. 2D-INADEQUATE spectrum of ethyl butyl ketone: (A) broad-band decoupled spectrum; (B) 2D-INADEQUATE spectrum with the couplings identified; (C) same as (B), but drawn as rows of matrix.

carbonyl group. It is very easy to correlate the satellite signals with the peaks in the 1D spectrum since the chemical shifts (on the horizonal axis) remain unchanged.

The main drawbacks in the procedure are that due to the low sensitivity of the experiment, several hundred milligrams of the compound are required and the spectra need to be recorded for many hours before good spectra can be obtained. However, the power of the method is obvious since the entire carbon framework can be worked out in one experiment by determining the carbon–carbon connectivities.

5.6.2 Two-Dimensional Double-Quantum Coherence Echo Correlated Spectroscopy (DECSY)

Two-dimensional double-quantum coherence NMR spectroscopy has been found useful for analyzing J_{CC} connectivity in ^{13}C spectra as well as J_{HH} connectivity in ^1H spectra. Double-quantum or multiple-quantum filter has also been used for unraveling complex, overlapping ^1H-NMR spectra. A new

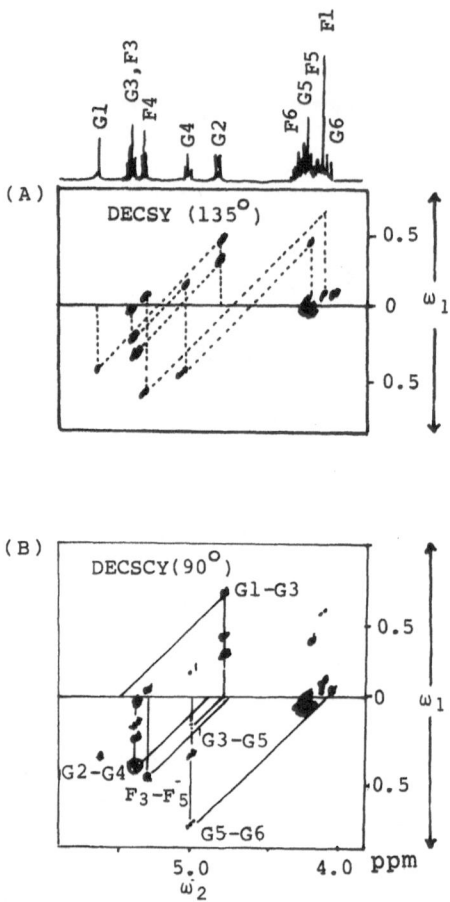

Figure 5.86. (A) DECSY spectrum of sucrose octaacetate with $\alpha = 135°$. The dotted lines show the connectivity between directly coupled protons. (B) DECSY spectrum of sucrose octaacetate with $\alpha = 90°$. The remote connectivities are shown by the diagonal lines which when drawn at an angle of 45° cross the central line ($\omega_1 = 0$) at the midpoint between the chemical shifts of the remotely connected protons.

method has recently been described for correlating coupled spectra which is denoted as DECSY (double-quantum coherence echo correlated spectroscopy) and employs quadrature detection in both dimensions. The spectra resemble the SECSY spectra described earlier, but in addition they provide information on remote connectivity and magnetic equivalence of protons.* The DECSY spectrum of sucrose octaacetate with $\alpha = 135°$ is shown in Figure 5.86A. The dotted lines establish the direct connectivities. Figure 5.86B shows the DECSY spectrum recorded with $\alpha = 90°$, exhibiting a greater number of cross peaks. These additional cross peaks establish remote

* M. Ikura and K. Hikichi, Two dimensional double quantum coherence echo correlated spectroscopy (DECSY), *J. Am. Chem. Soc.*, **106** 4275 (1984) and references therein.

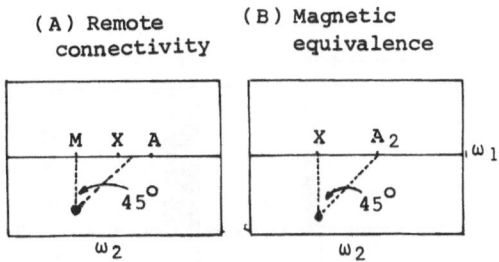

Figure 5.87. (A) Schematic representation of 2D DECSY spectrum showing remote connectivity in an AMX system. (B) Connectivity in a magnetically equivalent (A_2X) system.

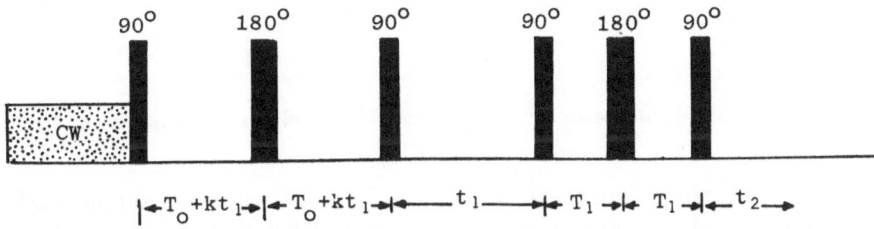

Figure 5.88. Pulse sequence for recording two-dimensional double-quantum 2QT NMR spectra. This results in cross peaks between directly coupled protons (i.e., protons on the α-carbons) as well as more remote protons (on the β-carbon atoms) by a relayed effect.

connectivities (i.e., connectivity between A and X through M in an AMX system) or magnetic equivalence (e.g., A_2X system). As shown in the schematic diagram of Figure 5.87, the connectivity of nucleus M is established by drawing a line at an angle of 45° from the cross peak to the ω_1-center line. The point at which it meets the central horizontal line is the *midpoint* between the chemical shifts of the two remote nuclei A and X. In the case of magnetically equivalent nuclei, this intersection is *at* the chemical shift of the equivalent (A_2) nuclei. The pulse sequence in DECSY is given by:

$$90°(0)-\tau-180°(\pi/2)-\tau-90°(0)-(t_1/2)-\alpha(\psi)-t_1-\text{acq.}(t_2)$$

5.6.3 Relayed Double-Quantum 2D-NMR Spectroscopy

Relayed transfer of two-dimensional double-quantum coherence (2QT) has recently been described in homonuclear and heteronuclear correlated spectral system.* The pulse sequence used in the relayed 2QT experiment is shown in Figure 5.88. The first three mixing pulses (90°, 180°, 90°) serve to mix double-

---

* S. Macura, N.G. Kumar, and L.R. Brown, Homonuclear relayed double quantum 2D NMR spectroscopy, *J. Magn. Resonance* **60**, 99 (1984) and references therein.

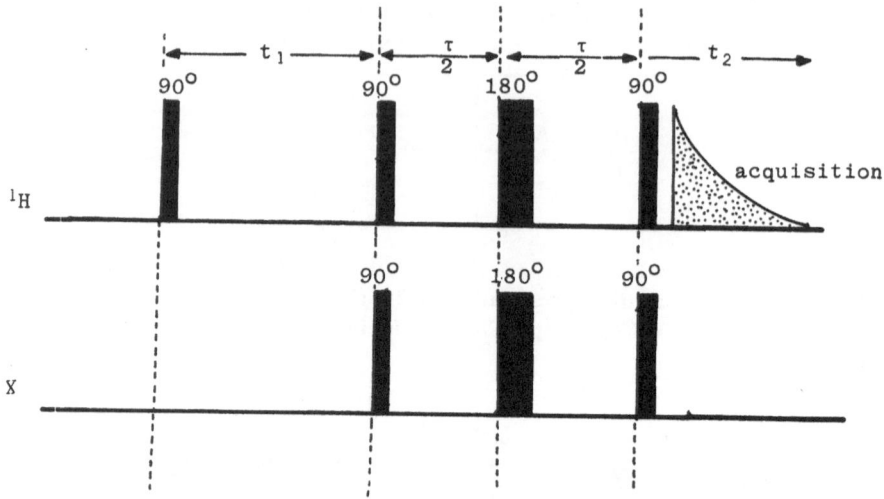

Figure 5.89. Pulse sequence for the X-relayed ^{1}H–^{1}H COSY experiment. From [140].

quantum coherence which evolves during t_1. Thus while the simple 2QT experiment results in cross peaks being observed between directly coupled protons, the *relayed* 2QT pulse sequence shown in Figure 5.88 results in cross peaks between directly coupled *and* remote protons. Thus in peptides the connectivity of the amide NH proton with the protons at both the α- and β-carbons can be seen by this experiment, and the intense water peak can be suppressed effectively.

5.6.4 X-Relayed ^{1}H–^{1}H Correlated Spectroscopy (X-Relayed ^{1}H-^{1}H COSY)

A new form of heteronuclear relayed coherence transfer spectroscopy has recently been described which can establish H–X–H connectivities directly.* The pulse sequence employed is shown in Figure 5.89. A ^{31}P-relayed ^{1}H–^{1}H COSY spectrum of triethylphosphonoacetate is shown in Figure 5.90. Protons A and B are both coupled to a common ^{31}P nucleus and this connectivity is clearly indicated by the ^{31}P-relayed ^{1}H–^{1}H COSY spectrum.

5.6.5 Carbon-Relayed ^{1}H–^{13}C Correlated Spectroscopy

Carbon–carbon connectivites for proton-bearing carbon atoms can be conveniently determined through a combination of ^{1}H–^{1}H COSY and

---

* D. Neuhaus, G. Wider, G. Wagner, and K. Wuthrich, X-Relayed ^{1}H–^{1}H correlated spectroscopy, *J. Magn. Resonance* **57**, 164 (1984) and references therein.

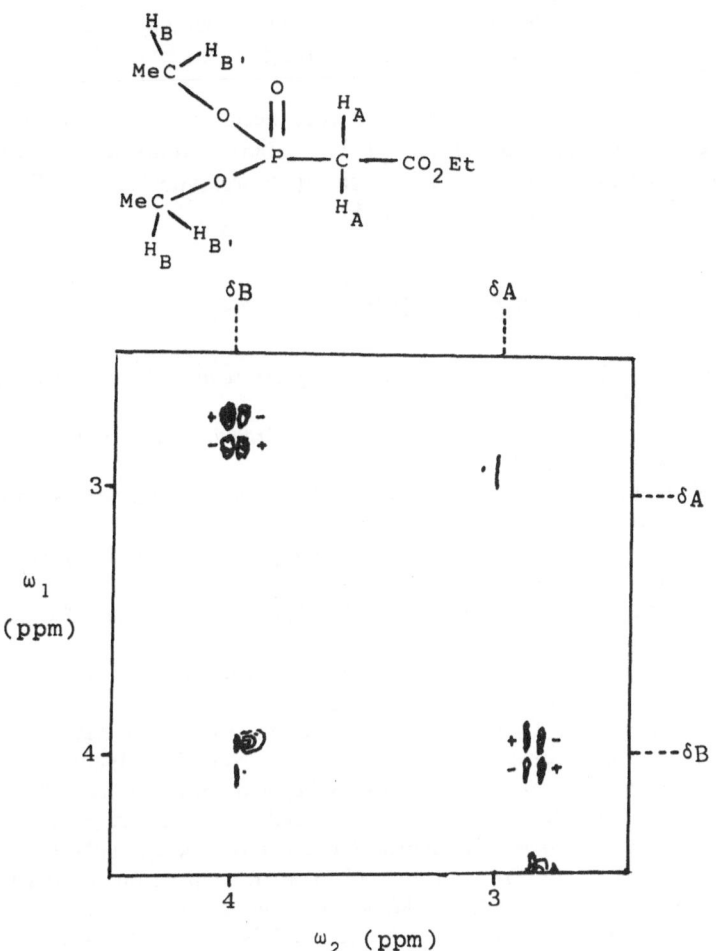

Figure 5.90. ^{31}P-relayed ^{1}H–^{1}H COSY spectrum of triethylphosphonoacetate. The cross peaks between the protons H_A and H_B establish their connectivity through the $H_A \leftrightarrow {}^{31}P \leftrightarrow H_B$ coupling pathway. From [140].

^{1}H–^{13}C COSY spectra. These experiments, however, do not allow assignments of quaternary carbon atoms, for which a [1]H–[13]C COLOC experiment is used* which allows long-range couplings between protons and carbons (protonated or quaternary) to be observed over two, three, or four bonds. It is sometimes difficult to distinguish between $^2J_{HC}$, $^3J_{HC}$, and $^4J_{HC}$ from the COLOC experiment. To overcome this, an INEPT-

---

* H. Kessler, C. Griesinger, and J. Lautz, Determination of the connectivites of weak proton–carbon couplings with a variation of the two-dimensional NMR technique, *Angew. Chem.*, **96**, 434 (1984).

INADEQUATE experiment was developed* which records couplings between directly connected carbon nuclei, but the sensitivity of the experiment is low.

A new pulse sequence has recently been described in which ^{13}C-relayed ^{1}H$-^{13}$C COSY spectra are obtained. The experiment allows assignments to quaternary carbon atoms.† A 1D-version (^{13}C-relayed ^{1}H$-^{13}$C INEPT) of the experiment has also been reported.

RECOMMENDED READING

1. G.C. Levy, *Topics in Carbon-13 NMR Spectroscopy*, Wiley-Interscience, New York (1984).
2. J.W. Cooper, *Spectroscopic Techniques for Organic Chemists*, John Wiley and Sons, New York (1980).
3. D.Shaw, *Pulsed Fourier Transform NMR Spectroscopy*, Elsevier Scientific Publishing Co., Amsterdam (1979).
4. F.W. Wehrli and T. Wirthlin, *Interpretation of Carbon-13 NMR Spectra*, Heyden and Son Ltd., London (1978).
5. E. Breitmaier and W. Voelter, *^{13}C-NMR Spectroscopy*, Verlag Chemie, Weinheim (1978).
6. C. Brevard and P. Granger, *Handbook of High Resolution Multinuclear NMR*, John Wiley and Sons, New York (1981).
7. G.C. Levy, R.L. Lichter, and G.L. Nelson, *Carbon-13 Nuclear Magnetic Resonance Spectroscopy*, John Wiley and Sons, New York (1980).
8. H. Günther, *NMR Spectroscopy*, John Wiley and Sons, New York (1980).
9. W.P. Aue, E. Bartholdi, and R.R. Ernst, Two-dimensional spectroscopy. Application to nuclear magnetic resonance, *J. Chem. Phys.* **64**, 2229 (1976).
10. G. Bodenhausen, R. Freeman, and D.L. Turner, Suppression of artifacts in two-dimensional *J* spectroscopy, *J. Magn. Resonance* **27**, 511 (1977).
11. G. Bodenhausen, R. Freeman, G.A. Morris, R. Niedermeyer, and D.L. Turner, A simple approach to single-channel quadrature detection, *J. Magn. Resonance* **25**, 559 (1977).
12. P. Bachmann, W.P. Aue, L. Muller, and R.R. Ernst, Phase separation in two-dimensional spectroscopy, *J. Magn. Resonance* **28**, 29 (1977).
13. W.P. Aue, P. Bachmann, A. Wokaun, and R.R. Ernst, Sensitivity of two-dimensional NMR spectroscopy, *J. Magn. Resonance* **29**, 523 (1978).
14. A. Kumar, Two-dimensional spin-echo NMR spectroscopy: a general method for calculation of spectra, *J. Magn. Resonance* **30**, 227 (1978).
15. L. Muller and R.R. Ernst, Coherence transfer: Heteronuclear 2-D NMR spectra, *Mol. Phys.* **38**, 963 (1979).

---

* O.W. Sorensen, R. Freeman, T.A. Frenkiel, T.H. Mareci, and R. Schuck, Observation of carbon-13–carbon-13 couplings with enhanced sensitivity, *J. Magn. Resonance*, **46**, 180 (1982).
† H. Kessler, W. Bermel, and C. Griesinger, Determination of carbon–carbon connectivities, assignment of quaternary carbons, and extraction of carbon–carbon coupling constants of carbon-relayed hydrogen–carbon spectroscopy, *J. Magn. Resonance*, **62**, 573 (1985), and references therein.

16. M.H. Levitt and R. Freeman, Phase adjustment of two-dimensional NMR spectra, *J. Magn. Resonance* **34,** 675 (1979).

17. G. Bodenhausen, S.P. Kempsell, and R. Freeman, Spinning sidebands in two-dimensional spin-echo spectra, *J. Magn. Resonance* **35,** 337 (1979).

18. A. Bax, A.F. Mehlkopf, and J. Smidt, A fast method for obtaining 2D *J*-resolved absorption spectra, *J. Magn. Resonance* **40,** 213 (1980).

19. L.D. Hall and S. Sukumar, A versatile strategy for the generalized acquisition of proton spin-echo data: measurement of partially relaxed two-dimensional *J* spectra, *J. Magn. Resonance* **40,** 405 (1980).

20. G. Bodenhausen and D.L. Turner, Artifacts in two-dimensional *J* spectra, *J. Magn. Resonance* **41,** 200 (1980).

21. S. Brownstein, Slicing and projection in 2-D arrays, *J. Magn. Resonance* **42,** 150 (1981).

22. R.L. Vold and R.R. Vold, Separation of single quantum spin echoes by two-dimensional Fourier transform techniques, *J. Magn. Resonance* **42,** 173 (1981).

23. R. Freeman and J. Keeler, Suppression of artifacts in two-dimensional *J* spectra, *J. Magn. Resonance* **43,** 484 (1981).

24. D.L. Turner, The measurement of coupling constants from two-dimensional NMR spectra, *J. Magn. Resonance* **39,** 391 (1980).

25. A.D. Bain, A superspin analysis of two-dimensional FT NMR experiments, *J. Magn. Resonance* **39,** 335 (1980).

26. A. Bax, R. Freeman, and G.A. Morris, A simple method for suppressing dispersion-mode contributions in NMR spectra: the "pseudo echo", *J. Magn. Resonance* **43,** 333 (1981).

27. D.L. Turner, Two-dimensional spin-echo spectroscopy of oriented systems, *J. Magn. Resonance* **46,** 213 (1982).

28. W.P. Aue, J. Karhan, and R.R. Ernst, Homonuclear broad band decoupling and two-dimensional *J*-resolved NMR spectroscopy, *J. Chem. Phys.* **64,** 4226 (1976).

29. G. Bodenhausen, R. Freeman, G.A. Morris and D.L. Turner, A simple pulse sequence for selective excitation in Fourier transform NMR, *J. Magn. Resonance* **31,** 75 (1978).

30. K. Nagayama, P. Bachmann, K. Wuthrich, and R.R. Ernst, The use of cross-sections and of projections in two-dimensional NMR spectroscopy, *J. Magn. Resonance* **31,** 133 (1978).

31. A. Bax, A.F. Mehlkopf, and J. Smidt, Homonuclear broad band-decoupled absorption spectra, with linewidths which are independent of the transverse relaxation rate, *J. Magn. Resonance* **35,** 167 (1979).

32. K. Nagayama, A. Kumar, K. Wuthrich, and R.R. Ernst, Experimental techniques of two-dimensional correlated spectroscopy, *J. Magn. Resonance* **40,** 321 (1980).

33. M.S. Broido and D.R. Kearns, Proton NMR evidence for an unusual conformation of poly C in solution, *J. Magn. Resonance* **41,** 496 (1980).

34. G. Wider, R. Baumann, K. Nagayama, R.R. Ernst, and K. Wuthrich, Strong spin-spin coupling in the two-dimensional *J*-resolved 360-MHz proton NMR spectra of the common amino acids, *J. Magn. Resonance* **42,** 73 (1981).

35. K. Nagayama and K. Wuthrich, Systematic application of two-dimensional proton nuclear magnetic resonance techniques for studies of proteins. 1. Combined use of spin-echo correlated spectroscopy and *J*-resolved spectroscopy for the identification of complete spin systems of non-labile protons in amino acid residues, *Eur. J. Biochem.* **114,** 365 (1981).

36. G. Wagner, A. Kumar, and K. Wuthrich, Systematic application of two-dimensional proton nuclear magnetic resonance techniques for studies of proteins. 2. Combined use of correlated spectroscopy and nuclear Overhauser spectroscopy for sequential assignments of backbone resonances and elucidation of polypeptide secondary structures, *Eur. J. Biochem.* **114,** 375 (1981).

37. C. Bosch, A. Kumar, R. Baumann, R.R. Ernst, and K. Wuthrich, 1-D and 2-D NOE in proteins, *J. Magn. Resonance* **42,** 159 (1981).

38. A. Bax, R. Freeman, and G. Morris, Correlation of proton chemical shifts by two-dimensional Fourier transform NMR, *J. Magn. Resonance* **42,** 164 (1981).

39. M. Ohuchi, M. Hosono, K. Matushita, and M. Imanari, A new pulse sequence in two-dimensional FT NMR. Solvent elimination using a Hahn echo, *J. Magn. Resonance* **43,** 499 (1981).

40. A. Kumar, R.R. Ernst, and K. Wuthrich, A two-dimensional nuclear Overhauser enhancement (2D NOE) experiment for the elucidation of complete proton–proton cross-relaxation networks in biological macromolecules, *Biochem. Biophys. Res. Commun.* **95,** 1 (1980).

41. A. Kumar, G. Wagner, R.R. Ernst, and K. Wuthrich, Studies of *J*-connectivities and selective proton–proton Overhauser effects in aqueous solutions of biological macromolecules by two-dimensional NMR experiments, *Biochem. Biophys. Res. Commun.* **96,** 1156 (1980).

42. J. Jenner, B.H. Meier, P. Bachmann, and R.R. Ernst, 2-D NOE, *J. Chem. Phys.* **71,** 4546 (1979).

43. B.H. Meier and R.R. Ernst, Elucidation of chemical exchange networks by two-dimensional NMR spectroscopy: the heptamethylbenzenonium ion, *J. Am. Chem. Soc.* **101,** 6441 (1979).

44. S. Macura and R.R. Ernst, Elucidation of cross relaxation in liquids by two-dimensional NMR spectroscopy, *Mol. Phys.* **41,** 95 (1980).

45. S. Macura, Y. Huang, D. Suter, and R.R. Ernst, Two-dimensional chemical exchange and cross-relaxation spectroscopy of coupled nuclear spins, *J. Magn. Resonance* **43,** 259 (1981).

46. A. Bax and R. Freeman, COSY, COSY-45, F_1 decoupling, *J. Magn. Resonance* **44,** 542 (1981).

47. R. Baumann, G. Wider, R.R. Ernst, and K. Wuthrich, Improvement of 2D NOE and 2D correlated spectra by symmetrization, *J. Magn. Resonance* **44,** 402 (1981).

48. G. Bodenhausen and R.R. Ernst, The accordion experiment, a simple approach to three-dimensional NMR spectroscopy, *J. Magn. Resonance* **45,** 367 (1981).

49. R. Baumann, A. Kumar, R.R. Ernst and K. Wuthrich, Improvement of 2D NOE and 2D correlated spectra by triangular multiplication, *J. Magn. Resonance* **44,** 76 (1981).

50. S. Macura, K. Wuthrich, and R.R. Ernst, Separation and suppression of coherent transfer effects in two-dimensional NOE and chemical exchange spectroscopy, *J. Magn. Resonance* **46,** 269 (1982).

51. L. Muller, A. Kumar, and R.R. Ernst, Two-dimensional carbon-13 NMR spectroscopy, *J. Chem. Phys.* **63,** 5490 (1975).

52. G. Bodenhausen, R. Freeman, and D.L. Turner, Two-dimensional *J* spectroscopy: Proton-coupled carbon-13 NMR, *J. Chem. Phys.* **65,** 839 (1976).

53. G. Bodenhausen, R. Freeman, R. Niedermeyer, and D.L. Turner, High-resolution NMR in inhomogeneous magnetic fields, *J. Magn. Resonance* **24,** 291 (1976).

54. R. Freeman, G.A. Morris, and D.L. Turner, Proton-coupled carbon-13 J spectra in the presence of strong coupling, I, *J. Magn. Resonance* **26**, 373 (1977).
55. G. Bodenhausen, R. Freeman, G.A. Morris, and D.L. Turner, Proton-coupled carbon-13 J spectra in the presence of strong coupling, II, *J. Magn. Resonance* **28**, 17 (1977).
56. A.A. Maudsley and R.R. Ernst, Indirect detection of magnetic resonance by heteronuclear two-dimensional spectroscopy, I, *Chem. Phys. Lett.* **50**, 368 (1977).
57. D.L. Turner and R. Freeman, The proton-coupled carbon-13 spectrum of cholesterol, *J. Magn. Resonance* **29**, 587 (1978).
58. R. Freeman and G.A. Morris, Experimental chemical shift correlation maps in nuclear magnetic resonance spectroscopy, *J. Chem. Soc., Chem. Commun.*, 684 (1978).
59. G. Bodenhausen and R. Freeman, Correlation of chemical shifts of protons and carbon-13, *J. Am. Chem. Soc.* **100**, 320 (1978).
60. A. Hohener, L. Muller, and R.R. Ernst, ^{13}C 2-D in liquid crystals, *Mol. Phys.* **38**, 909 (1979).
61. P.H. Bolton and G. Bodenhausen, Heteronuclear two-dimensional NMR as a conformational probe of cellular phosphates, *J. Am. Chem. Soc.* **101**, 1080 (1979).
62. R. Freeman, S.P. Kempsell, and M.H. Levitt, Elimination of dispersion-mode contributions from two-dimensional NMR spectra, *J. Magn. Resonance* **34**, 663 (1979).
63. L.D. Hall, G.A. Morris, and S. Sukumar, Resolution and assignment of the 270-MHz proton spectrum of cellobiose by homo- and heteronuclear two-dimensional NMR, *J. Am. Chem. Soc.* **102**, 1745 (1980).
64. G. Bodenhausen, Heteronuclear J spectroscopy, *J. Magn. Resonance* **39**, 175 (1980).
65. L.D. Hall and G.A. Morris, Measurement of carbon-13 proton coupling constants in oligosaccharides by two-dimensional carbon-13 NMR spectroscopy, *Carbohyd. Res.* **82**, 175 (1980).
66. A.A. Maudsley, L. Muller, and R.R. Ernst, Cross-correlation of spin-decoupled NMR spectra by heteronuclear two-dimensional spectroscopy, *J. Magn. Resonance* **28**, 463 (1977).
67. G. Bodenhausen and R. Freeman, Correlation of proton and carbon-13 NMR spectra by heteronuclear two-dimensional spectroscopy, *J. Magn. Resonance* **28**, 471 (1977).
68. R. Niedermeyer and R. Freeman, Carbon–carbon spin-spin coupling studied by two-dimensional Fourier transformation, *J. Magn. Resonance* **30**, 617 (1978).
69. L. Muller, High-sensitivity in ^{13}C 2-D NMR, *J. Magn. Resonance* **36**, 301 (1979).
70. L. Muller, Off-resonance decoupling in ^{13}C 2-D NMR, *J. Magn. Resonance* **38**, 79 (1980).
71. A.G. Avent and R. Freeman, NMR spin-lattice relaxation studied by magnetization transfer, *J. Magn. Resonance* **39**, 169 (1980).
72. G. Bodenhausen and P.H. Bolton, Elimination of flip-angle effects in two-dimensional NMR spectroscopy. Application to cyclic nucleotides, *J. Magn. Resonance* **39**, 339 (1980).
73. M.H. Levitt and R. Freeman, Simplification of NMR spectra by masking in a second frequency dimension, *J. Magn. Resonance* **39**, 533 (1980).
74. A. Bax and G.A. Morris, An improved method for heteronuclear chemical shift correlation by two-dimensional NMR, *J. Magn. Resonance* **42**, 501 (1981).

75. D.M. Thomas, M.R. Bendall, D.T. Pegg, D.M. Doddrell, and J. Field, Two-dimensional carbon-13-proton polarization transfer J spectroscopy, *J. Magn. Resonance* **42**, 298 (1981).

76. G.A. Morris, Indirect two-dimensional J spectroscopy: measurement of proton multiplet structure via carbon-13 signals, *J. Magn. Resonance* **44**, 277 (1981).

77. P.H. Bolton, Investigation of phospherine and cytidine 5'-phosphate by heteronuclear two-dimensional spectroscopy: samples with strong proton coupling, *J. Magn. Resonance* **45**, 539 (1981).

78. A. Bax and R. Freeman, Relative signs of couplings, ^1H–^{13}C 2-D, *J. Magn. Resonance* **45**, 177 (1981).

79. P.H. Bolton and G. Bodenhausen, Resolution enhancement, coherence transfer echoes. *J. Magn. Resonance* **46**, 306 (1982).

80. M.R. Bendall, D.T. Pegg, D.M. Doddrell, and D.M. Thomas, A superior pulse sequence for two-dimensional chemical shift correlation spectroscopy, *J. Magn. Resonance* **46**, 43 (1982).

81. P.H. Bolton and G. Bodenhausen, Double resonance in heteronuclear two-dimensional spectroscopy, *J. Magn. Resonance* **43**, 339 (1981).

82. S. Vega and A. Pines, Operator formalism for double quantum NMR, *J. Chem. Phys.* **66**, 5624 (1977).

83. A. Wokaun and R.R. Ernst, Selective detection of multiple quantum transitions in NMR by two-dimensional spectroscopy, *Phys. Lett.* **52**, 407 (1977).

84. L. Muller, Detection of weak nuclei via MQ coherence, *J. Am. Chem. Soc.* **101**, 4481 (1979).

85. A. Bax, R. Freeman, and S.P. Kempsell, Natural abundance carbon-13–carbon-13 coupling observed via double-quantum coherence, *J. Am. Chem. Soc.* **102**, 4849 (1980).

86. G. Bodenhausen, R.L. Vold, and R. R. Vold, Multiple quantum spin-echo spectroscopy, *J. Magn. Resonance* **37**, 93 (1980).

87. A. Minoretti, W.P. Aue, M. Reinhold, and R.R. Ernst, Coherence transfer by radiofrequency pulses for heteronuclear detection of multiple quantum transitions, *J. Magn. Resonance* **40**, 175 (1980).

88. A. Bax, S.P. Kempsell, and R. Freeman, Investigation of carbon-13–carbon-13 long range couplings in natural abundance samples, *J. Magn. Resonance* **41**, 349 (1980).

89. A. Bax and R. Freeman, Investigation of carbon-13–carbon-13 couplings in natural abundance samples: the strong coupling case, *J. Magn. Resonance* **41**, 507 (1980).

90. O.W. Sorensen, R. Freeman, T.A. Frenkiel, T.H. Mareci, and R. Schuck, Observation of carbon-13–carbon-13 coupling with enhanced sensitivity, *J. Magn. Resonance* **46**, 180 (1982).

91. W. Ammann, R. Richarz, T. Wirthlin, and D. Wendisch, Proton and carbon-13 chemical shifts and coupling constants of lupane. Application of two-dimensional NMR techniques, *Org. Magn. Resonance* **20**(4), 260 (1982).

92. G.A. Gray, Multinuclear two-dimensional NMR: Assignments of natural abundance polypeptide ^{13}C, ^1H and ^{15}N chemical shifts and demonstration of isomer interconversion, *Org. Magn. Resonance* **21**(2), 111 (1983).

93. R. Richarz and W. Ammann, Practical Tips for Homonuclear 2D-NMR Experiments, Varian Application Note No. Z-81, July 1981.

94. 2D NMR of Rotenone, Jeol Application Note No. NM 16 (1984).

95. R. Richarz, W. Ammann, and T. Wirthlin, XL-200/300 in DEPT (h): A New Experiment for the ADEPT Spectroscopist, Varian Application Note No. Z-15, Aug. 1982.

96. W. Ammann, R. Richarz and T. Wirthlin, COSMIC: A New Approach to Automatic Structure Elucidation. A Pascal Programme for the XL-200 Data System, Varian Application Note No. Z-12, April 1981.

97. A. Bax, R. Freeman, T.A. Frenkiel and M.H. Levitt, Assignment of carbon-13 NMR spectra via double quantum coherence, *J. Magn. Resonance* **43,** 478 (1981).

98. G. Bodenhausen, R. Freeman, R. Niedermeyer, and D.L. Turner, Double Fourier transformation in high resolution NMR, *J. Magn. Resonance* **26,** 133 (1977).

99. R. Freeman and G.A. Morris, Two-dimensional Fourier transformation in NMR, *Bull. Magn. Resonance* **1**(1), 5 (1979).

100. D.H. Williams, M.P. Williamson, D.W. Butcher, and S.J. Hammond, Detailed binding sites of the antibiotics vancomycin and ristocetin A: Determination of intermolecular distances in antibiotic/substrate complexes by use of the time-dependent NOE, *J. Am. Chem. Soc.* **105,** 1332 (1983).

101. D. Wenhaus, R.N. Sheppard, and I.R.C. Bick, Structural and conformational study of repanduline using long-range nuclear Overhauser effect difference spectroscopy, *J. Am. Chem. Soc.* **105,** 5996 (1983).

102. J.C. Steffens, J.L. Roark, D.G. Lynn, and J.R. Riopel, Host recognition in parasitic angiosperms: Use of correlation spectroscopy to identify long-range coupling in an haustorial inducer, *J. Am. Chem. Soc.* **105,** 1669 (1983).

103. A.C. Pinto, M.L.A. Goncalves, R.B. Filho, A. Neszmelyi, and G. Lukacs, Natural abundance $^{13}C-^{13}C$ coupling constants observed via double quantum coherence: Structural elucidation of velloziolide, a diterpene with a novel skeleton, *J. Chem. Soc., Chem. Commun.,* 293 (1982).

104. A. Bax, *Two Dimensional Nuclear Magnetic Resonance in Liquids*, Delft University Press, Delft, Holland (1982).

105. D.A. Aikens, S.C. Bunce, O.F. Onasch, H.M. Schwartz, and C. Hurwitz, Two dimensional NMR investigation of the protonation sequence in spermidine, *J. Chem. Soc., Chem. Commun.,* 43 (1983).

106. E. Haslinger and H. Kalchhauser, 2D-NMR of natural products, Part III, Homo- and heteronuclear NMR-spectroscopy of a cyclic tetrapeptide related to chlamydocin, *Tetrahedron Lett.* **24**(25), 2553 (1983).

107. M.L. Martin, G.J. Martin, and J.J. Delpuech, *Practical NMR Spectroscopy*, Heyden and Sons Ltd., London (1980).

108. Two-dimensional NMR, Bruker Application Note, April 1982.

109. D.L. Foxall, New Water Suppression Experiments, Varian Application Note No. NMR-23, January 1984.

110. R. Richarz and T. Wirthlin, CCCP: Carbon–carbon Connectivity Plots on the XL-200, Varian Application Note No. 2-13, April 1983.

111. C. LeCocq and J.-Y. Lallemand, Precise carbon-13 NMR multiplicity determination, *J. Chem. Soc., Chem. Commun.,* 150 (1981).

112. A.C. Pinto, S.K. Do Prado, R.B. Filho, W.E. Hull, A. Neszmelyi, and G. Lukacs, Natural abundance $^{13}C-^{13}C$ coupling constants observed via double quantum coherence. Structural elucidation by the one-and the two-dimensional NMR experiments of velloziolone, a new seco-diterpene, *Tetrahedron Lett.* **23**(50), 5267 (1982).

113. E.L. Ulrich, W.M. Westler, and J.L. Markley, Reassignments in the ^{1}H-NMR spectrum of flavin adenine dinucteolide by two-dimensional homonuclear chemical shift correlation, *Tetrahedron Lett.* **24**(5), 473 (1983).

114. C.J. Turner, INEPT on the XL-200, Varian Application Note No. Z-11, September 1980.

115. R. Richarz, W. Ammann, and T. Wirthlin, Relayed Coherence Transfer in 2D-NMR, Varian Application Note No. Z-17, September 1982.

116. A. Bax, Broadband homonuclear decoupling in heteronuclear shift correlation NMR spectroscopy, *J. Magn. Resonance* **53**, 517 (1983).

117. M.L. Levitt, G. Bodenhausen, and R.R. Ernst, The illusions of spin decoupling, *J. Magn. Resonance* **53**, 443 (1983).

118. G. Lukacs and A. Neszmelyi, Computer-assisted determination of carbon connectivity patterns based on natural abundance one-bond ^{13}C–^{13}C coupling constants: Terpenes, *Tetrahedron Lett.* **22**(50), 5053 (1981).

119. D.J. Cookson and B.E. Smith, Improved method for assignment of multiplicity in ^{13}C-NMR spectroscopy with application to the analysis of mixtures, *Org. Magn. Resonance* **16**(2), 111 (1981).

120. M.R. Bendall and D.T. Pegg, ^{1}H–^{13}C two-dimensional chemical shift correlation spectroscopy using DEPT, *J. Magn. Resonance* **53**, 144 (1983).

121. G. King and P.E. Wright, Application of two-dimensional relayed coherence transfer experiments to ^{1}H-NMR studies of macromolecules, *J. Magn. Resonance* **54**, 328 (1983).

122. J.N. Shoolery, Recent developments in ^{13}C- and proton-NMR, *J. Nat. Prod.* **47**(2), 226 (1984).

123. H. Kessler and D. Ziessow, Zweidimensionale NMR-spektroskopie, *Nachr. Chem. Tech. Lab.* **30**(6), 448 (1982).

124. D.L. Foxall, Broadband Decoupled Heteronuclear Correlation and Semiselective Heteronuclear 2D *J* Spectroscopy, Varian Application Note, 1984.

125. A. Bax, Two-dimensional heteronuclear relayed coherence transfer spectroscopy, *J. Magn. Resonance* **53**, 149 (1983).

126. R. Benn and H. Günther, Modern pulse methods in high-resolution NMR spectroscopy, *Angew. Chem., Int. Ed. Engl.* **22**, 350 (1983).

127. J.M. Bulsing, W.M. Brooks, J. Field, and D.M. Doddrell, Polarisation transfer via an intermediate multiple quantum state of maximum order, *J. Magn. Resonance* **56**, 167 (1984).

128. M.J. Gidley and S.M. Bociek, Selective 2D-heteronuclear *J*-resolved NMR spectroscopy, *J. Chem. Soc., Chem. Commun.*, 220 (1985).

129. J.A. Wilde and P.H. Bolton, Suppression of homonuclear couplings in heteronuclear two-dimensional spectroscopy, *J. Magn. Resonance* **59**, 343 (1984).

130. T.T. Nakashima, B.K. John, and R.E.D. McClung, Selective 2D DEPT heteronuclear shift correlation spectroscopy, *J. Magn. Resonance* **59**, 124 (1984) and references therein.

131. H. Kessler, C. Griesinger, J. Zarbock, and H.R. Looslie, Assignment of carbonyl carbons and sequence analysis in peptides by heteronuclear shift correlation via small coupling constants with broad-band decoupling in t_1(COLOC), *J. Magn. Resonance* **57**, 331 (1984).

132. L. Muller, Mapping of spin–spin coupling via zero-quantum coherence, *J. Magn. Resonance* **59**, 326 (1984).

133. P.J. Hore, Two-dimensional chemical shift correlation using water suppression pulses, *J. Magn. Resonance* **56**, 535 (1984) and references therein.
134. C. Yu and G.C. Levy, *J. Am. Chem. Soc.* **106**, 6533 (1984) and references therein.
135. G. Wagner, Two dimensional relayed coherence transfer-NOE spectroscopy, *J. Magn. Resonance* **57**, 497 (1984)
136. C.A.G. Haasnoot, F.J.M. van de Ven, and C.W. Hilbers, COCONOSY, combination of 2D correlated and 2D nuclear Overhauser enhancement spectroscopy in a single experiment, *J. Magn. Resonance* **56**, 343 (1984).
137. M.A. Delsuc, E. Guittet, N. Trotin, and J.Y. Lallemand, Two dimensional correlation spectroscopy with heteronuclear relay, *J. Magn. Resonance* **56**, 163 (1984).
138. M. Ikura and K. Hikichi, Two dimensional double quantum coherence echo correlated spectroscopy (DECSY), *J. Am. Chem. Soc.* **106**, 4275 (1984) and references therein.
139. S. Macura, N.G. Kumar, and L.R. Brown, Homonuclear relayed double quantum 2D NMR spectroscopy, *J. Magn. Resonance* **60**, 99 (1984).
140. D. Neuhaus, G. Wider, G. Wagner, and K. Wuthrich, X-Relayed $^1H-^1H$ correlated spectroscopy, *J. Magn. Resonance* **57**, 164 (1984) and references therein.
141. E. Breitmaier, Die Kohlenstoff-13-NMR spektroskopie, *Pharmazie in unserer Zeit* **13**(4), 102 (1984).
142. H. Günther and P. Schmitt, Zweidimensionale Messtechniken der Hochauflosenden Kernresonanzspektroskopie (Teil I): J, δ Spektren, *Kontakte* (2), 3 (1985).
143. H. Kessler, A. Muller, and H. Oschkinat, Differences and sums of traces within COSY spectra (DISCO) for the extraction of coupling constants: decoupling after the measurement, *Magn. Resonance in Chemistry*, **23**(10), 844 (1985).
144. H. Kessler, W. Bermel, and C. Griesinger, Recognition of NMR proton spin systems of cyclosporin A via heteronuclear proton–carbon long range couplings, *J. Am. Chem. Soc.*, **107**, 1083 (1985).
145. H. Kessler, W. Bermel, and C. Griesinger, Determination of carbon–carbon connectivities, assignments of quaternary carbon atoms, and extraction of carbon–carbon coupling constants by carbon relayed hydrogen–carbon spectroscopy, *J. Magn. Resonance*, **62**, 573 (1985).
146. M.J. Gidley and S.M. Bociek, Long-range $^{13}C-^1H$ coupling in carbohydrates by selective 2D heteronuclear J-resolved NMR spectroscopy, *J. Chem. Soc., Chem. Commun.*, 220 (1985).

Appendix A

Problems

PROBLEMS IN ¹H-NMR SPECTROSCOPY

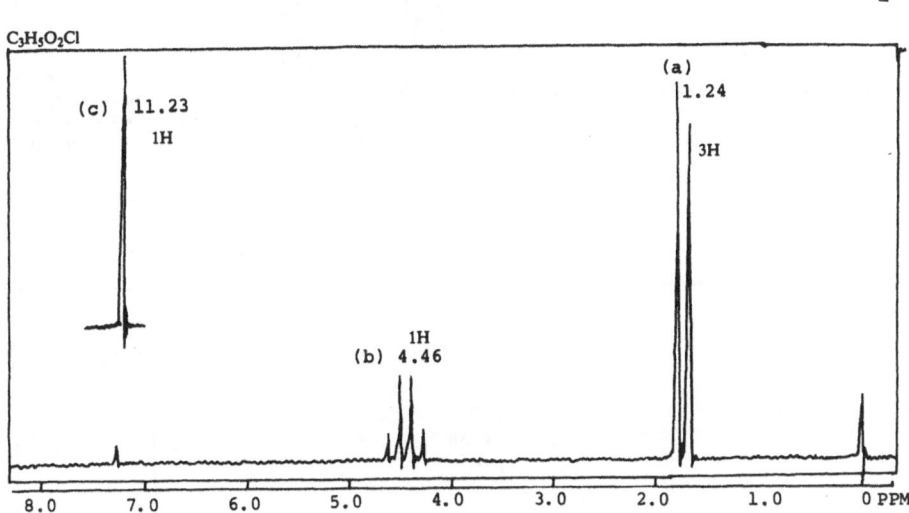

$C_3H_5O_2Cl$

(c) 11.23
1H

(b) 4.46 1H

(a) 1.24 3H

8.0   7.0   6.0   5.0   4.0   3.0   2.0   1.0   0 PPM

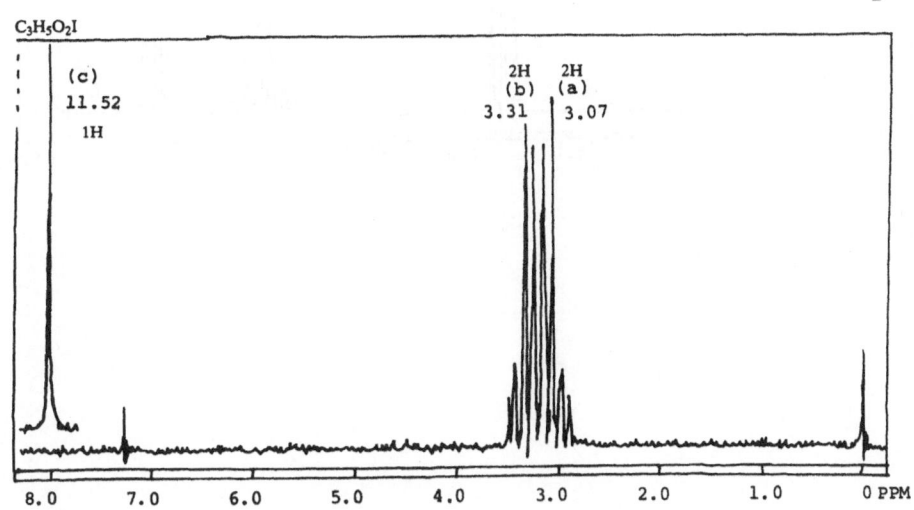

$C_3H_5O_2I$

(c) 11.52
1H

2H (b) 3.31
2H (a) 3.07

8.0   7.0   6.0   5.0   4.0   3.0   2.0   1.0   0 PPM

3

C₃H₆ClBr

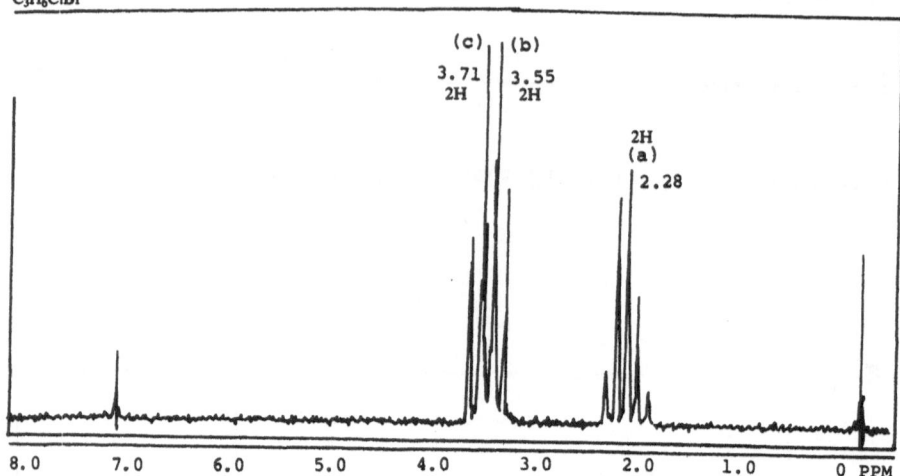

4

C₃H₇O₂N

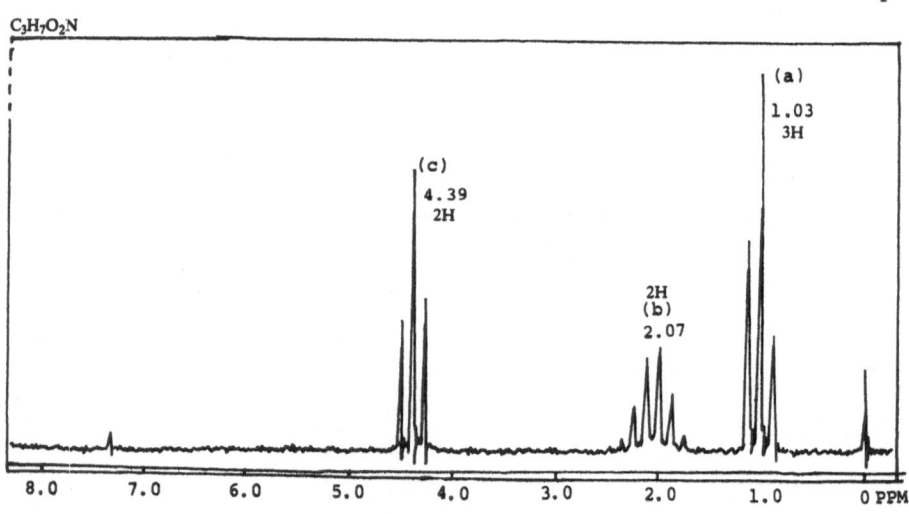

5

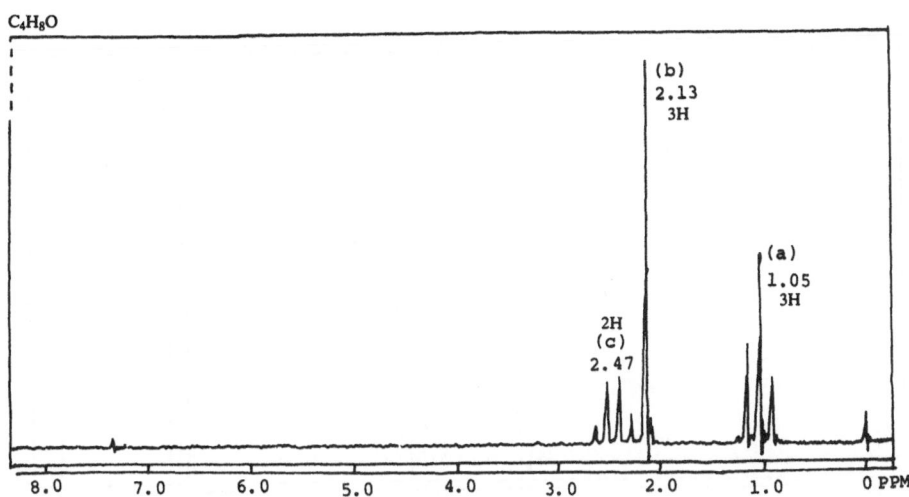

C₄H₈O

(b)
2.13
3H

2H
(c)
2.47

(a)
1.05
3H

8.0   7.0   6.0   5.0   4.0   3.0   2.0   1.0   0 PPM

6

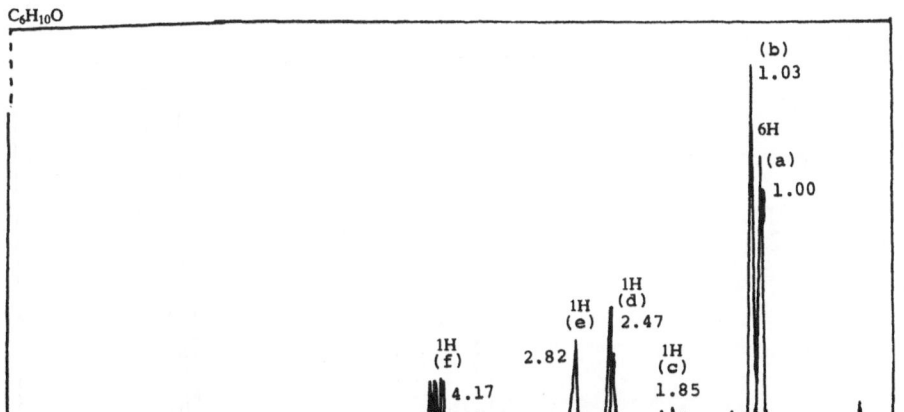

C₆H₁₀O

(b)
1.03

6H

(a)
1.00

1H
(d)
2.47

1H
(e)

1H
(c)
1.85

2.82

1H
(f)
4.17

8.0   7.0   6.0   5.0   4.0   3.0   2.0   1.0   0 PPM

7

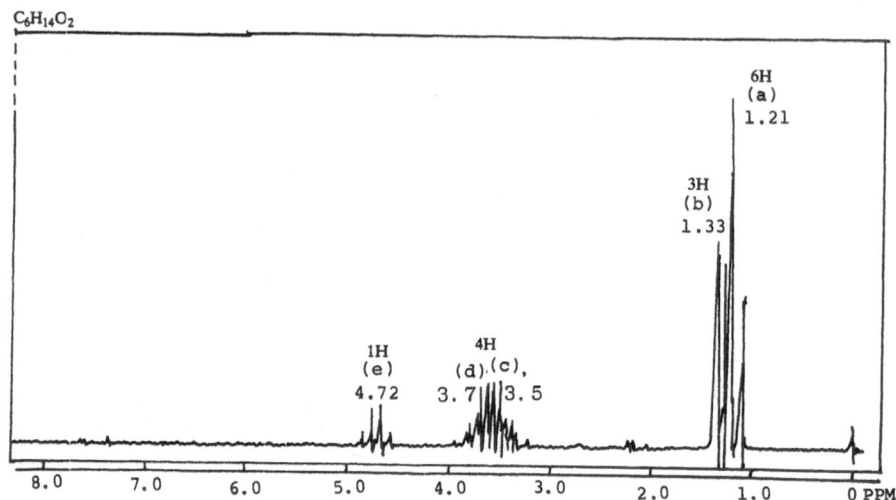

$C_6H_{14}O_2$

8

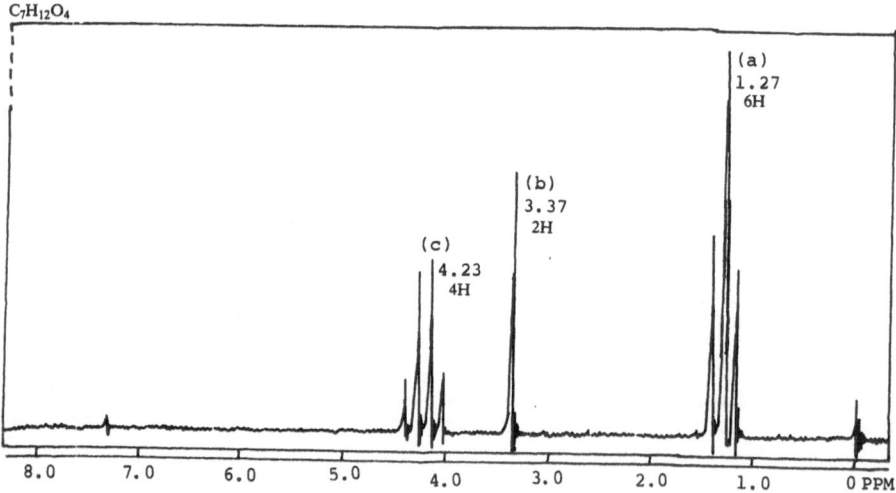

$C_7H_{12}O_4$

9

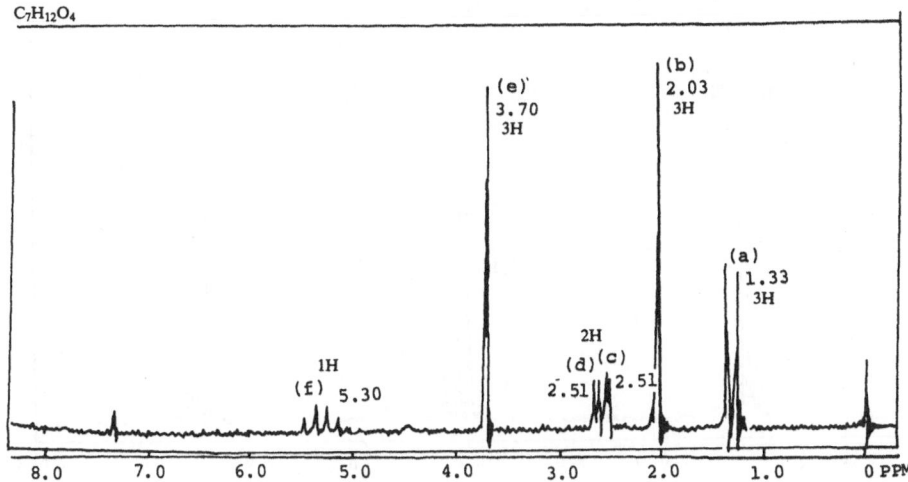

10

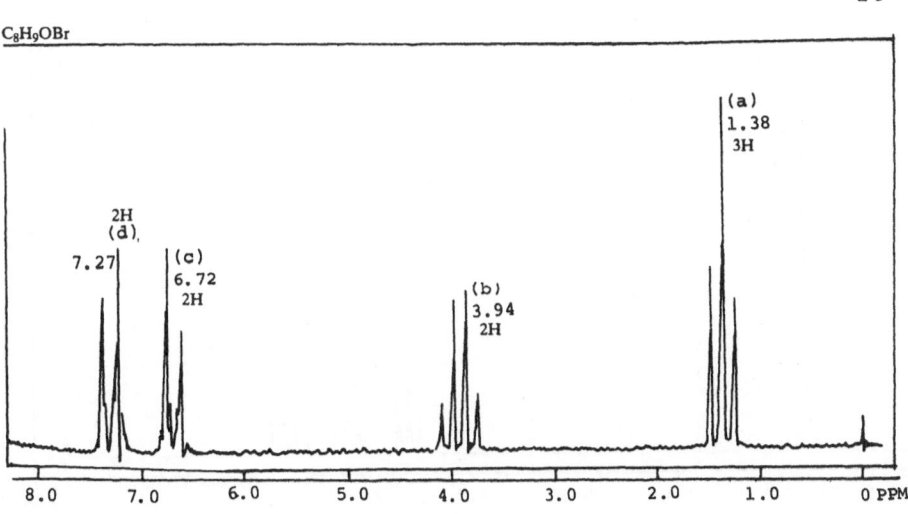

11

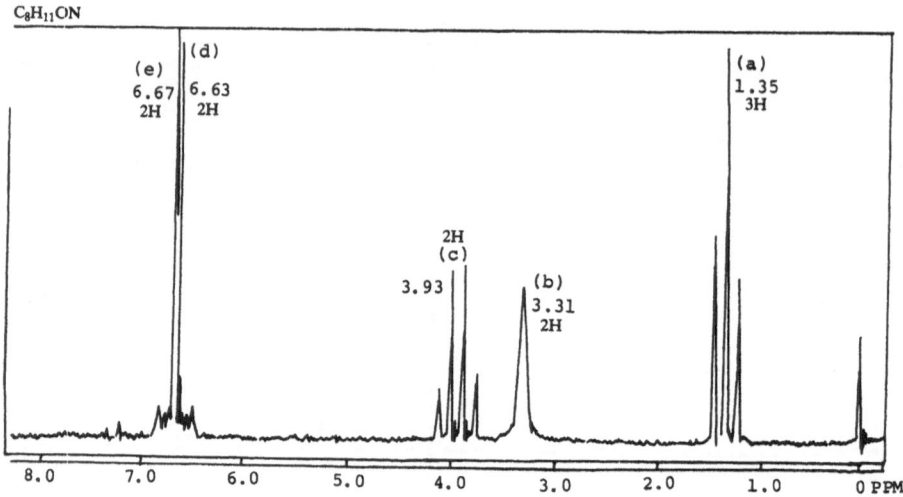

C₈H₁₁ON

12

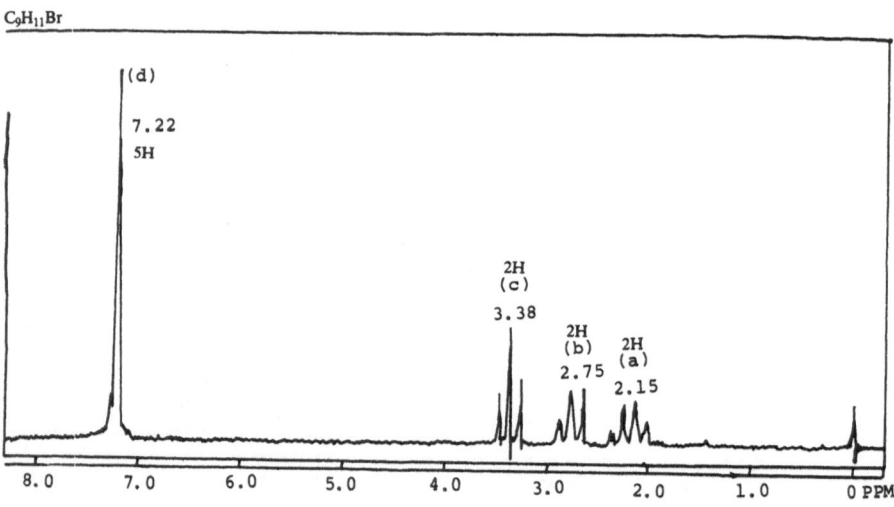

C₉H₁₁Br

13

$C_{10}H_{13}O_2N$

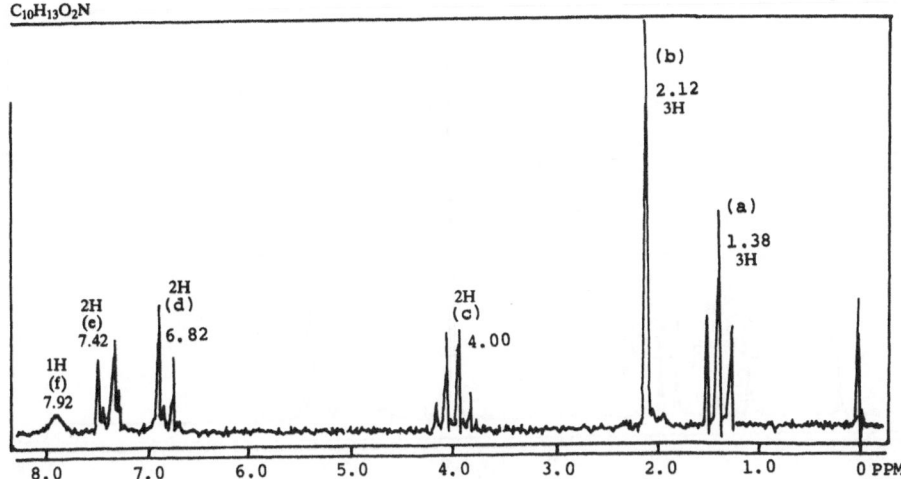

14

$C_{10}H_{16}O_5$

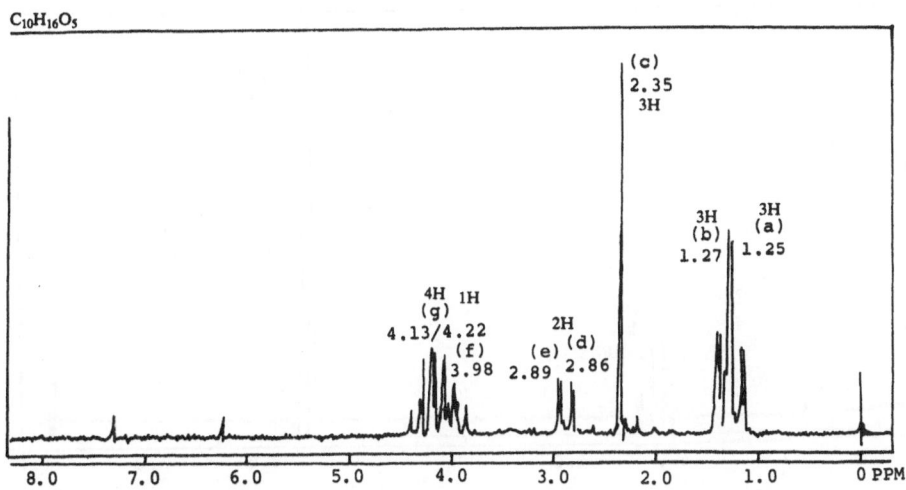

15

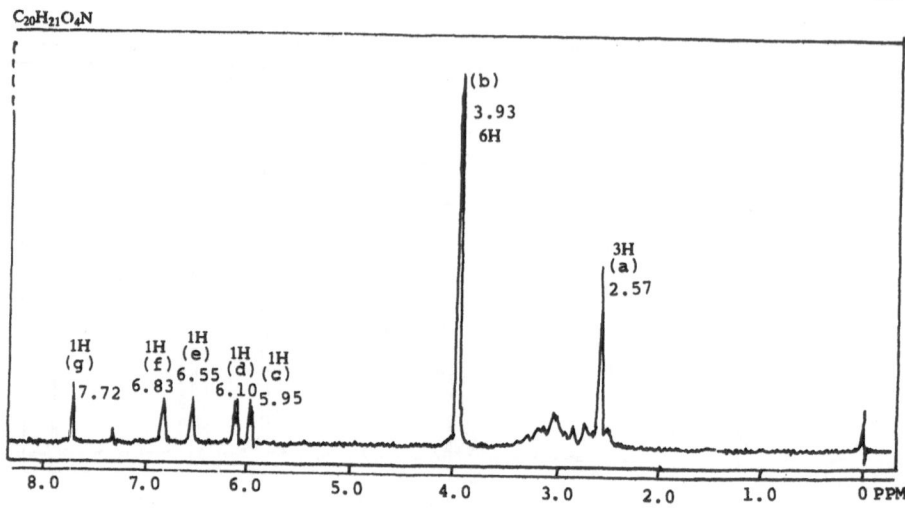

16

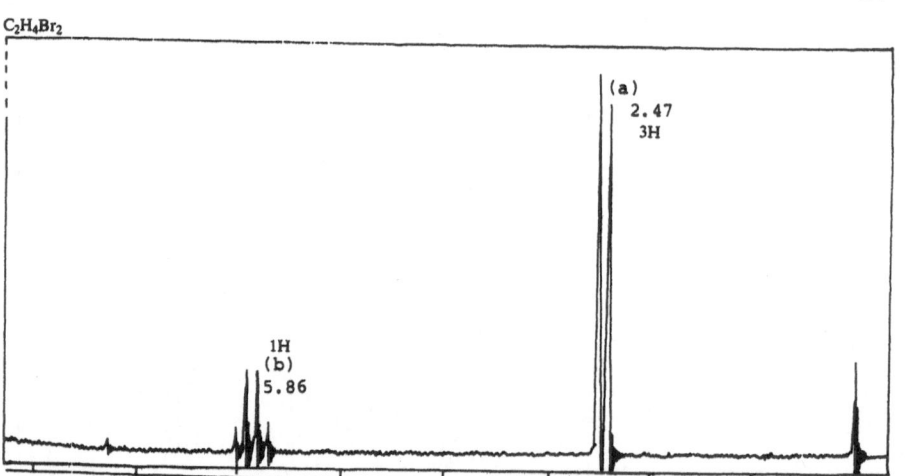

17

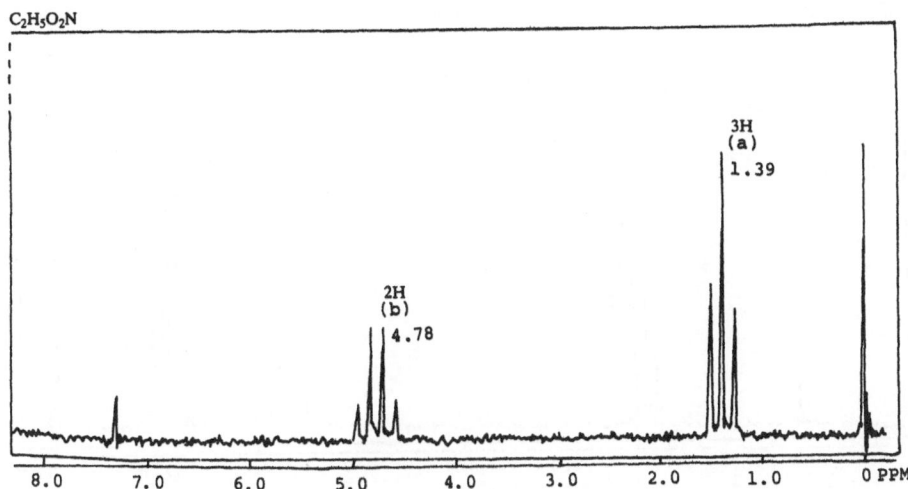

18

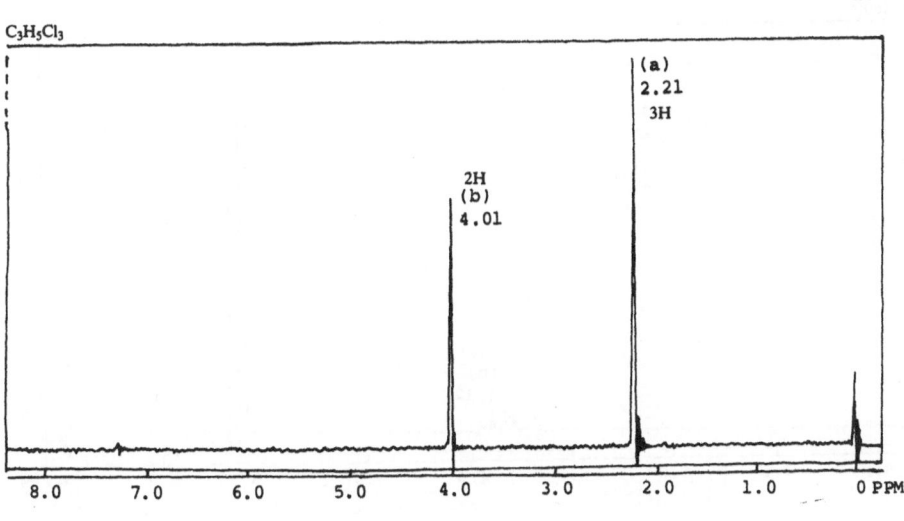

19

C₃H₆O₂NCl

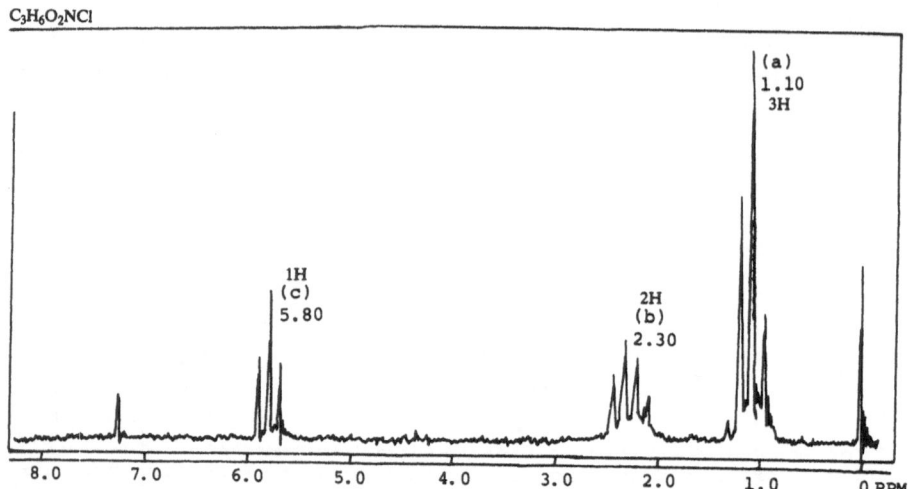

20

C₃H₇Br

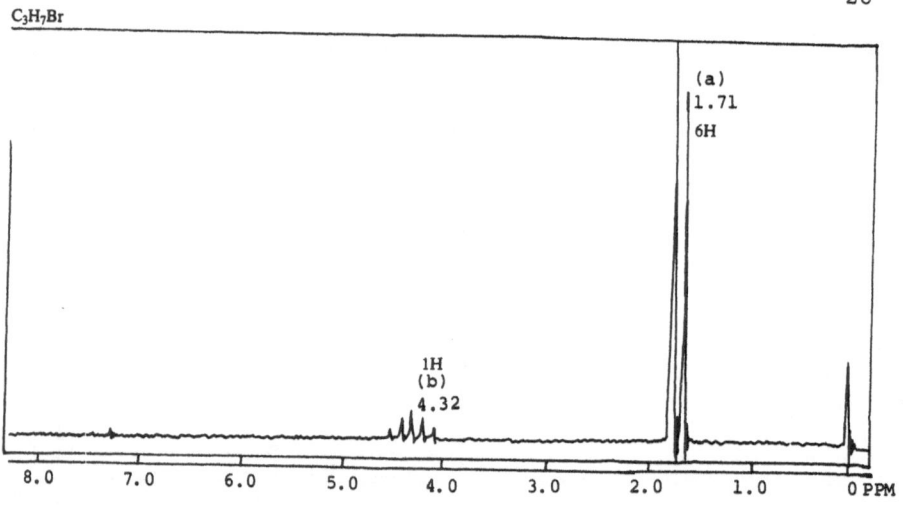

21

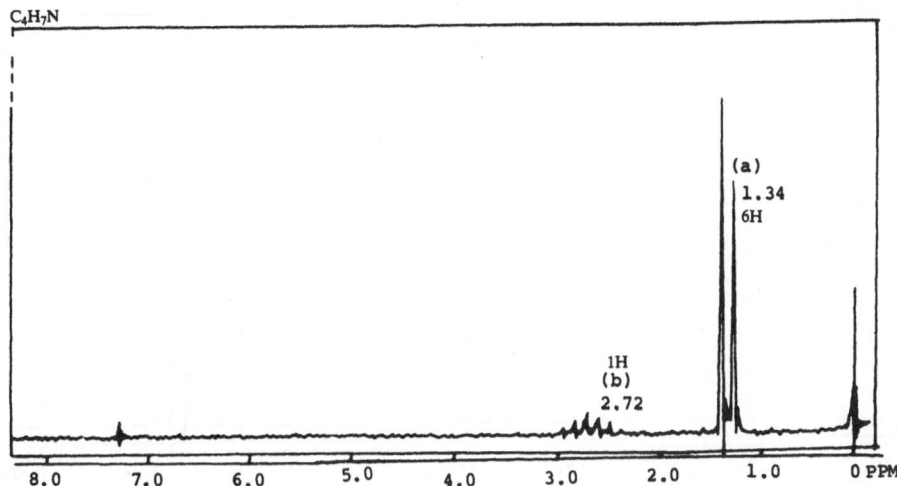

C_4H_7N

(a)
1.34
6H

1H
(b)
2.72

8.0   7.0   6.0   5.0   4.0   3.0   2.0   1.0   0 PPM

22

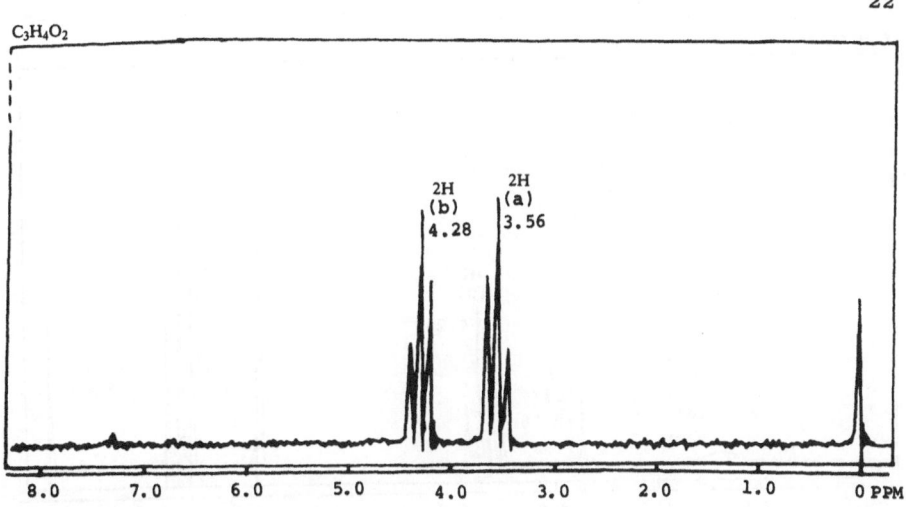

$C_3H_4O_2$

2H
(b)
4.28

2H
(a)
3.56

8.0   7.0   6.0   5.0   4.0   3.0   2.0   1.0   0 PPM

23

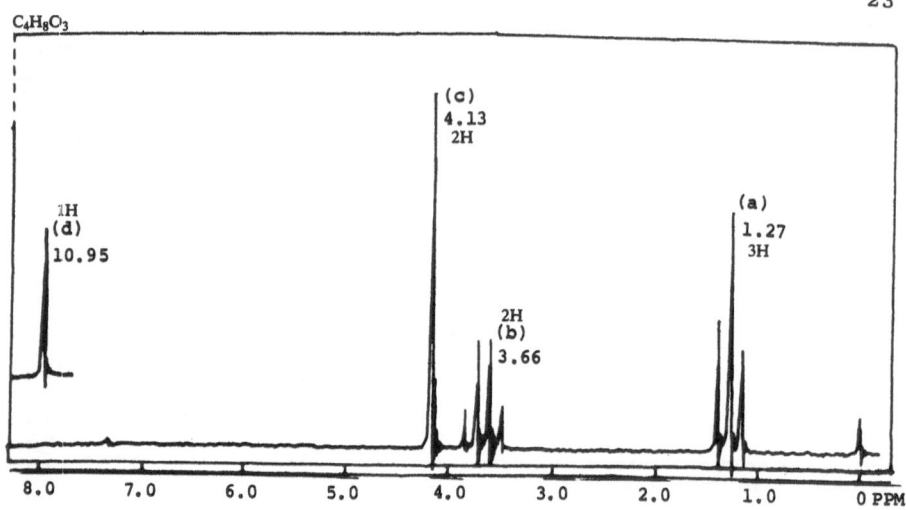

24

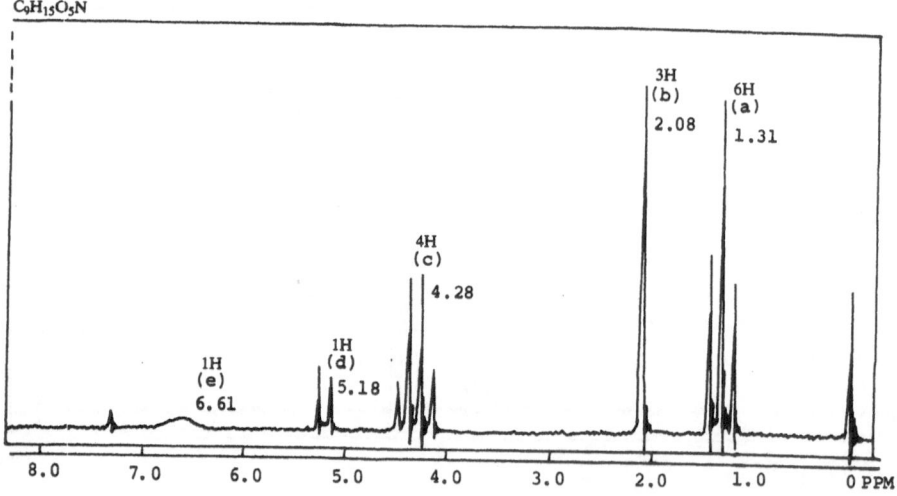

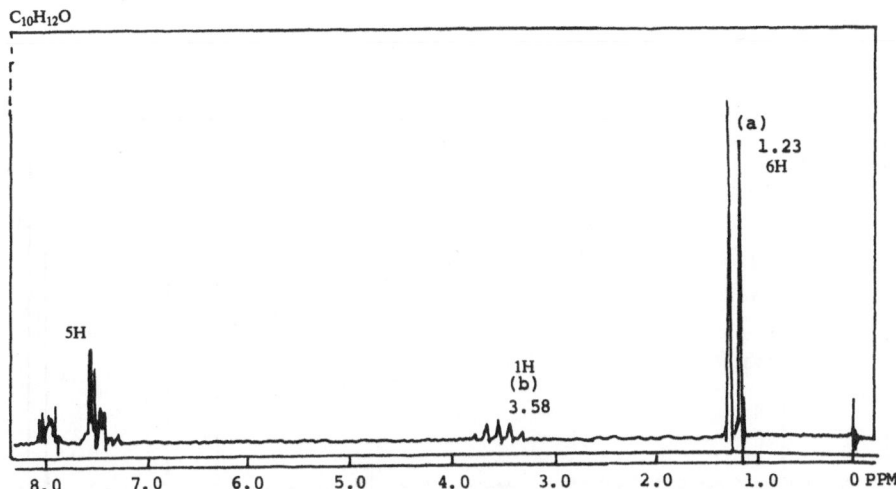

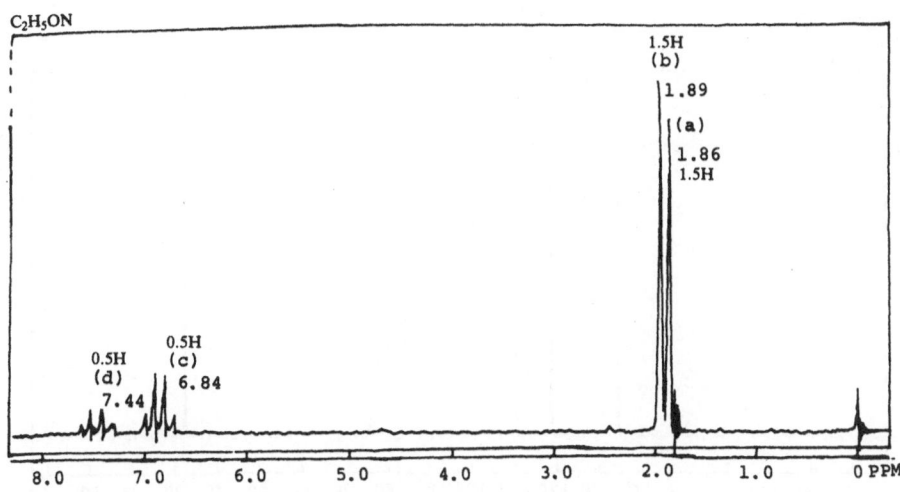

PROBLEMS IN ¹³C-NMR SPECTROSCOPY

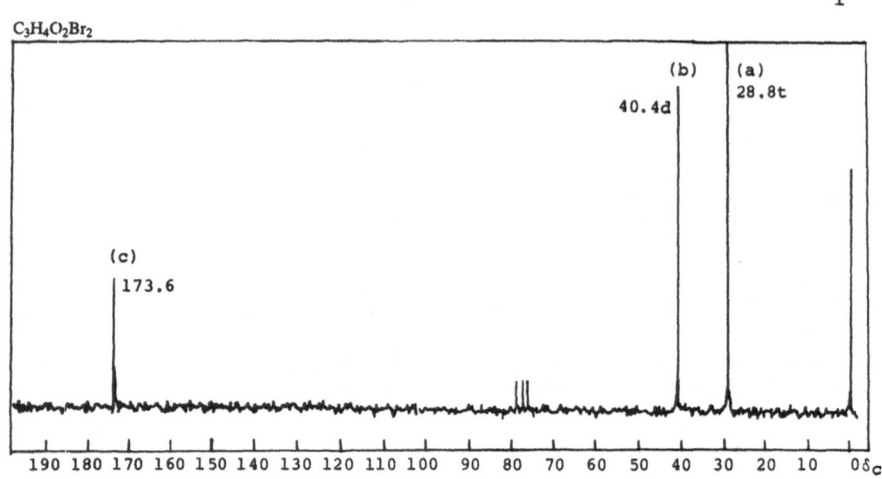

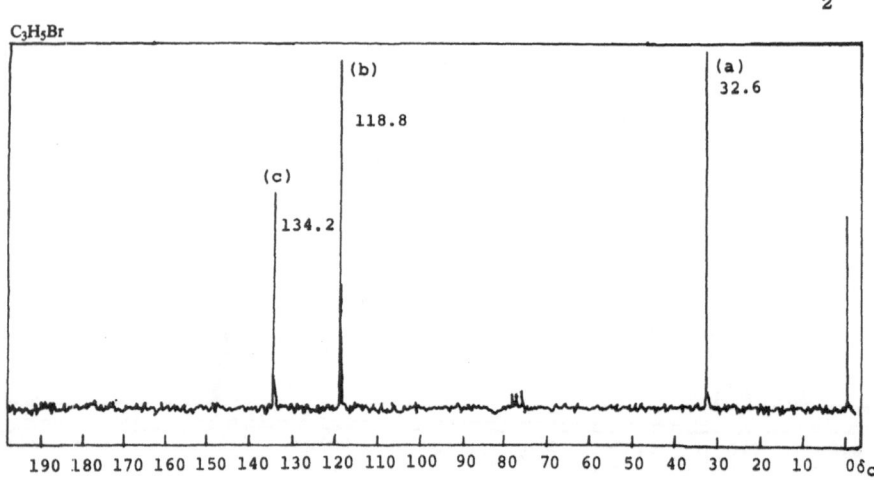

3

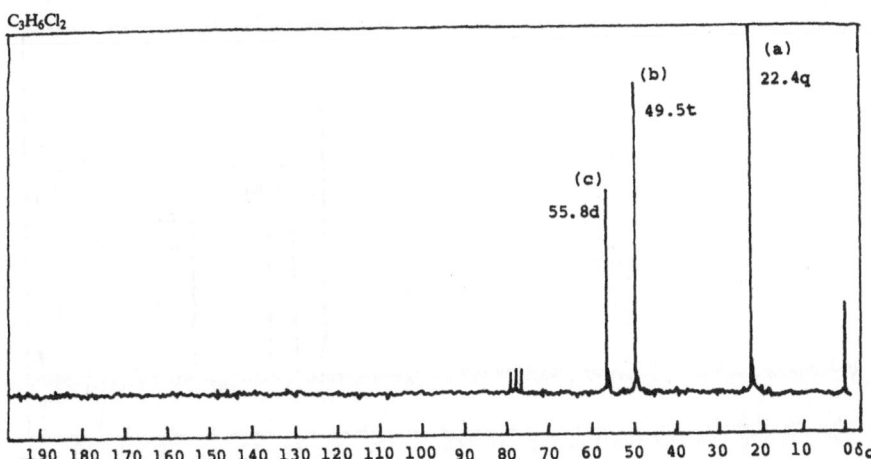

C₃H₆Cl₂

(a) 22.4q

(b) 49.5t

(c) 55.8d

4

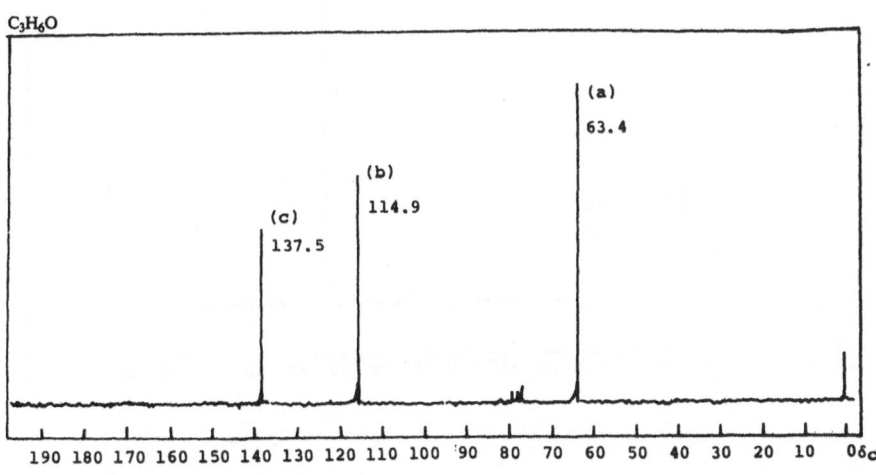

C₃H₆O

(a) 63.4

(b) 114.9

(c) 137.5

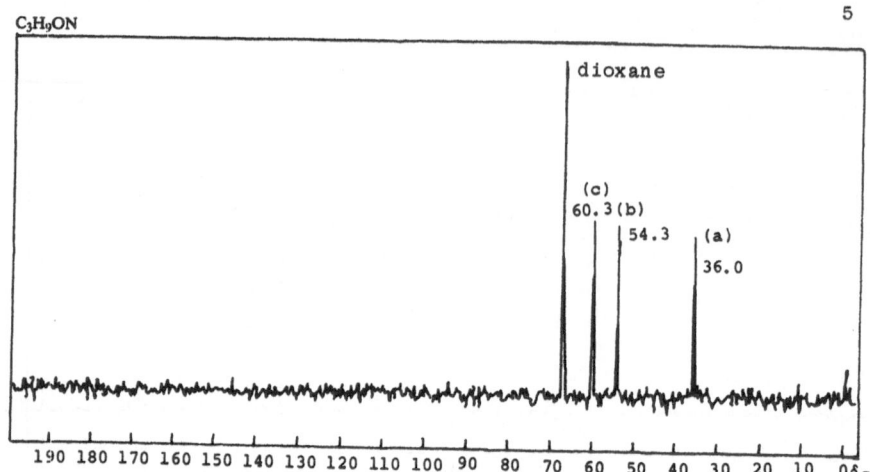

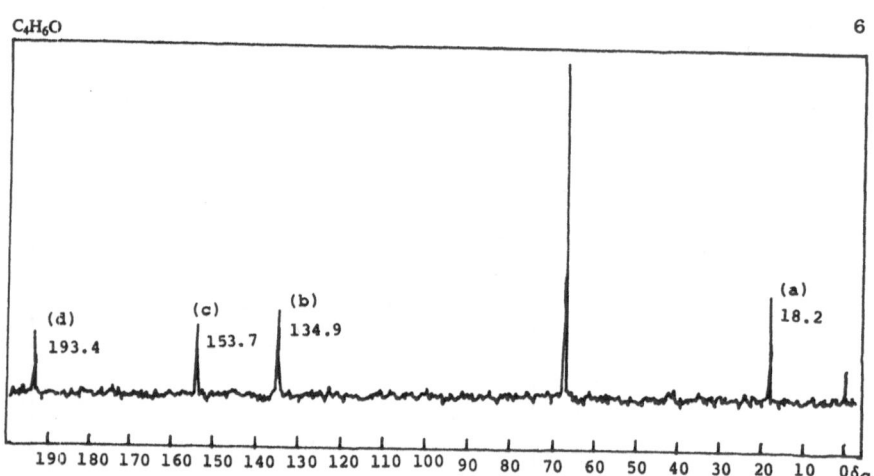

7

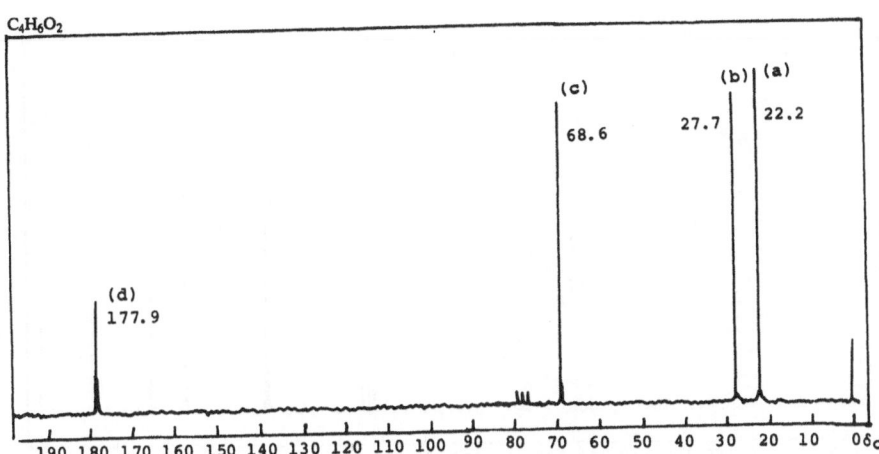

8

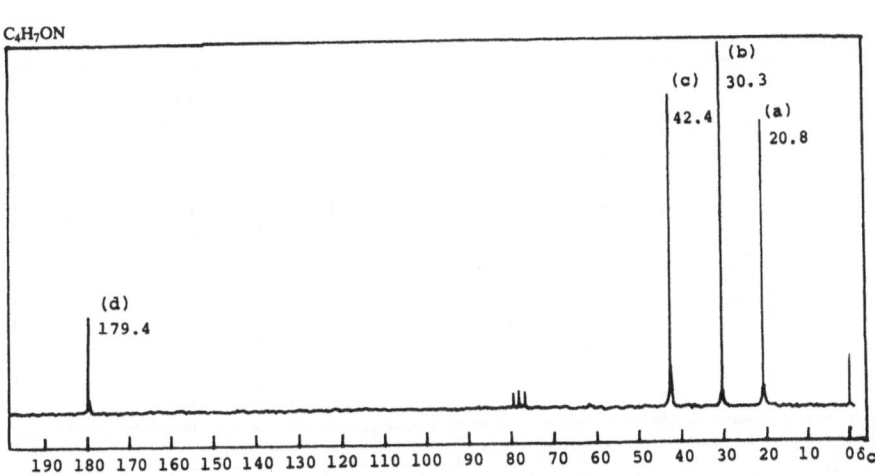

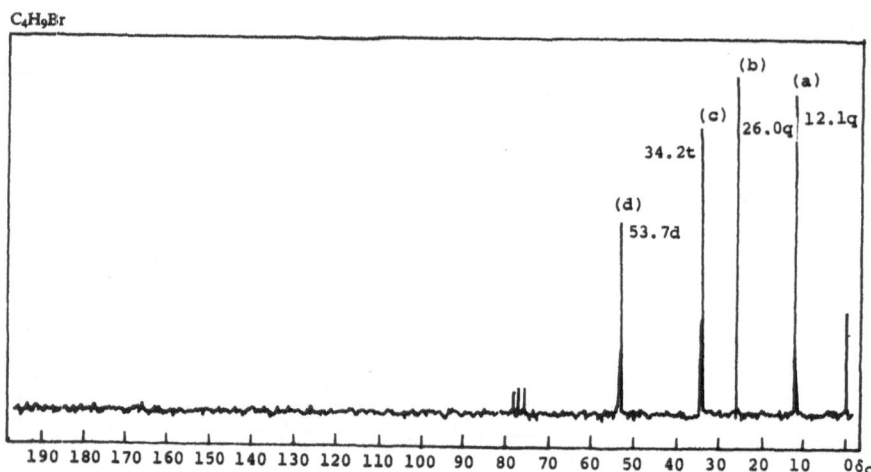

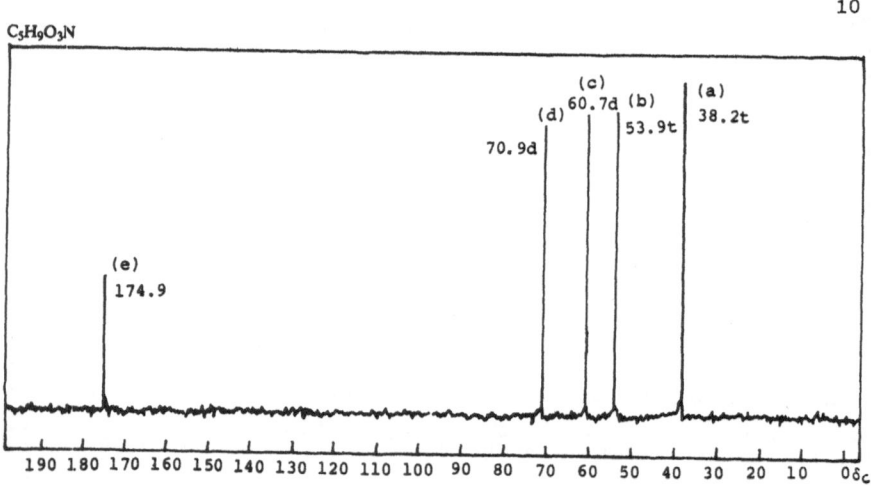

11

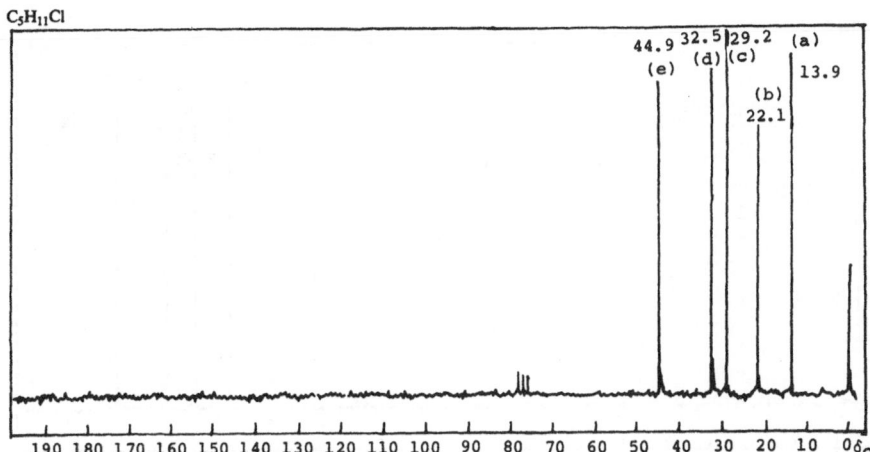

C₅H₁₁Cl

12

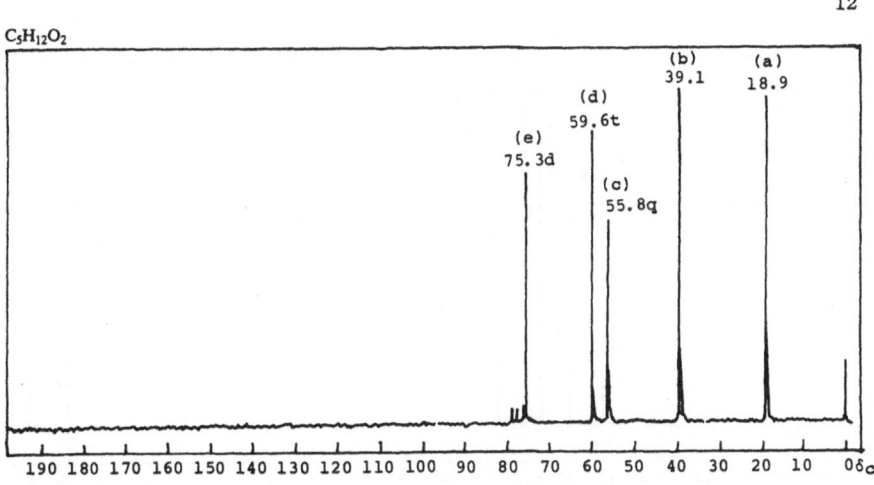

C₅H₁₂O₂

13

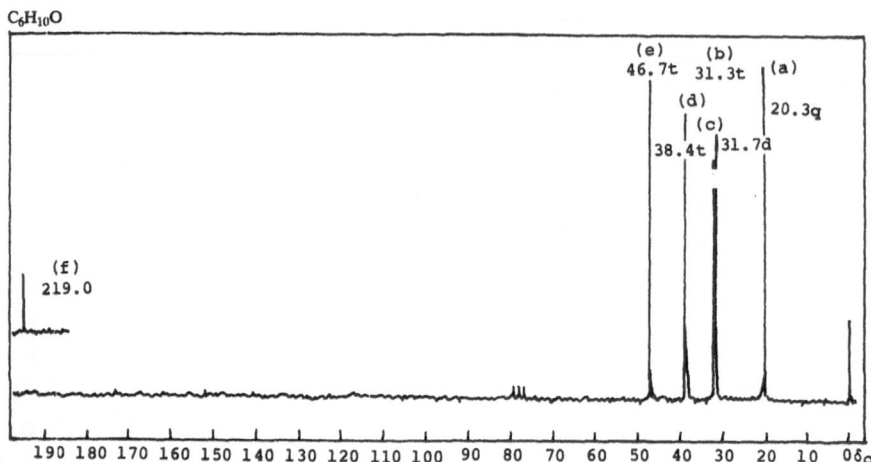

$C_6H_{10}O$

(e) 46.7t (b) 31.3t (a) 20.3q
(d) 38.4t (c) 31.7d
(f) 219.0

190 180 170 160 150 140 130 120 110 100 90 80 70 60 50 40 30 20 10 0 δ_C

14

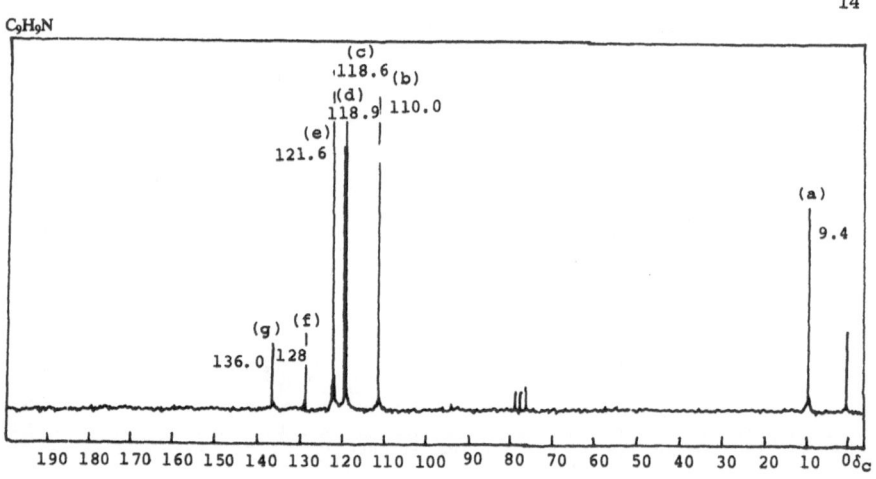

C_9H_9N

(c) 118.6 (b) 110.0
(d) 118.9
(e) 121.6
(a) 9.4
(g) 136.0 (f) 128

190 180 170 160 150 140 130 120 110 100 90 80 70 60 50 40 30 20 10 0 δ_C

15

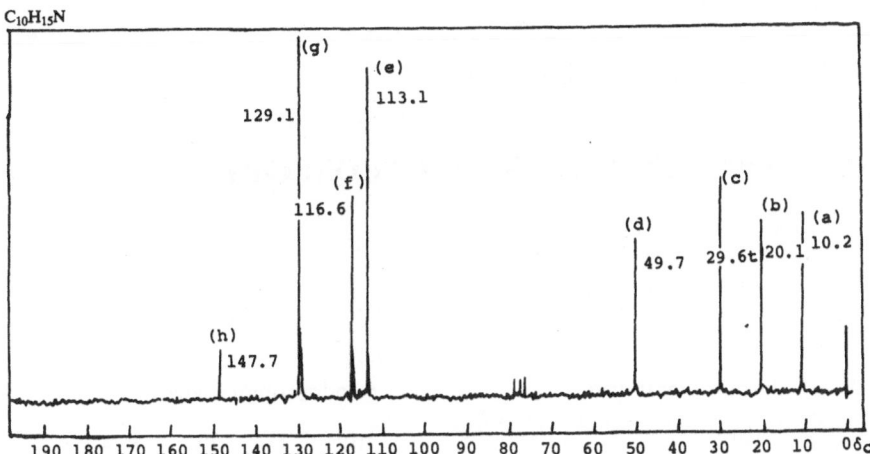

16

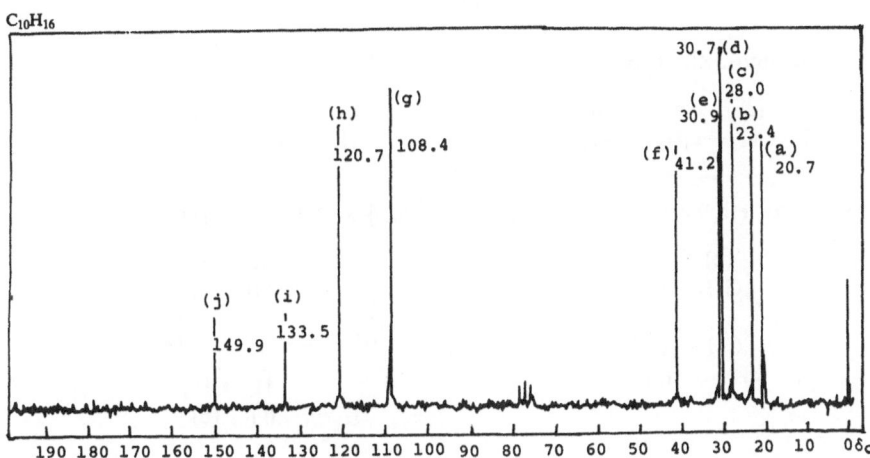

Appendix B

Answers to Problems

PROBLEMS IN ¹H-NMR SPECTROSCOPY

1. α-Chloropropionic acid

$$
\begin{array}{c}
\text{(b)} \\
\text{H} \\
\text{(a)} \quad | \quad \text{(c)} \\
\text{CH}_3\text{—C—COOH} \\
| \\
\text{Cl}
\end{array}
$$

2. β-Iodopropionic acid

$$
\begin{array}{ccc}
\text{(b)} & \text{(a)} & \text{(c)} \\
\text{ICH}_2 & \text{—CH}_2 & \text{—COOH}
\end{array}
$$

3. 1-Bromo-3-chloropropane

$$
\begin{array}{ccc}
\text{(c)} & \text{(a)} & \text{(b)} \\
\text{Cl—CH}_2 & \text{—CH}_2 & \text{—CH}_2\text{—Br}
\end{array}
$$

4. 1-Nitropropane

$$
\begin{array}{ccc}
\text{(a)} & \text{(b)} & \text{(c)} \\
\text{CH}_3 & \text{—CH}_2 & \text{—CH}_2\text{—NO}_2
\end{array}
$$

5. Methyl ethyl ketone

$$
\begin{array}{c}
\text{(b)} \\
\text{CH}_3 \\
\diagup \\
\text{O=C} \\
\diagdown \\
\text{CH}_2\text{—CH}_3 \\
\text{(c)} \quad\quad \text{(a)}
\end{array}
$$

6. 4-Methyl-1-pentyne-3-ol

$$
\begin{array}{c}
\text{(e)} \quad \text{(a)} \\
\text{OH CH}_3 \\
\text{(d)} \quad | \quad | \quad \text{(c)} \\
\text{HC}\!\equiv\!\text{C—C—C—H} \\
| \quad | \\
\text{H} \quad \text{CH}_3 \\
\text{(f)} \quad \text{(b)}
\end{array}
$$

7. Diethyl acetaldehyde acetal

$$
\begin{array}{c}
\text{(c)} \quad\quad \text{(a)} \\
\text{H} \diagdown \quad \diagup \text{CH}_3 \\
\text{C} \\
\text{(b)} \quad\quad \diagup\text{O} \quad \diagdown \text{H (d)} \\
\text{CH}_3\text{—CH (e)} \\
\diagdown\text{O} \quad \diagup\text{H (d)} \\
\text{C} \\
\text{H} \diagup \quad \diagdown \text{CH}_3 \\
\text{(c)} \quad\quad \text{(a)}
\end{array}
$$

8. Diethyl malonate

$$
\begin{array}{c}
\text{(c)} \quad\quad \text{(a)} \\
\text{O} \diagdown \quad \diagup\text{O—CH}_2\text{—CH}_3 \\
\text{C} \\
| \\
\text{CH}_2 \text{ (b)} \\
| \\
\text{C} \quad\quad \text{(c)} \quad\quad \text{(a)} \\
\text{O} \diagup \quad \diagdown \text{O—CH}_2\text{—CH}_3
\end{array}
$$

9. β-Acetoxy methyl butyrate

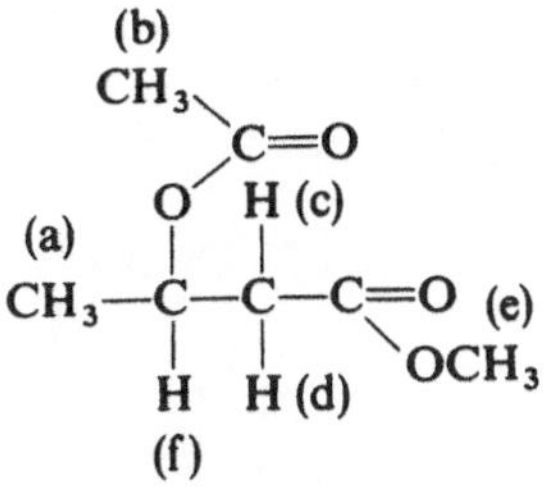

(b)
CH₃
C=O
O H (c)
(a)
CH₃—C—C—C=O (e)
H H (d) OCH₃
(f)

10. p-Bromophenetole

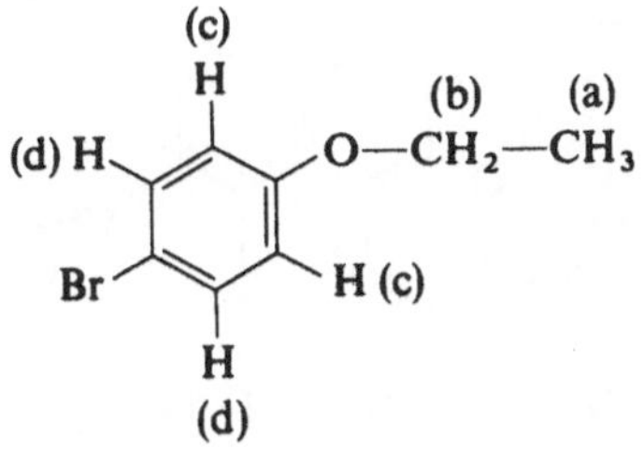

(c)
H
(b) (a)
(d) H
O—CH₂—CH₃
Br
H (c)
H
(d)

11. p-Phenetidine

(e)
(d) H
H
(a) (c)
(b)
CH₃—CH₂—O
NH₂
H H
(d) (e)

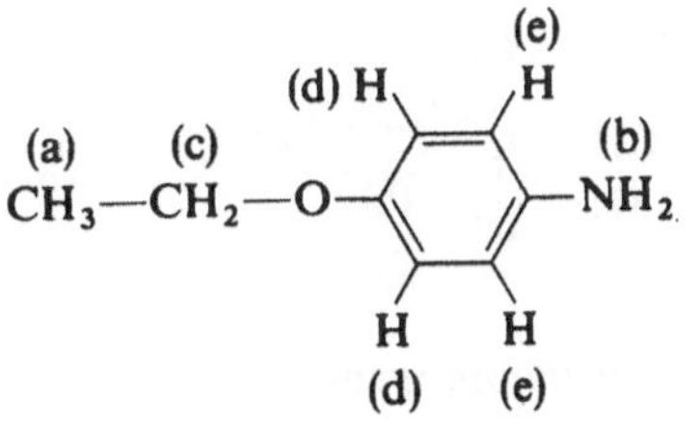

12. 1-Phenyl-3-bromopropane

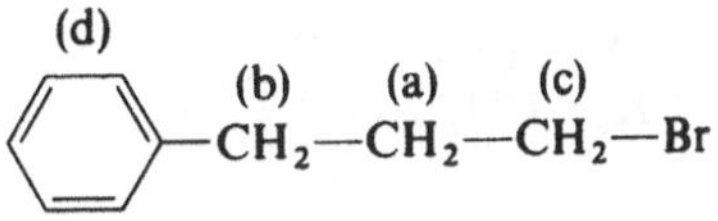

(d)
(b) (a) (c)
—CH₂—CH₂—CH₂—Br

13. Phenacetin

(d) (e)
H H
(f)
(a) (c)
H (b)
CH₃—CH₂—O
N
CH₃
C
H H
O
(d) (e)

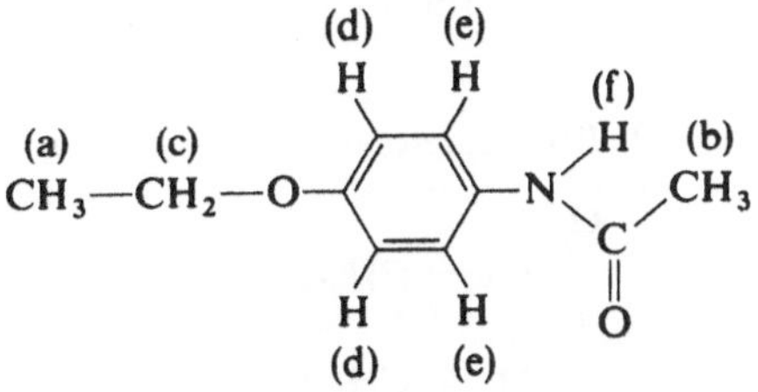

14. Diethyl acetyl succinate

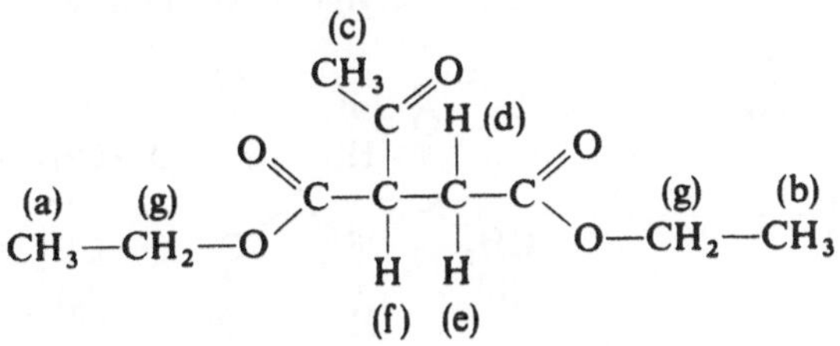

(c)
CH₃ O
C
H (d)
O
O
(a) (g)
C—C—C—C
(g) (b)
CH₃—CH₂—O
O—CH₂—CH₃
H H
(f) (e)

15. Dicentrine

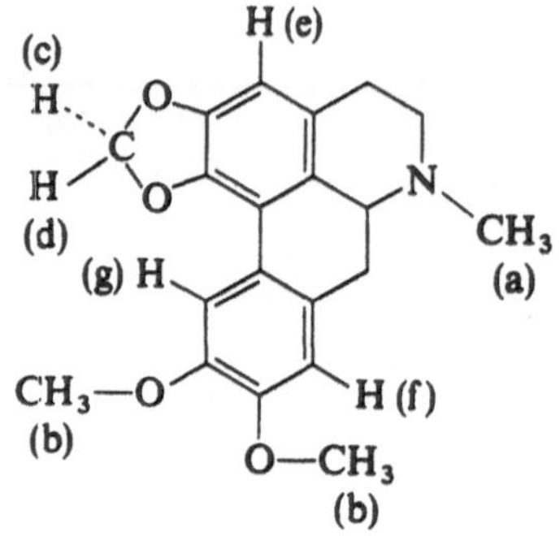

16. 1,1-Dibromoethane

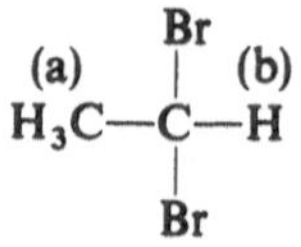

$$\underset{\text{(a)}}{H_3C}-\underset{\underset{Br}{|}}{\overset{\overset{Br}{|}}{C}}-\underset{\text{(b)}}{H}$$

17. Ethyl nitrite

$$\underset{\text{(a)}}{CH_3}-\underset{\text{(b)}}{CH_2}-O\overset{N}{\diagdown}_O$$

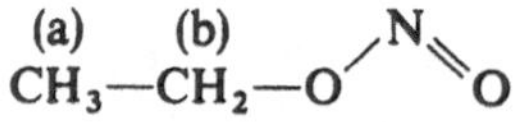

18. 1,2,2-Trichloropropane

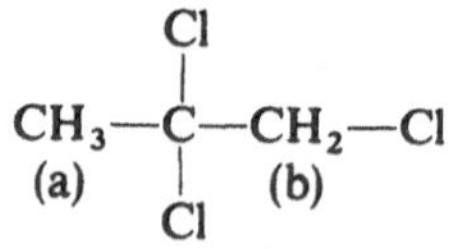

$$\underset{\text{(a)}}{CH_3}-\underset{\underset{Cl}{|}}{\overset{\overset{Cl}{|}}{C}}-\underset{\text{(b)}}{CH_2}-Cl$$

19. 1-Chloro-1-nitropropane

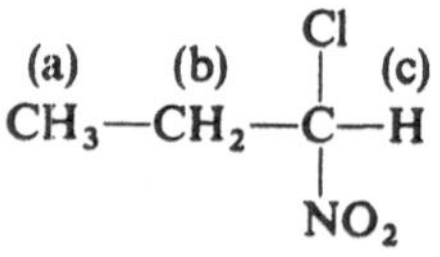

$$\underset{\text{(a)}}{CH_3}-\underset{\text{(b)}}{CH_2}-\underset{\underset{NO_2}{|}}{\overset{\overset{Cl}{|}}{C}}\underset{}{\overset{\text{(c)}}{H}}$$

20. 2-Bromopropane

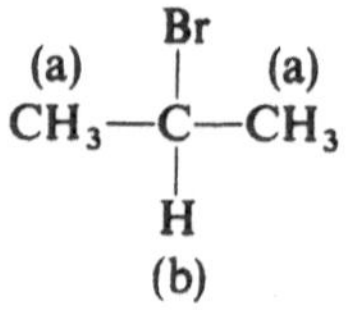

$$\underset{\text{(a)}}{CH_3}-\underset{\underset{\underset{\text{(b)}}{H}}{|}}{\overset{\overset{Br}{|}}{C}}-\underset{\text{(a)}}{CH_3}$$

21. Isopropyl cyanide

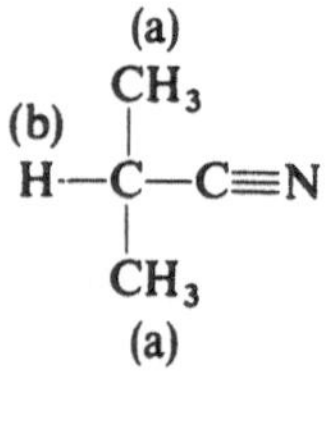

$$\underset{\text{(b)}}{}H-\underset{\underset{\underset{\text{(a)}}{CH_3}}{|}}{\overset{\overset{\overset{\text{(a)}}{CH_3}}{|}}{C}}-C\equiv N$$

22. Propiolactone

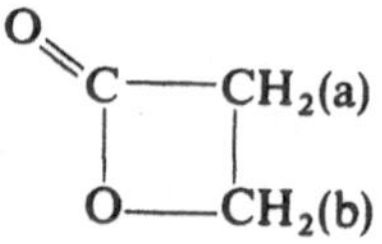

$$O\diagdown\diagup\begin{matrix}C-CH_2(a)\\ |\\ O-CH_2(b)\end{matrix}$$

23. Ethoxyacetic acid

$$\underset{\text{(a)}}{CH_3}\diagup\underset{\text{(b)}}{CH_2}\diagdown O\diagup\underset{\text{(c)}}{CH_2}\diagdown\underset{\underset{O}{\|}}{C}\diagup\underset{\text{(d)}}{OH}$$

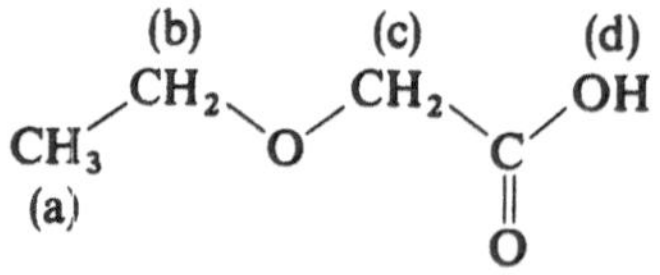

24. 2-(N-Acetyl)-diethylmalonate

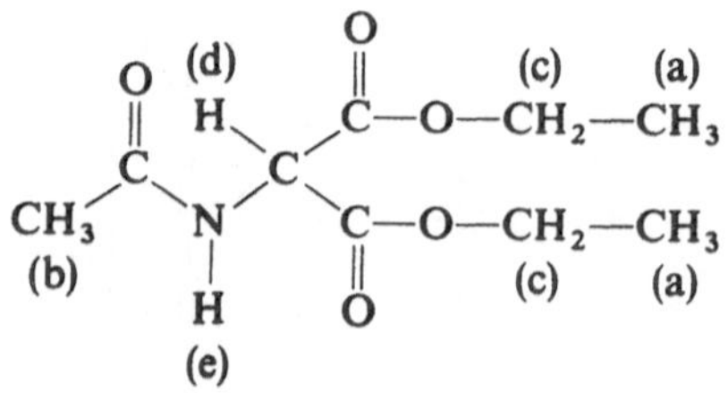

25. Isopropylphenone

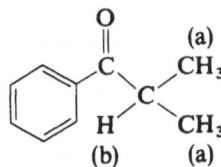

26. Acetaldoxime

(c) and (d)

H
\
C=N—OH
/
CH₃
(a) and (b)

PROBLEMS IN ¹³C-NMR SPECTROSCOPY

1. 2,3-Dibromopropionic acid

O
‖(c)
C (b)   (a)
HO   CH—CH₂—Br
|
Br

2. Allyl bromide

(a)
(b) (c) CH₂—Br
H₂C=C
\
H

3. 1,2-Dichloropropane

Cl
(b) | (a)
Cl—CH₂—CH—CH₃
(c)

4. 3-Hydroxy-1-propene

(a)
(b) (c) CH₂—OH
H₂C=C
\
H

5. 2-Methylaminoethanol

H
|
(a) N (b) (c)
H₃C   CH₂—CH₂—OH

6. Crotonaldehyde

(d) H
\
O=C (b) (c) H
\   \
C=C (a)
/   \
H   CH₃

7. 4-Butyrolactone

(a)   (b)
(c)   (d)
O   O

8. 2-pyrrolidone

(a)   (b)
(c)   (d)
N   O
|
H

9. 2-Bromobutane

Br
(b) | (c) (a)
CH₃—CH—CH₂—CH₃
(d)

10. Hydroxyproline

HO
(d) (a)
(b)   (c) OH
N   C (e)
| ‖
H   O

11. 1-Chloropentane

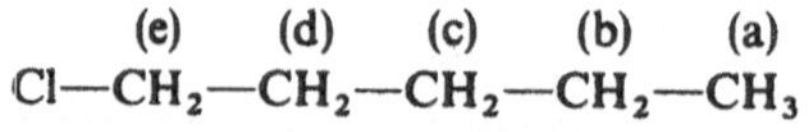

$$\underset{\text{Cl}}{\text{Cl}}-\overset{(e)}{\text{CH}_2}-\overset{(d)}{\text{CH}_2}-\overset{(c)}{\text{CH}_2}-\overset{(b)}{\text{CH}_2}-\overset{(a)}{\text{CH}_3}$$

12. 3-Methoxy-1-butanol

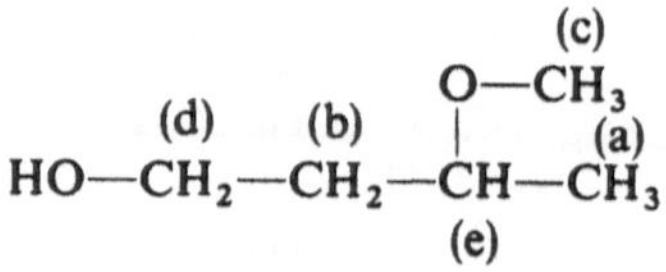

13. 3-Methylcyclopentanone

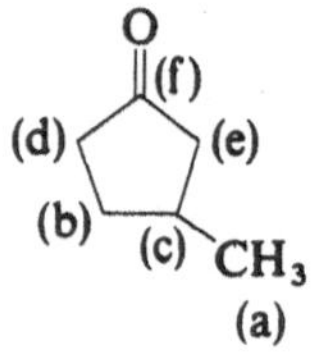

14. 3-Methylindole

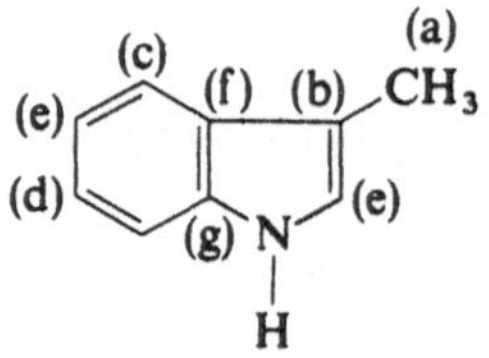

15. *N-sec*-Butylaniline

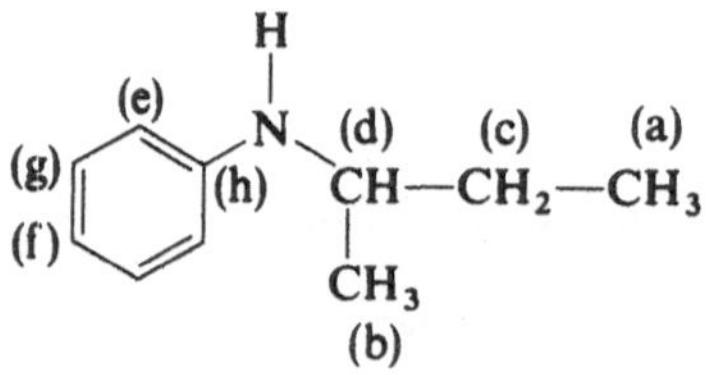

16. Limonene

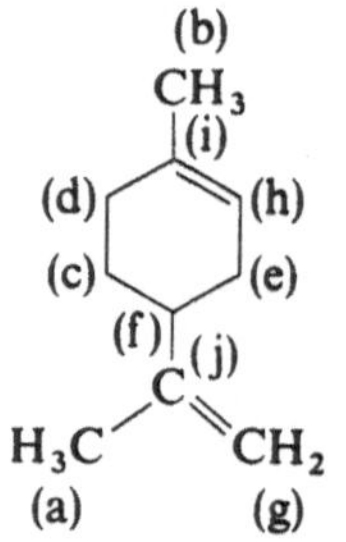

Index